ALCOHOL
Use and
Abuse in America

ALCOHOL
Use and
Abuse in America

Jack H. Mendelson
Nancy K. Mello

Little, Brown and Company
Boston   Toronto

FIRST EDITION

LIBRARY OF CONGRESS CATALOGING IN PUBLICATION DATA

Mendelson, Jack H. (Jack Harold), 1929–
    Alcohol: use and abuse in America.

    Bibliography: p.
    Includes index.
    1. Alcoholism—United States—History. 2. Alcohol—
Physiological effect. I. Mello, Nancy K. II. Title.
HV5292.M4   1985   362.2′92′0973   85-6959
ISBN 0-316-56663-2

*Acknowledgments for copyright material
appear on page ix.*

VB

DESIGNED BY JEANNE F. ABBOUD

*Published simultaneously in Canada
by Little, Brown & Company (Canada) Limited*

PRINTED IN THE UNITED STATES
OF AMERICA

Contents

Acknowledgments

The authors are grateful to the following for permission to use previously copyrighted material:

Jai Press, Inc. for the use of "Etiological Theories of Alcoholism" by Nancy Mello, from *Advances in Substance Abuse: Behavioral and Biological Research*, Volume III.

Alcoholics Anonymous World Services, Inc. for "The Twelve Steps" (Copyright 1939, 1955, 1976 by Alcoholics Anonymous World Services, Inc.) and for an excerpt from *Alcoholics Anonymous Comes of Age* (Copyright © 1957 by Alcoholics Anonymous World Services, Inc.).

Viking Penguin Inc. for an excerpt from "On Magic in Medicine" from *The Medusa and the Snail* by Lewis Thomas. Originally published in *The New England Journal of Medicine*. Copyright © 1978 by Lewis Thomas.

The Economic Research Department of the Wine Institute, the United States Brewers Association, Inc., the Distilled Spirits Council of the United States, Inc., and the Brewers Association of Canada, for tables and charts as indicated in the text.

Preface

The intent of this volume is to provide an overview of contemporary American drinking practices and problems within a historical context of cultural perceptions of alcohol use and abuse. We were stimulated to write this book following discussions with our students and colleagues about two seminal publications that have achieved a prominent position in American literature. The first, *Licit and Illicit Drugs*, a comprehensive analysis prepared by Edward M. Brecher for the Editors of *Consumer Reports* and published by Little, Brown and Company in 1972, highlighted the importance of alcohol use in American society. However, alcohol was discussed as part of a consideration of many abused drugs and substances, so the scope was necessarily limited. Since alcohol use and abuse has a major impact upon society, culture, politics and public health, we felt a more detailed exposition of contemporary alcohol-related issues and problems would be of value to the American consumer.

An earlier text, *The Neutral Spirit: A Portrait of Alcohol* by Berton Roueché, was a collection of essays, initially serialized in the *New Yorker* magazine and later published in book form by Little, Brown and Company in 1960. This fine example of science writing was among the first to present a clear description of the biological and behavioral aspects of alcohol use and abuse for the nonscientist reader. Our goal was to update information discussed by Roueché and to incorporate new findings, theories and thoughts that have emerged from alcohol-related research in medicine, biology, psychology and the social sciences.

It has been said that great discoveries in science do not represent individual contributions as much as the collective efforts of many scientists striving to achieve a common goal of understanding. Expositions of achievements in science, art and literature are also made possible by many writers whose collective efforts have brought a better appreciation of the significance of accomplishments in art and science to the public. As scientists, we welcome the opportunity to share with our readers information, ideas and undoubtedly a few prejudices about alcohol use and abuse in contemporary America. As writers about science, it has been a great joy to discover in history, art and literature, some facts and fancies that make science more comprehensible, and provide a new perspective on the problems and promises of humanity.

> Jack H. Mendelson, M.D.
> Nancy K. Mello, Ph.D.
> November, 1984
> Cambridge, Massachusetts

Part I

Alcohol Use in Perspective

CHAPTER 1

In the Beginning

The miracle of wine, a mysterious potion to the ancients, became richly interwoven with myths of creation, death and resurrection, gods and demons. Wine seemed to reflect the unpredictable forces of nature; intoxication, the raptures and furies of the gods. Early legends described the origins of wine as a god-given gift or affliction. The power of the gods to give joy and inflict pain seemed synonymous with the power of wine to create a cycle of ecstasy, sorrow and silent sleep. The myths of antiquity are linked to Judeo-Christian theology and contemporary beliefs through a remarkably similar array of common images and themes. The seasonal rhythms of the vine, growth, maturity, harvest, death and rebirth became an allegory of death and resurrection, and the wine itself, a ceremonial symbol of sin and salvation.

Some fundamental fears and illusions about wine and intoxication transcend cultures and time. The primeval origins of superstition reflected the unpredictability of the universe. The awesome forces of nature were feared and worshipped — until propitiatory rituals became somewhat systematized, any act might unleash a terrifying storm or a savage predatory attack. Since alcohol arose spontaneously from fermentation of fruits and grains, it also was a mysterious force of nature, which changed man's behavior in unprecedented and strange ways. We can only speculate about the experience of intoxication in early man. William Golding describes an episode of alcohol intoxication by primitive humans in his novel *The*

Inheritors. The prose style mimics early man's meager language in a time when acquisition of linguistic and communication skills were just beginning.

> *The old man . . . held a round stone thing in his hand which he put to his mouth every now and then and in between while he was singing. The other men and women were scattered round the clearing. They held more of these round stones and now Lok saw that they were drinking from them. His nose caught the scent of what they drank. It was sweeter and fiercer than the other water, it was like the fire and the fall. It was a bee-water, smelling of honey and wax and decay, it drew toward and repelled, it frightened and excited like the people themselves. There were other stones nearer the fire with holes in their tops and the smell seemed to come particularly strongly from them. Now Lok saw that when the people had finished their drink they came to these and lifted them and took more to drink. The girl Tanakil was lying in front of one of the caves, flat on her back as if she were dead. A man and a woman were fighting and kissing and screeching and another man was crawling round and round the fire like a moth with a burnt wing. Round and round he went, crawling, and the other people took no notice of him, but went on with their noise. . . . The old man sang and the people fought, the man crawled round the fire, Tuami burrowed at the fat women and all the time the clearing jumped back and forth, sideways [pp. 154–155].*

As man evolved and civilization developed, so did man's involvement with wine and its gods. A rudimentary viniculture and nascent enology were part of the earliest civilizations. Over six thousand years ago in Mesopotamia, the Sumerians worshipped Gestin, a goddess descended from the great earth-mothers, as protector of the vine. As early as 3000 B.C., the Egyptian god Osiris was worshipped as god of wine and lord of the dead. This curious amalgamation of titles was one of the earliest myths in which wine was involved in the symbolism of death and resurrection. According to Egyptian legend, Osiris taught wisdom, art and law to the people and was first to plant the vine, tread the grape and cultivate the soil. He was twice killed by his jealous brother and twice resurrected by Isis, his sister-wife. When the harvest was past and the land lay dry and parched, the tears of Isis, sorrowing for Osiris, caused the Nile to flood and rejuvenate the soil. The burial of Osiris coincided with the planting season. Osiris also presided over the harvest festival, a celebration of his resurrection and a feast for the dead. These feasts were similar to the Dionysian feasts of Anthesteria in ancient Greece, which were also attuned to the seasonal rhythms of growth, decay and rebirth.

Inconsistent and contradictory effects of intoxication have been chronicled in many legends; wine is ever benefactor and malefactor, harbinger of both good and evil. The Chinese emperor Yu is credited with both the discovery of rice wine during the third millennium B.C. and its subsequent prohibition. He came to believe that wine produced such powerful effects on behavior that it would lead to the fall of his kingdom. A Persian legend describes how King Dshemshid became convinced that the juice of fermented grapes was poison. He changed his opinion when a beautiful woman suffering from chronic headaches attempted suicide by drinking the "poison" and enjoyed a miraculous cure. The Persian court then called wine "a delightful poison." The king named his serendipitous discovery "royal medicine."

Greece was the source of the most lyrical and ambivalent myths and legends about wine. Dionysus, god of wine, was a very old deity in Greek folklore by the time Homer composed the *Iliad* and the *Odyssey*. The worship of Dionysus probably originated in very early agrarian communities near the Aegean Sea, long before viniculture and wine production began. The original Dionysus controlled mysterious forces of the forest and the prowess of animals and was worshipped as god of the trees. When the intoxicating product of the fermented grape was discovered, it was logical that control of this new powerful force of nature also be attributed to Dionysus.

Through time, the informal and individual worship of Dionysus evolved into ritualized customs and celebrations. The most elaborate of these ceremonies were the feasts of Anthesteria, a four-day period of preparation and distribution of wine, intoxication and revelry, sobering up and atonement and finally, celebration of the resurrection and return of Dionysus. This Dionysian feast contained the germ of ideas that later appeared in various forms in Judeo-Christian theology. Yom Kippur is the day of atonement; Rosh Hashanah, the festive celebration of the new year; Easter, the festival of rejoicing at the resurrection of Christ.

Drama had its origins in the formalization of expressions of atonement during Dionysian worship and comedy emerged from the spontaneous revelry and merrymaking during intoxication. In Athens, the ruins of the theater of Dionysus on the south side of the Acropolis are a mute witness to the emergence of Greek dramatic arts at the feasts of Anthesteria. By the fifth century B.C., poets competed to offer a tragedy, a comedy and a dithyramb or choral ode to Dionysus at the festival. The classic plays of Sophocles, Aeschylus, Euripides and Aristophanes were written and performed for the Dionysian feasts. Dionysus later became the Roman god Bacchus; *bacca* is the Latin word for grape. Bacchus, like Dionysus, was both saint and satyr, miscreant and venerated deity. The Bacchanalian feasts, orgies of intoxication, were also seasonal celebrations of the harvest, death and rebirth of the vine. With the development of Judeo-Christian theology, these ancient festivals of the planting and the harvest gradually became more identified with the sins of intoxication.

Ambivalence about the salutory and dangerous effects of alcohol has persisted in art, literature, philosophy and science throughout the ages. We can only speculate how intoxication of the mythologists may have shaped the intricate and conflicting stories about the spirits of man.

Today, we celebrate many of the best events in life, and life itself, with a toast, a prayer and a drink. Birth, good fortune, marriage and communion with God may sanction alcohol use. We drink in celebration and in sorrow. We use alcohol as a beverage, as naturally as we ingest food. We abuse alcohol under conditions of stress, provocation and for no good reason at all. Transition from accepted drinking traditions to alcohol abuse may be precipitated by a myriad of private, social and cultural stimuli. Today, as in the past, the use and abuse of alcohol has a powerful influence upon our health, culture and societal evolution. In this book, we examine alcohol yesterday and alcohol today in relation to drinking and alcoholism in contemporary America.

CHAPTER 2

Alcohol Use
in Colonial Times

Colonial America evokes images of a tranquil, pastoral society, enviable in its simplicity, celebrating the spring planting and the autumn harvest in communal harmony. This bucolic idyll, if it ever existed, was achieved under great hardships and deprivations that are almost inconceivable today. We can only imagine the contrast between seventeenth-century rural England and Cape Cod in November of 1620. The survivors of the *Mayflower*'s first voyage confronted the alien winter wilderness with resolute courage, and an instantly constructed framework for social order — the Mayflower Compact. Puritans seeking religious freedom were only a minority of that first colony — the rest were laborers and servants searching for better opportunities in the New World. The ominous prospect of anarchy was evaded by the Puritans' political astuteness. They prevailed in this first exercise in self-governance in America and set the pattern for future Colonial settlements.

Almost nothing is known about alcohol use during the first years, but in a time of constant scarcity it is unlikely that alcohol was readily available in any form. However, the Colonists' concern with social order was soon extended to include alcohol intoxication. By 1633, in Plymouth Colony, a John Holmes was censured for drunkenness. His penalty — "to sitt in the stocks, and was amerced forty shillings." In 1629, the governor of the Massachusetts colony was advised by his English superior that: "We pray yow endeavor, though there be much *strong waters* sent for sale, yett see to order it, as that the salvages may not for or lucre sake bee induced to excessive use or rather abuse of it, and, at any hand, take care or people give us ill example; and if any shall exceed in that inordinate kind of drinking as to become drunck, wee hope you will take care his punishment be made exemplary for all others."

The religious Puritans believed alcohol was God's gift to man. Alcohol was considered a restorative with broad medicinal powers and nutritional value. Alcohol was used to ease pain from childbirth to terminal illness. Fevers, fatigue, indigestion, insomnia, hypertension and nervous breakdown are only a few of the common disorders for which alcohol was prescribed. Marginal water supplies and dried and salted food further accentuated the need for a potable beverage. Festive drinking and revelry were part of the Colonists' British heritage. The prospect of convivial drinking was probably not unattractive to the hardworking farmers, struggling to wrest a crop from the rocky New England soil before winter, ever vigilant for hostile "savages" and marauding animals.

Ambivalent attitudes about alcohol are a common thread that links all societies where drinking occurs and Colonial America was no exception. The many virtues and medicinal necessities of alcohol were accepted by all, on biblical authority. The evils of inebriety were deplored by the clergy and penalized by the law. Yet consider the inventiveness, time and labor devoted to making alcohol in Colonial times.

COLONIAL SPIRITS

The production of alcoholic beverages was just one of many tasks that strained the ingenuity and resourcefulness of the early Colonists. The development of the ciders, wines and beers that became colonial staples required apples, grapes and berries and hops, and perhaps most important, a supplementary supply of sugar to enhance fermentation. Apples were not native to New England, and the orchards had to be grown from English seed. The first orchard was planted about 1625 on Beacon Hill in Boston by a Reverend Blaxton and probably required over ten years to reach maturity. More orchards were planted as the land was cleared, and apples became abundant in New England and New York. Within two hundred years, about two thousand varieties of apples were grown throughout America.

Cider was the most common beverage in the Colonies. Crushed apples were weighted with stones or logs to release their juices and aged in barrels over the winter. Some cider fermented into hard cider by spring, then was racked off into storage barrels to age further. The addition of ever scarce honey or cane sugar increased the potency and provided some natural carbonation. An "apple champagne" was a special treat in Colonial times.

Cider was served to every member of the family at breakfast, dinner and supper. Cider was consumed in the fields between meals, and was a regular staple at all the communal social functions. Cider's popularity was eventually overshadowed by the distilled beverages, rum and whiskey, that came into vogue during the eighteenth century.

Beer accompanied the pilgrims who sailed on the *Mayflower* and remained a favorite beverage of the early colonists. Hops grew wild in New England, and cultivated hops eventually supplanted the sparse natural supply. Hop seeds were ordered from England by the Massachusetts Bay Co. as early as 1629. As an interim measure, Colonists devised a beer made from red and black spruce twigs, boiled in water, as well as a ginger beer. Once the intricacies of brewing were mastered, beers became readily available, and many farmers made their own with the help of a malster who malted their barley, or more often, corn.

The designations of beers as X, XX or XXX referred to their alcohol content. The weakest used only sugar from the malt and such "small beers" were most commonly drunk. "Ships beers" were somewhat stronger and also readily available. The strongest beer, brewed with malt and supplementary sugar, was available only to the affluent few.

Colonial farmers learned to make an astonishing variety of fruit and vegetable wines. Beets, carrots, celery, corn silk, dandelions and goldenrod, even spinach, tomatoes, onions and squash were mashed, boiled, sweetened and fermented into mildly alcoholic wines. Strawberries, blackberries, cranberries, currants, elderberries and gooseberries were popular sources of fruit wine. Four quarts of berries could be processed to yield about a gallon of wine. Flowers, herbs and oak leaves were also converted to wine, and that list by no means exhausts the colonial wine repertoire.

Honey was valuable as a source of sugar for fermentation and as an alcoholic beverage itself. Mead was known in the Middle Ages and consisted of honey diluted with water, then fermented. The Colonists flavored fermented honey with herbs and spices into a drink called Old Metheglin. Honey was sometimes fermented with fruits to make melomel. Each honey-based drink could be served hot or cold and these were favored to combat winter chills.

Rum resembles mead in that it also involves fermentation of a sweet base, molasses, with sugar, but it is a distilled spirit. Rum was rarely available to the Colonists until after 1650, when trade with Barbados began and West Indian rum was imported. Soon molasses and sugar cane were imported directly, and enterprising New Englanders manufactured the rum at a lower cost. Rum distilleries emerged after 1700 and were common within forty years. Imported rum also continued to flow into the colonies because it had become a standard currency for exchange. By the time of the Revolution, it was estimated that the Colonies imported four million gallons and distilled five million gallons, of which seven-eighths were consumed by the Colonists. The rum industry grew dramatically in Boston, Newport and Providence. Almost every other important town from Massachusetts to the Carolinas possessed a still house to meet the local demand. Rum was often consumed in the form of mixed drinks like punch and flip. Flip was a winter favorite made from rum and beer sweetened with sugar and warmed by plunging a red-hot poker from the fireplace into the mug as it was served. Rum soon challenged cider as the favored beverage of Colonial times.

ATTITUDES TOWARD DRINKING

Throughout the Colonies, alcoholic beverages were considered necessary and beneficial. No serious attempt was made to restrict the normal use of fermented and distilled beverages. The intimate role that alcohol played in all aspects of Colonial society seems contrary to our stereotypes about the stern, self-denying Puritan and the stoic Quaker. But it was the descendants of the early Puritans who later condemned drinking and inveighed against the "Demon Rum." In 1673, Increase Mather praised alcohol, saying that "Drink is in itself a creature of God, and to be received with thankfulness." Careful study of the Puritan legacy of attitudes about alcohol shows that they made an important distinction between drinking and drunkenness. A distant echo of contemporary ambivalence is expressed succinctly in Mather's pronouncement: "The wine is from God, but the drunkard is from the Devil."

Although there was no formalized opposition to drinking, intoxication was treated with rapid and sometimes severe punishment. Habitual drinkers were whipped or forced to wear a mark of shame. Once so labeled, they could be refused the right to purchase liquor. During the seventeenth century, all of the Colonies specified a fine or prescribed the stocks for the first drunkenness offense. Repeated offenders often received sentences to hard labor or corporal punishment.

Although there is ample evidence for the need to prosecute drinking-related infractions, the general sobriety of the first Colonists is remarkable given the abundant availability and widespread use of alcoholic beverages. In all likelihood, it was only the strict sanctions against blatant intoxication, the discipline expected in all aspects of daily living and the close-knit nature of Colonial towns that prevented the development of more serious drinking problems. One example of an early measure taken to control alcohol abuse was the designation of a tithingman. The tithingman was commissioned to oversee approximately ten families and to report any liquor infractions to the minister. Little is known about the effects of this type of surveillance, but its official tenure was mercifully short. Soon, after 1675, only a few New England towns used tithingmen.

A somewhat less intrusive effort to control drinking was a limit on when alcohol could be sold. In New England taverns, a patron could spend only a limited time "tippling." Guests were not allowed to drink beyond 9:00 P.M. Liquor was not served during hours of worship or during the entire Sabbath. In New York and Massachusetts there were prohibitions on gaming, dancing and excessive drinking.

Despite the general acceptance of drinking as a normal, even integral part of the Colonial social fabric, this privilege was not extended to all. Many legislatures prohibited the sale of intoxicants to servants, apprentices, slaves, American Indians, debtors and habitual drunkards. But these restrictions were often violated by the illegal tippling houses that were opened clandestinely to accommodate these groups.

Control of the use of alcohol for economic, as well as moral purposes, was an early and pervasive concern. All thirteen Colonies adopted statutes to guarantee the quality of inns and taverns, to raise revenues, to suppress drunkenness, and to "protect" the American Indian. Many statutes framed by Colonial legislatures were clearly intended to encourage innkeeping, brewing and distilling. Inns were needed to promote the expansion of travel and commercial activity. Domestic production of spiritous and malt liquors was encouraged because they were considered an integral part of the daily diet and of social intercourse.

Licensing was adopted as the main vehicle of control. The authority to license was assumed early by Colonial legislatures, and licensing power was delegated to either the governor, county court or the selectmen. It was expected that public houses would be established only by reputable citizens, and that these would be located near places frequented by the justices holding county court, along traveled roads, near important bridges and ferries, thereby serving the traveler and those engaged in legal activity or trade. Licensing fees were kept low to promote the development of retail liquor trade.

Tavern owners were socially prominent members of their communities. Securing a license to run a public house was often the end point of a career in public service — a sinecure unrivaled in that era. Sons of tavern owners preceded sons of clergymen in an early Harvard listing of students according to their relative social rank. Alcohol as well as public drinking had become an accepted part of Colonial society.

DRINKING PATTERNS IN TRANSITION

As the Colonies evolved from a simple rural society into a network of towns and cities, there was a gradual change in drinking patterns. The ciders, beers and fruit wines produced locally for family use were increasingly replaced by commercially distilled rums and liquors. The infant urbanization also brought changes in family and community cohesiveness. The early Colonists often emigrated with families and religious communities intact, rapidly established stable social groups, and believed devoutly in the divine authority of the Bible. To them the use of wine, cider and beer was a healthy alternative to unsafe drinking water, a nutritious part of their diet and a traditional complement to communal celebrations.

As Colonial society became more complex, there was a gradual erosion of many traditional values. The moral stigma once associated with acute intoxication was reduced as public drunkenness became almost commonplace. Rum and whiskey were increasingly popular and far more intoxicating than cider.

During the decade before the Revolution, Americans drank an estimated 3.7 gallons of alcohol per year, most of it New England rum. Rum was cheap, available and copiously imbibed by all at work, at meals and at leisure. In sickness, liquor became a common part of the pharmacopoeia of every colonial physician. In health, spirits were believed to be a useful source of energy to manual laborers, farmers and sailors, who were typically provided with a free daily ration of rum as a fringe benefit of their employment. Communal gatherings such as weddings, house raisings or the ordination of a minister were occasions for the liberal provision of cider, punch and rum. Whether it was a meeting of the town selectmen, local merchants or Colonial militia, social drinking was often a common denominator of daily business. Church leaders were expected to serve alcoholic refreshment at ordinations and ecclesiastical meetings, and the same was expected of town authorities at public ceremonies, Colonial musters and militia training.

Even in death, drinking assumed an important role as a common activity at wakes and funerals. In a society whose population was periodically decimated by cholera epidemics and infant mortality, funerals were a constant reminder of the hazards of frontier living. On the night before the funeral, mourners were treated to generous servings of liquor at the home of the deceased, and alcoholic beverages were freely dispensed at the funeral service. At the death of a minister or a pauper, the town itself often provided several gallons of rum or cider, suggesting that alcohol must have been regarded as one of the prime requisites of a decent burial. One indication of the extent to which alcohol was served at funerals can be found in the mortuary expense bills of the period. One such bill lists the expenses incurred at the funeral of David Porter, who was drowned near Hartford in 1678. As can be seen from the following list, alcohol accounts for the major part of the expenses:

By a pint of liquor for those who dived for him .... 1s.
By a quart of liquor for those who bro't him home .... 2s.
By 2 quarts of wine & 1 gallon of cyder to jury of inquest . 5s

By 8 gallons and 3 qts. wine for funeral .... 1 & 15s.
By barrel cyder for funeral ....16s.
1 coffin ....12s.
Windeing sheet ....18s.

Yet even the custom of funeral drinking was soon to change. By 1742, the General Court of Massachusetts forbade the use of wine and rum at funerals.

Another indication of the importance of alcohol in Colonial society is suggested by the rich vocabulary used to describe drinking and intoxication. No less an authority than Benjamin Franklin devoted a piece in his January 6, 1737, edition of the *Pennsylvania Gazette* to the rich vocabulary of Colonial drinking. Gathered "wholly from the modern Tavern-Conversations of Tipplers," Franklin's "Drinker's Dictionary" was presumably constructed from the first-hand experience of the Pennsylvania sage himself. Among the terms offered "to surprise as well as divert the sober Reader," the following, selected from more than 235, give a taste of the Puritan drinker's salty vocabulary and how it was used to describe the drunkard: "He's drunk as a Wheelbarrow, As Drunk as a Beggar, Got Corns in his Head, Loaded his Cart, Cock Ey'd, Fox'd, Frozen, Been at an Indian Feast, As Dizzy as a Goose, Got the Glanders, Juicy, Merry, Moon-ey'd, Mellow, Oil'd, Got the Night Mare, Like a Rat in Trouble, In the Sudds, As Stiff as a Ring-bolt, Soak'd, He carries too much Sail, Topsy Turvey, Tipsey, he's Wet." Considering that the familiar terms like "three sheets to the wind, high, half shot, bent, loaded and pickled" were added during the nineteenth century to describe "rummies," "lushes" and "souses," it is evident that many of the popular metaphors now used to describe the experience of drinking and intoxication have their origin in America's early drinking history.

Given the pervasive use of alcohol in all aspects of Colonial society, it is not surprising that alcohol-related problems were also in evidence. In eighteenth-century sermons, pamphlets, personal diaries and newspaper articles, some New Englanders noted with alarm the changes in personal conduct and public morals seemingly associated with increased liquor consumption. Puritan divine Cotton Mather, whose father had called wine "the good creature of

God," expressed his fear that "the flood of excessive drinking" was about to "drown Christianity." In Pennsylvania, Anthony Benezet, a French Quaker known for his early opposition to slavery, published a vehement tract in 1774. It was entitled, "The mighty Destroyer Displayed, in some account of the Dreadful Havock made by the mistaken Use as well as Abuse of DISTILLED SPIRITUOUS LIQUORS, by a Lover of Mankind." With a logic that surely inspired his more illustrious student, Temperance leader Benjamin Rush, Benezet identified spiritous liquors as the primary source of alcohol-related problems.

This was to become a dominant theme of the American Temperance Movement in the early 1800s, but prophetic chords were struck well before the Revolutionary War. Between 1730 and 1770, a myriad of social problems beset the Colonies. Economic development and urbanization were accompanied by increasing poverty, unemployment and crime. Many of these emerging social problems were ascribed to drunkenness and most particularly to rum. The period of development of the rum industry in New England (1720-1770) was paralleled by a "gin plague" among the urban poor in London. Gin was hailed as the "vampyre of the Nation" and was also blamed for poverty and crime. In America, "Demon" rum was seen as a "most terrible scourge" of society. This simplistic scapegoating of an intoxicant, or a minority group, now seems a predictable accompaniment of social unrest and economic problems. The basic scenario has been repeated often — opium, cocaine, marijuana, alcohol, each takes its turn as demon for a day.

RUM, REVOLUTION AND WHISKEY

By the eve of the Revolution, rum was a significant and profitable part of colonial commerce. Rum was traded extensively for flour from the Middle Colonies, tobacco from Virginia, pitch and turpentine from North Carolina. In many respects, rum was the currency of the time. The extent to which New England rum was important for export is unknown. It is widely believed that after 1730, New England rum was exchanged for West African slaves, who in turn were sold for West Indian molasses. Benezet, self-labeled "lover of mankind," also lamented the nefarious influence of the rum trade on the inhabitants of West Africa. This practice, known as the "triangle trade," may have been somewhat exaggerated by the rhetoric of the times and by later historians. More recent analysts of the rum trade conclude that since the Colonists consumed about 87 percent of the available rum themselves, it was not important as an export.

The Revolutionary War helped to popularize another distilled spirit, whiskey. In the early Colonial period, grain-based whiskey was distilled solely for home use or local consumption. The British blockade during the Revolution cut off the molasses and sugar cane imports, essential for making rum. Domestic production of whiskey was stimulated to meet the demand for distilled spirits and for provisions for the Revolutionary Army. The commercial value of whiskey as a competitor to rum was realized even before this period of national crisis and shortage of rum. In the frontier settlements beyond the Appalachians, whiskey production became a preferred way to use surplus grains. Whiskey traveled well over the poor mountain roads and could be bartered more easily than grain.

Many vignettes illustrate the importance of rum and whiskey in the Revolutionary War. The rigorous winter at Valley Forge was described by Washington's laconic troops in this refrain: "No pay, no clothes, no provisions, no rum." Medicinal and moral considerations aside, soldiers without rum may have been dispirited in several ways. It is now known that alcohol affects several hormones believed to be important in the expression of sexual and aggressive behaviors (Chapter 11). We can only speculate about the extent to which intoxication was a conscious plan by the military strategists of the time. Drinking apparently did not seriously interfere with the battle, and may have aided victory. An account of a voyage of the U.S.S. *Constitution*, probably apocryphal, illustrates some prevailing patterns of alcohol use.

The following is taken from the United States Naval Institute proceedings and is an account of a voyage taken on 23 August 1779, when the U.S.S. *Constitution* set sail from Boston:

She left with 475 officers and men, 48,600 gallons of fresh water, 7,400 cannon shot, 11,600 pounds of black powder, and 79,400 gallons of rum on board. Her mission was to destroy and harass English shipping.

Making Jamaica on 6 October, she took on 826 pounds of flour and

68,300 gallons of rum. Then she headed for the Azores, arriving there on 12 November. She provisioned with 550 pounds of beef and 64,300 gallons of Portuguese wine. On the 18th of November she set sail for England.

In the ensuing days, she defeated five British Men-of-War and captured and scuttled twelve English merchant men, salvaging only their rum. By 27 January her powder and shot were exhausted.

Unarmed, she made a raid up the Firth of Clyde. Her landing party captured a whiskey distillery and transferred 40,000 gallons to board by dawn. Then she headed home.

The Constitution arrived in Boston harbor on 20 February 1780, with no cannon shot, no powder, no food, no rum, no whiskey but with 48,000 gallons of stagnant water.

A disdain for water, stagnant or otherwise, may well have been endemic. In any event, the popularity of whiskey did not decline after the War for Independence. In fact, the economic power of the distillation industry was so great that Treasury Secretary Alexander Hamilton saw the whiskey makers as a good potential source of revenue to pay the enormous debt inherited by the young Republic. In 1791, Congress enacted an excise tax on distilled spirits, a tax that fell heavily on the mountain distillers of western Pennsylvania. The ensuing violence was known as the Pennsylvania Whiskey Rebellion. Federal agents who attempted to collect the tax were resisted with vigor. Some were tarred and feathered, others' homes were burned. It was not until 1794 that President Washington decided to intervene by mustering a federal army of twelve thousand troops (rather more than he had led in the Revolution). Washington's army, accompanied by a determined Alexander Hamilton, advanced into Pennsylvania and was victorious. Not a shot was fired by either side. According to some accounts, a ragtag army of seven thousand intoxicated whiskey rebels met the federal troops. According to others, the rebels vanished before the army arrived. Jefferson wryly commented that "an insurrection was announced and proclaimed and armed against, but never could be found." Jefferson recommended repeal of the hated excise tax, and Congress so acted in 1802.

CHAPTER 3

Loss of Innocence
and the Call to Reform

In the wake of the Revolution, tumultuous social change and turmoil affected every segment of America. The new Republic gradually shed the unwanted trappings of British colonialism and progressed steadily toward the establishment of a new social order. The path of progress was often meandering and erratic, sometimes misdirected, but never static. This period of political, social and economic transition was accompanied by changes in drinking customs and attitudes toward alcohol. Increased public drunkenness was met with mounting public disapproval. The struggle toward an orderly society took many forms, and the temperance reform movement was one dimension of that process.

By 1790, the estimated alcohol consumption by each adult had reached nearly six gallons *absolute* alcohol per year, more than twice the estimated level of consumption in America today.

Distilled spirits accounted for a greater proportion of America's total consumption than fermented beverages for the first time in the nation's history. As the demand for whiskey constantly increased and rum pushed the national consumption beyond seven gallons per adult ten years later, America was aptly described as a "nation of drunkards." Accounts by physicians, travelers, ministers and newspaper correspondents indicated that the young Republic had somehow lost its innocence in the social turmoil following the Revolutionary War. Alcohol abuse was increasingly denounced as a cause of poverty and disease. The national drinking binge, which continued into the 1830s, was an important catalyst to one of the most powerful and enduring social movements in the history of America. To understand the origins of the American Temperance Movement and how it irrevocably modified our national drinking patterns, it is important to review the influence of prominent moral leaders like Benjamin Rush and the Reverend Lyman Beecher.

Universally acknowledged as the father of the American Temperance Movement, Rush epitomized many of the qualities that were later to characterize the early wave of temperance reform that grew progressively up to the 1850s: moderation, patriotism and a genuine concern for public health. Born near Philadelphia in 1745, educated at Princeton, London and Paris, Rush was considered one of America's foremost physicians at the time of the Revolutionary War. In 1774, he helped organize the first antislavery society in America, and the following year he signed the Declaration of Independence as a member of the Continental Congress. As Army Surgeon-General during the Revolution, he published a pamphlet that noted the destructive influence of alcohol. In 1784, he published another essay, "An Inquiry into the Effects of Ardent Spirits Upon the Human Minde and Body," which had a far-reaching influence on public opinion. Rush distinguished between the relative evils of fermented beverages and distilled liquors. As to the pernicious effects of the latter on his whiskey- and rum-drinking contemporaries, he spoke in no uncertain terms: "In folly it causes him to resemble a calf; in stupidity, an ass; in roaring, a mad bull; in quarreling and fighting, a dog; in cruelty, a tiger; in fetor, a skunk; in filthiness, a hog; and in obscenity, a he-goat."

Rush was one of the first to label intemperance a disease, and to enumerate the symptoms of tolerance and alcohol withdrawal that now form the basis of the syndrome of alcohol dependence. To combat the physical and social effects of intemperance, Rush advocated a plan of action that served as a blueprint for Temperance reform long after his death in 1813: active involvement of church leaders, petitions to limit the number of taverns, heavy taxes on liquor, and consistent sanctions against public intoxication. Reprinted almost continuously throughout the nineteenth century, Rush's pamphlet served as an inspiration to subsequent reformers like Lyman Beecher, and as the scientific foundation of the first Temperance organizations.

EARLY TEMPERANCE SOCIETIES

One sympathetic student of Rush's advice and admonitions, country doctor Billy J. Clark of Moreau, New York, was so inspired by his reading of the venerable pamphlet that in 1808 he founded what is thought to be the first Temperance organization in America. Clark conceived of a Temperance organization that would involve both a social compact and a moral covenant. With the moral support and assistance of his minister, Clark founded the Union Temperance Society. Reacting in part to the seasonal binges of local woodcutters and lumberjacks, this voluntary organization applied Rush's theories for the first time to alter the prevailing drinking customs of a community. In the organization's charter its members merely pledged to abstain for one year from rum, gin, whiskey and wine "except by the advice of a physician, or in case of actual disease, also excepting wine at public dinners." The penalty for drinking was twenty-five cents; for intoxication, fifty cents. When the trial period had ended, the abstainers met again to voice their surprise that houses could be built, barns could be raised, grain could be harvested, christenings could be enjoyed and the dead could be mourned without the constant accompaniment of distilled spirits. Members were particularly taken by the belief that labor efficiency could be improved by renouncing the daily alcohol rations given on the job in the morning and afternoon, and it was decided to emphasize this point in a public education campaign initiated by means of pamphlets and broadsides. Appealing to the scientific authority of Dr. Rush, the members urged farmers to abandon the practice of providing free liquor to their workers.

From these meager beginnings one can discern the prototype of action that was soon to revolutionize American drinking patterns: local citizens organized into a voluntary society, public commitment of each member to a pledge of partial or total abstinence from alcohol, regular meetings to plan action and reinforce solidarity, and appeals to the public through speeches and pamphlets.

It appears that the clergy played a somewhat limited role in the early Temperance societies until about 1815. What is alleged to be the first Temperance sermon was given in Washington, Connecticut, in 1805, and it is a stirring prelude to the later integration of Temperance and the Protestant revival movement. The sermon was prompted by the death of a drunken vagrant who froze in the snow only a mile from the church. The Reverend Ebenezer Porter warned his congregation of the evils of drink, citing Isaiah V: 11: "Woe unto them that rise up early in the morning, that they may follow strong drink; that continue until night, till wine inflame them." He concluded with a pessimistic reflection on the drunkenness of the time, saying, "Probably this infant country has reached a maturity in this shameful vice which is without parallel in the history of the world. Probably no nation, ancient or modern, in proportion to its whole population, ever had so many male and female drunkards as this. Certainly in no other have the means of intoxication been procured with so much facility and used with so little restraint by all sorts of people." Over the ensuing decade, these opinions came to be more generally shared.

The principles and policies of Temperance reform did not have an immediate effect on American drinking problems. But Temperance activity did seem to take root wherever the destructive effects of excessive drinking were most visible. It is also likely that the social disruption, economic vicissitudes, and political turmoil that followed the Revolution and the War of 1812 contributed to the increase in drinking and drunkenness and also prompted efforts to reestablish the forces of public order. At a time when public morals were in decline, respect for the clergy was low, and the power of the New England Federalists was eroding, Temperance reform may have provided a convenient instrument to reassert conventional values.

It is therefore not surprising that the Temperance Movement in its early form was received most sympathetically in the more conservative Eastern seaboard cities like Boston. It was here in 1813 that an eminent group of Congregational clergy, Federalist politicians, and local business leaders met at the State House to form the Massachusetts Society for the Suppression of Intemperance (MSSI). As stated in the organization charter, their purpose was to "discountenance and suppress the too free use of ardent spirits, and its kindred vices, profaneness and gaming; and to encourage and promote temperance and general morality." To accomplish these goals the Massachusetts Society focused on the drinking habits of the "labouring classes." Believing that the "too free use of ardent spirits" was encouraged by employers who supplied liquor to their workers, the

MSSI sought to eliminate the liquor ration by appealing directly to merchants, sea captains, farm owners, mechanics and manufacturers. Pamphlets addressed to employers were careful not to condemn the moderate use of alcoholic beverages, but argued that immoderate drinking reduced worker productivity and caused accidents and absenteeism. It was pointed out that contrary to popular opinion alcohol was not a stimulant, but rather a dangerous depressant, which produced fatigue and idleness.

Another tactic employed by the MSSI was legal suasion. Convinced that existing restrictions on the sale of alcohol were not enforced adequately, Temperance reformers pressured local authorities to act against innkeepers who sold liquor to minors and habitual drunks. Another target was the illegal dramshops, small retail stores that sold liquor by the drink in addition to their usual merchandise. To gain support for their appeals, the first Temperance literature blamed spiritous liquors for increases in crime, unemployment, poverty and urban decay that characterized many port cities, particularly following the War of 1812. In addition to the Congregational clergy and Federalist politicians impressed by these arguments, the first Temperance organizations also attracted lawyers, physicians, merchants and country landowners. Membership grew to its highest level during the five years after the War of 1812, when numerous local auxiliaries were established throughout Massachusetts and Connecticut. Similar societies were formed in Rhode Island, New Hampshire, Vermont, Maine, New York and Pennsylvania.

In some respects, early Temperance reform can be seen as an attempt to minimize or prevent social problems associated with the use of alcohol, much like contemporary attempts to deal with teenage alcohol abuse or drunken driving. However, Temperance reform also can be interpreted as a moral crusade conducted by political and religious elitists who were prejudiced against an increasingly disorderly and irreligious segment of society. While there may be elements of truth in both interpretations, it is important to note the social context of drinking in the early Republic.

DRINKING IN THE NEW CENTURY

During the decade before and after 1800, alcohol consumption continued to increase beyond the post-Revolution levels. Estimates of consumption derived from production statistics, tax reports and customs receipts suggest that by 1810, American adults over age fifteen were drinking a yearly average of approximately seven gallons of absolute alcohol. This high level of consumption remained stable until 1830. Seven gallons per year is equivalent to about 2.5 ounces of alcohol per adult per day, which equals two whiskey cocktails or a quart of hard cider. These estimates are close to that given in 1829 by the Secretary of War, who wrote that 75 percent of the country's laborers drink four ounces or more of spirits a day.

Rum, whiskey and cider remained the most popular alcoholic beverages, but their relative contributions to what has been aptly termed the "national binge" were changing rather dramatically during this period. Consumption of both cider and rum declined significantly, only to be replaced by an even stronger taste for whiskey. By 1830, distilled liquors accounted for approximately 65 percent of the absolute alcohol consumed by Americans. Since wine and beer accounted for only a very small proportion of the alcohol consumed, and cider contained relatively little absolute alcohol, Temperance reformers were probably correct in identifying liquor as the major cause of alcohol problems.

What accounted for this significant change in national beverage preferences? Historians have devoted relatively little attention to the drinking patterns of this period, and expressed more interest in the motives of the reformers than of the reformed. Those who have ventured to explain the national switch to whiskey have typically favored some version of tension-reduction theory. Like its psychological counterpart so often used to explain individual drinking behavior, tension reduction theory sees societal stress as the major cause of the early-nineteenth-century whiskey binge. Unfortunately, there is very little evidence to suggest that living in the early Republic was any more stressful than it was in Colonial times or than it is today.

A more likely interpretation lies in the dramatic changes that were taking place in American society and in the alcoholic beverage industry as a consequence of technological advances, improved transportation and changing agricultural markets. These economic changes occurred at a time when American dietary habits favored the more "hygienic" alcoholic beverages over impure water, and social customs openly sanctioned the free use of alcoholic beverages at both work and leisure. These several factors made possible the mass marketing of cheap whiskey to a young and restless population. Distillation, long a profitable, locally based industry in the East, became increasingly popular on the Midwestern frontier because it provided a ready solution to an embarrassing problem: agricultural surplus. By transforming cumbersome fruit and grain into more easily transportable distilled spirits, farmers could find new markets for their surplus and increase their profits enormously. During the early part of the nineteenth century, there were significant advances in the technology of portable stills, which made it possible for every frontier farmer to set up a cheap, efficient home distillery. Jefferson's general trade embargo in 1807, intended to avert war with Britain, also helped to cripple the Eastern rum industry. With the development of the canal system, steamboats and the railroad, Western whiskey was able to penetrate the Eastern markets at a price that made it highly competitive with domestic rum and brandy. By 1820 there was a whiskey glut on the frontier. Prices were so low that even the poorest agricultural laborer could afford to drink liquor daily, if it was not already provided free by his employer. Distillation became concentrated in the areas like southwest Ohio, upstate New York, and southwestern Pennsylvania, where distance from the population centers and proximity to the grain belt made distillation the most economically feasible alternative. As new markets were found and demand increased, there was a concomitant economic concentration of the industry into larger and larger producers. Ironically, the same economic conditions that made possible this second "whiskey rebellion" also served as an essential prerequisite for the New England industrial revolution, a development which was to signal the necessity for a more disciplined and sober work force.

THE SOCIAL IMPACT OF DRUNKENNESS

Given a thirst for alcohol unparalleled in the nation's history, it is important to inquire about the consequences of this drinking tradition on public health and public order. Was Temperance reform necessary to proceed with the national agenda of industrial development, westward expansion and assimilation of the immigrant hoards? To what extent was the liquor consumption of the period the "cause" of poverty, crime, insanity and a host of other social problems? Finally, were the vigorous and at times extreme measures of the Temperance reformers in any way instrumental in reducing the prevalence of alcohol abuse and alcoholism?'

Unfortunately, the evidence needed to answer these questions is fragmentary at best. Organized police departments capable of maintaining reliable records and of dealing with widespread public intoxication did not emerge until the 1820s, and even then only in a few of the largest cities. The first asylums for the insane capable of treating alcoholics date from the same period, while public charity organizations devoted to serving the poor and homeless were not established until the 1830s. Nevertheless, the early records of these first public institutions do suggest a high prevalence of alcohol-related problems. In Boston, New York and Philadelphia, drunkenness accounted for more criminal convictions than any other category of offense. At Philadelphia's Pennsylvania Hospital, and at Boston's McLean Asylum, the category of "insanity caused by intemperance" accounted for more admissions than any other type of mental disorder. An analysis of Boston's bills of mortality for the two decades preceding 1832 revealed that only consumption, "lung fever" and dysentery were considered to have caused more deaths than intemperance. Many physicians of the period believed that intemperance contributed to an individual's susceptibility to yellow fever, cholera and consumption. This was a consistent theme in letters and articles published in the medical journals of the time. Physicians also expressed concern about the drinking habits of many members of their own profession, whose reputation for insobriety rivaled that of the clergy. "He is an excellent doctor if called when sober," was a common caveat heard during informal discussions of a physician's competence.

If intemperance among physicians and the clergy constituted a national scandal, the heavy drinking habits of other occupations provided further support for the Temperance reformers' argument. Stage drivers, teamsters, fur traders, lumberjacks and sailors were also known for heavy drinking sprees, which usually followed long periods of social isolation and relative abstinence. Unlike the communal celebrations of Colonial society, drinking now took on the character of an end in itself. To be sure, the practice of daily drams, election day treating and Fourth of July frolics continued unabated in the agricultural towns and villages where the majority of the population still resided. But early in the nineteenth century, the rapid economic developments were accompanied by conditions that were more conducive to binge drinking. These included a more mobile work force, the dissolution of traditional religious institutions, rapid urban growth and the breakdown of Colonial social cohesion. Combined with the increasing popularity of whiskey, these conditions contributed to a more explosive, disruptive and visible style of drinking; one which invited increased private concern and public censure.

A fascinating example of the way in which binge drinking and pathological intoxication came to be viewed as a national disgrace can be found in a curious essay published in 1818, entitled "The Drunkard's Looking Glass." It consists of a series of case histories, each a short vignette designed to show "that there is no folly, no madness ever yet committed out of Bedlam, which a drunken man is not perfectly capable of." One story depicts the adventures of a group of young students, with "their noodles full of whiskey" out for a night of wild "capers." First they steal the tavern sign which reads, "Cakes and Beer for Sale Here," and place it on the parson's door. Then they push a "squadron of Carts" over the wharves and into the river. Threatened with arrest, the boys are forced to pay hundreds of dollars in restitution to the car-men.

Another story describes the misadventures of a clergyman, who, getting "a drop in his eye," misses his chance to become the parson of a wealthy congregation. It seems that the night before his first sermon, some "wicked young fellows" took him to the tavern under pretense of an oyster supper. Instead, they all got drunk, "the man of God, as gay and frisky as the best among them." After a variety of excesses, a game of follow-the-leader commenced, each one of the party throwing articles of clothing into the fire until all of them were "naked as the day they were born." As the conspirators retire to their rooms, the clergyman is left to explain the commotion to the tavernkeeper and his wife. Too embarrassed to face the congregation, he leaves the village the next morning, "accompanied with reflections."

As the cases continue they become more tragic. A young lawyer in a "drunken frolic" is disinherited by his father. A young doctor makes himself a patient after challenging another drunk to a duel. Youthful revelers are involved in fatal riding accidents, a drunken sailor mistakes a powder keg for a candle stand and blows up his ship, and a number of other unfortunates are driven insane during drunken episodes.

As absurd as they seem, these stories were used to great effect by

Temperance reformers in their campaign to portray the effects of drunkenness as something far more insidious than wholesome fun. By showing how crime, insanity, accidents and inappropriate behavior were the direct result of drinking, the author struck a sensitive chord in an audience that usually had firsthand experience with the consequences of alcohol abuse.

THE ASCENDANCE OF TEMPERANCE AND THE PROTESTANT REVIVAL

Despite the early popularity of Temperance reform, and the apparent social need, by the 1820s there was a waning of interest in the Temperance Movement. Social elitism, ineffective tactics and the internal squabbling among political and religious factions have been blamed for the apparent failure of the early Temperance Movement. It is also possible that the nation was not yet ready for reform, and the reformers were not yet equal to the challenge before them. Yet only ten years later, Temperance reemerged as a major social movement, its power and influence increased almost in direct proportion to its companion religious reform movement, the Great Awakening. This Protestant revival movement began about 1815 and continued unabated throughout the next two decades. The Great Awakening spread throughout the country, through the vehicles of enthusiastic revival meetings, home missionary activities and mass circulation of printed propaganda. Just as these religious evangelists saw alcohol as a major cause of immorality and religious backsliding, Temperance reformers saw in revivalism a major way to fuel the engines of their waning movement. Nowhere is this confluence of historical forces more apparent than in the life of the Reverend Lyman Beecher.

Born in 1775, Beecher studied divinity at Yale before embarking on a long and controversial career that included published tracts on the menace of Catholicism, public censure for his opposition to slavery, charges of heresy, leadership of a dissident Presbyterian group, and the incitement of an evangelical crusade against alcohol. In his *Autobiography* Beecher traces his opposition to liquor to his experience with a drunken "grog-seller" who had defrauded and corrupted the Montauk Indians through the "trafficking of alcohol." In 1810, he became pastor of the Congregational Church at Litchfield, Connecticut, and two years later he drafted a seminal report for a state committee on the question: "How can drunkenness be prevented?" His answer included the following modest proposals: abstention from liquor at ecclesiastical meetings, and in family settings; cessation of unlawful purchase and sale of spirits by church members; ministers' discourses on intemperance; and circulation of Dr. Rush's pamphlet; substitution of nutritious drinks and other compensation for the liquor employers typically provided for their employees; formation of voluntary associations to enforce existing laws.

However, fourteen years passed before Lyman Beecher's message fell upon sympathetic ears. A chance encounter at the home of his most esteemed parishioner, whom he found drunk in bed, led to the preparation of his famous "Six Sermons on the Nature, Occasions, Signs, Evils and Remedy of Intemperance." Delivered in 1825 on six successive Sabbaths and published formally one year later, Beecher's rhetoric ignited New England into one of the most significant social movements of the century. He was one of the first to urge the outright prohibition of distilled beverages. "Let it be engraven upon the heart of every man," he declared, "that the daily use of ardent spirits, in any form, or in any degree, is intemperance." In his discussion of the "Signs of Intemperance," Beecher, like Benjamin Rush before him, characterized intemperance in both physical and moral terms: "Intemperance is a disease as well as a crime, and were any other disease as contagious, of as marked symptoms, and as mortal, to pervade the land, it would create universal consternation ... to drink daily, at stated times, any quantity of ardent spirits is intemperance, or to drink periodically, as often as days ... is intemperance ... the beginning of a habit, which cannot fail to generate disease, and will not be pursued by one hundred men without producing many drunkards." Here Beecher seems to be suggesting that regular drinking and frequent intoxication, if not a disease in itself, is at least the kind of habitual behavior pattern that can cause disease, especially among a significant proportion of those who drink. In his final three sermons on "The Remedy of Intemperance," Beecher again outlined the tactics that were soon to become the modus operandi of the new Temperance revival: 1) "the banishment of ardent spirits from the list of lawful articles of commerce, by a correct and efficient public sentiment; such as has turned slavery out of half our land;" 2) the dissemination of information on intemperance; 3) the formation of an organization to carry on this work; 4) the support of churches and of physicians; 5) renunciation of spirits as a medicinal drug.

Beecher's sermons were widely circulated by both Temperance organizations and by the churches. His appeal for "the banishment of ardent spirits from the list of lawful articles of commerce" was soon taken up by the Reverend Justin Edwards, pastor of the Park Street Church in Boston. In 1826, Edwards convened a state convention to study ways of combatting intemperance. Represented largely by Massachusetts clergymen, the group adopted a constitution and baptized their organization the American Society for the Promotion of Temperance. Thus began a new and significant stage in the history of alcohol control in America. It differed from previous reform efforts in three major respects. First, the fight against intemperance took on the mantle of divine inspiration. Second, missionary tactics were used to preach the gospel and convert the unbeliever. Finally, there was to be no compromise with the source of evil. Total abstinence from all *distilled* beverages was established as the formal badge of the temperate.

In 1827 the General Assembly of the Presbyterian Church passed resolutions encouraging support of the society, whose name had now been abbreviated to the American Temperance Society. The following year the Methodist General Conference advised its members to abandon any involvement in the manufacture and sale of spirits, and to discontinue the practice of giving liquor gratuitously to employees. Within the next five years, the Congregationalists and Southern Baptists also took firm positions on Temperance.

With such strong involvement of the nation's Protestant churches it is not surprising that the "Temperance crusade" took on many of the characteristics of a religious revival, both in its ideology and in its tactics. Convinced that ardent spirits were a major obstacle to the success of the church's mission, Temperance was preached as a logical corollary to the gospel. And at a time when evangelical American missionaries were spreading the "good news" of the Bible worldwide, the Temperance message was carried to sympathetic listeners in almost all of the European nations. By the 1830s Temperance organizations had been established in England, Norway, Sweden, Finland and Ireland. While Temperance reform in each nation had its own particular origins and evolution, the example and influence of Americans were acknowledged as an important catalyst.

Unlike previous reform movements, the attention of these evangelical Temperance workers was directed, not so much at the drunkard or alcoholic, as at themselves and those involved in the liquor trade. Church members involved in the distillation, transportation or sale of alcohol were singled out for particular censure. To support their arguments that the liquor business was incompatible with Christian morality, Temperance workers presented an endless array of statistics and case histories designed to prove that spiritous liquors were directly or indirectly responsible for the majority of crime, poverty, disease and insanity in the land. In many cases, the data presented were a truthful representation of official statistics, although the conclusions drawn from these statistics were often based on a rather crude logic. For example, the claim that drunkenness was a major cause of crime was true only to the extent that public intoxication accounted for the majority of "crimes" committed in the nation's major cities, such as New York, Philadelphia and Boston. And while intemperance was correctly cited as a major cause of commitment to the nation's early insane asylums, the "insanity" of the intemperant patient was often no more than a brief episode of severe alcohol withdrawal or, in some instances, delirium tremens. Finally, the growing problem of urban poverty was thought to be caused by liquor, not so much because respectable citizens were brought to ruin by drink, but rather because many of the heaviest drinkers were unemployed, unmarried, foreign-born males who congregated in urban drinking establishments for want of better alternatives.

TEMPERANCE AND THE TACTICS OF SPECIAL-INTEREST POLITICS

While it would be a mistake to credit a single organization with the accomplishments of a major social movement, in many respects the activities initiated by the American Temperance Society and its affiliates constituted a major innovation in the politics of special interest groups. By organizing large numbers of energetic, middle-class men and women into locally based voluntary interest groups, Temperance reformers were able to raise funds and apply pressure directly on the political process. Exploiting the power of the printed word, a massive propaganda campaign was initiated through the distribution of millions of cheap Temperance tracts. After a large special interest network was established, the literature of protest continued through the mailing and distribution of annual reports, Temperance almanacs, special circulars, weekly newspapers, and Temperance journals. This literature served to communicate Temperance ideology in attractive, pithy terms while at the same time informing its readers of the "progress of Temperance reform" throughout the land.

Eschewing direct political involvement at first, Temperance leaders wisely focused on the reform of drinking habits within their own ranks, and on moral suasion outside their ranks. Surprisingly, the master strategy of the American Temperance Society did not include the reform of alcoholics. Instead, the reformers attempted to reduce the demand for alcohol and to restrict its availability by attending to the supply of liquor to the community and the moderate drinking habits of its members.

By 1835 more than thirty-five thousand ministers had signed the abstinence pledge. But rather than a sign of progress of Temperance reform, some critics saw the enthusiastic involvement of the churches as a self-serving effort to recruit members. While there may be an element of truth to this notion, the enormous degree of popular support evidenced during this period suggests that Temperance reform reached far beyond the Sunday revival meeting. In 1829 approximately one thousand societies were affiliated with the American Temperance Society. These locally autonomous voluntary organizations claimed to have more than a hundred thousand active members. Two years later membership had doubled to twenty-two hundred affiliates with a total of one hundred seventy thousand adherents. By 1835 membership had surpassed a million men, women and children. New England, with one-sixth of the U.S. population, supported one-third of the Temperance societies. In growing industrial towns like Worcester, Massachusetts, as much as fifty percent of the adult population subscribed to the local Temperance organization. By this time Temperance had become independent of the churches in its appeal to various special-interest groups. Legislative abstinence societies were established both in the Congress and in many of the state houses. Workingman's abstinence societies were formed in the emerging factory towns. Young people's Temperance societies and women's auxiliaries were also involved in the political arena for the first time. Although women were not allowed to assume positions of leadership in the organizational hierarchy, they were admitted on an equal footing and quickly demonstrated their ability to lead boycotts and organize support for the families of alcoholics.

As the Temperance Movement gained momentum during the 1830s, its support broadened to include a social base much more diverse than evangelical Christians. University groups were established at Yale, Brown and Dartmouth. Medical Temperance societies involved the most prominent physicians of the period in research, writing and public speaking, thereby conferring an aura of scientific respectability and academic prestige on the movement. The support of the medical profession was particularly strong, although it has been suggested that Temperance was useful to the "regulars" as a means of consolidating their professional position and improving their tarnished image.

But if women, the clergy and the medical profession were the only supporters of the new Temperance crusade of the 1830s, it is likely that the movement would have merely repeated the failures of its predecessor, the Massachusetts Society for the Suppression of Intemperance. What distinguished the new Temperance reformers from the old was that religious evangelists were able to combine forces with an ambitious and upwardly mobile group of industrial promoters who were leading the crest of the industrial revolution. By enforcing total abstinence on their workers, encouraging their participation in Temperance agitation, and by assuming positions of leadership alongside the clergy and physicians, these newly successful entrepreneurs, manufacturers and commercial farmers saw Temperance as a means of improving the productivity and living conditions of the working classes.

During the 1830s when Temperance partisans entered the political arena to oppose liquor licensing laws, they did not conform to the traditional stereotype of the conservative, elitist, religious fanatic conveyed in superficial historical accounts. In the emerging factory towns like Worcester, Massachusetts, opposition to the issuing of further liquor licenses came from manufacturers, mechanics, skilled tradesmen, farm managers as well as the clergy. Support for the existing license system, on the other hand, was limited to a more specialized group of taverners, hoteliers, grocers and merchants whose businesses were linked directly or indirectly to the liquor trade.

One interpretation of this alignment of interest groups lies in its relationship to the emerging economic order. The rapid development of the new industrial economy had prompted an important change in attitudes toward work and leisure. With the introduction of new and more sophisticated machinery on the farms and in the factories, drunken and inattentive workers could no longer be tolerated. Drunkenness compromised the employee's safety and the employer's profits. Combining the economic liabilities of drinking with the image of liquor as a destroyer of family life and morality was enough to make Temperance attractive to a wide spectrum of society.

Although a significant and well-organized segment of the population was behind the Temperance cause, the degree to which the great Temperance reform of the 1830s accomplished its goals remains a controversial question. A belief in the perfectability of man and the power of truth led Temperance activists to undertake a massive public education campaign as the primary instrument of moral suasion. The most striking result of this campaign is the dramatic change in per capita alcohol consumption, which began about 1830. Consumption of both liquor and cider plunged to levels so low that by 1849 American adults were drinking 75 percent less absolute alcohol than they had drunk just two decades before. This apparent decline in consumption coincided with important changes in drinking customs. By this time manufacturing establishments throughout New England and New York had stopped providing free liquor to manual workers, and this could account for a significant reduction in yearly per capita consumption. On the farms, haying, harvesting, barn-raising, plowing and planting were being conducted without the assistance of rum, whiskey and brandy for the first time. In cities, coffee and mild wines replaced distilled spirits on holidays and New Year's social visits, an innovation first organized in 1834 by education reformer Horace Mann.

Under pressure from the churches and their Temperance allies, and abetted by the reduction in demand for spirits, numerous distilleries and grog shops closed their doors. In 1830 the town of Canton, Connecticut, boasted forty distilleries. By 1840 their number had been reduced to two. Similarly the city of Boston supported one hundred eighteen distilleries in 1834. By 1838 this number had declined to forty-six. While competition from Western distillers and the concentration of the industry into larger units undoubtedly contributed to the decline in local distilleries, it is likely that pressures within the local communities were in part responsible for the waning of small-scale distillation.

Another sign of the influence of Temperance agitation was seen in the annual reports of hospital administrators and city officials. At the McLean Asylum and the Worcester State Hospital the proportion of persons admitted for reasons connected with intemperance dropped significantly during this period. In the City of Boston the number of arrests for drunkenness declined from the alarming levels of the 1820s.

THE TEETOTALERS: FROM TEMPERANCE TO TOTAL ABSTINENCE

By many accounts, the Temperance Movement had reached its peak in enthusiasm and effectiveness by 1835. At about this time, there was increasing dissension within the movement over the issue of *fermented* beverages. One argument frequently advanced by the advocates of total abstinence was that the moral example of the temperate was the only way to reform the intemperate. In addition, the availability of fermented beverages would constitute a constant source of temptation to those who could not control their drinking. Another argument against fermented beverages was framed in terms of the central role of alcohol in the addiction process. Far from providing a healthy substitute for spiritous liquors, fermented beverages were suspected of generating an insidious need for stronger "stimulants." As the debate raged, those advocating total abstinence from all alcoholic beverages began to sign their Temperance pledges with the addition of the letters "T.A.," thereby gaining the now familiar sobriquet "teetotaler." Gradually the teetotalers gained ascendancy, a shift that led to the departure of their companion reformers who loved wine, beer and cider.

Total abstinence brought with it another shift in tactics; moral suasion gave way to legal suasion as a primary instrument in Temperance reform. In 1838 state legislatures in New England and the Middle Atlantic states were deluged by petitions asking state governments to withdraw approval from the liquor business. In that same year, Massachusetts Temperance advocates succeeded in passing the "15 Gallon Law," the first attempt at total prohibition. This statute forbade the sale of spiritous liquors in quantities less than fifteen gallons and effectively prohibited the sale of distilled beverages in taverns and retail shops. Opposition to the law was framed in terms of its class bias — it could be interpreted as favoring the rich who could afford to stockpile their liquor. Tavern-keepers were particularly incensed by this infringement on their ability to conduct business. Many supporters of Temperance reform were ambivalent about the law, first because it was difficult to enforce, second because it was too far in advance of public sentiment. While many retailers were convicted and fined for violating the law, the majority operated with impunity in defiance of a law the authorities in Boston were loath to enforce. By 1840, the law was regretted by both the public and legislature, and it was repealed on the recommendation of a Democratic governor known for his sympathy for Temperance.

Despite major attention devoted by Temperance advocates to statewide prohibition laws, most significant restrictions on drinking were imposed through passage of "local option" laws. These laws passed by most of the state legislatures allowed local town meetings to decide whether alcoholic beverages could be sold. During the 1840s Temperance sentiment was translated into local prohibition. The majority of state legislatures, reluctant to pass statewide prohibition laws, voted to pass this responsibility to the local governments. Local option laws were then passed in towns and cities to such an extent that the majority of jurisdictions in Pennsylvania, Rhode Island, Connecticut, Wisconsin, Michigan and Iowa went dry. These victories proved to be fleeting, however, as many towns, cities and counties reversed their votes from one year to the next.

Temperance agitation on the legal front reached its peak in the early 1850s with the onset of what has been termed the "first great wave" of statewide prohibition. It began in Maine under the able leadership of Neal Dow, a prominent Portland merchant who had long been active in Temperance work. Convinced that liquor was the major cause of labor inefficiency and social problems, Dow allied himself with the Washingtonians to enforce prohibition in Portland in 1842. Realizing that local boundaries provided a poor defense against the liquor traffic and traffickers, Dow initiated a ten-year campaign to pass a statewide prohibition statute. On June 2, 1851, such a law was enacted, forbidding the manufacture and sale of intoxicating liquors within the confines of Maine.

The passage of the Maine law electrified the nation. Dow, who personified this new turn in the Temperance Movement, was called the Prophet of Prohibition by some, and the Prince of Fanatics by others. Dow epitomized a new type of moral stewardship in the Temperance Movement: men who were actively opposed to liquor, slavery and the immigration of Roman Catholics, not so much from a sense of moral superiority, as from an inner sense of duty to correct the evils of an imperfect but perfectable society.

The triumph of the Maine law was followed by similar statewide victories in Minnesota, Rhode Island, Massachusetts, Vermont, Michigan, Connecticut, Indiana, Delaware, New York, Iowa, Nebraska and New Hampshire. By 1855 a total of thirteen states had adopted "Maine laws." In many states amendments permitted the use of beer and wine, and nowhere was the personal use of alcohol prohibited. The intent of these laws was primarily to abolish public intoxication, not private drinking, and thereby to protect the American family from the kind of lower-class drunken disorder long identified with the infamous grog shops. As with the local option laws, however, statewide prohibition was maintained at the mercy of a fickle electorate. Many laws were repealed, most were found to be unconstitutional, and all failed to correct the social problems they were designed to address.

ALCOHOLICS AND THE TEMPERANCE MOVEMENT

It seems curious that alcoholics were excluded from Temperance reform until the later phases of the movement. The prevailing belief was "that there was little hope of the reformation of drunkards while the rum traffic continued in public and enticing forms." It was not until 1840 that reformed alcoholics became an active and powerful force in Temperance reform. At the same time as the teetotalers were pursuing prohibition through legal measures and Temperance reformers were evangelizing the middle class with moral suasion and temperance tales, the "Washingtonians" were reforming the nation's alcoholics through moral example.

The Washingtonian Revival began in a Baltimore tavern when a curious group of barroom patrons decided to send two representatives to attend a Temperance lecture. The Temperance message proved so inspiring to the group that it decided to form a new society devoted to the reform of drunkards and heavy drinkers. Just as

General Washington had delivered the Republic from the op-pression of King George III, so were the Washingtonians embarking upon a campaign to deliver America from another oppressor, King Alcohol. Unlike their more pious and bourgeois allies in Temper-ance reform, however, the Washingtonians employed an entirely new set of tactics and appealed to a different segment of the popula-tion. While their principles appealed to both alcoholics and ab-stainers from all social classes, the movement was dominated by reformed alcoholics of the lower-middle and working classes.

As the new society worked to convert drunkards to the cause of teetotalism, news of its activity spread to New York and Boston. There, speakers were invited to address public meetings which were enthusiastically promoted by sympathetic Temperance workers. The New York rallies were acclaimed as a huge success, with thou-sands of putative alcoholics signing their names to the pledge. Quickly, the organizational blueprint emerged for this new phase of Temperance reform. First, local Temperance societies would invite the Washingtonian speakers to their towns, carefully preparing the way with advance publicity and a covenient speaking forum. Once the rally had been conducted and pledges had been signed, the new adherents were organized into local societies that operated accord-ing to principles remarkably similar to those popularized by Alco-holics Anonymous a century later. Meetings were held weekly with the express goal of providing encouragement, support and advice to fellow alcoholics. This was accomplished primarily through the nar-ration of personal experiences, which highlighted their unhappy drinking careers, their conversion experience to total abstinence, and the benefits derived from sobriety.

In addition to rendering direct assistance to sick and dependent alcoholics, the Washingtonians actively spread their novel message of hope to the general population. The notion that moral and reli-gious persuasion could reform even the common drunkard was per-fectly consistent with the ideology of Temperance, a fact which explains why the Washingtonians were so eagerly received wherever they went. Inspired by dynamic platform speakers like John Gough, the "Apostle of Cold Water," and John Hawkins, who delivered as many as 2,500 addresses during the 1840s, the movement claimed the support of more than a million adherents by 1843. If the figures of the times can be believed, one hundred thousand habitual drunkards and half a million temperate drinkers had been persuaded to sign the total abstinence pledge by the Washingtonians. Although these figures may be unreliable and inflated, contemporary newspaper accounts and Temperance reports suggest that the Washingtonians may have had a salutory effect on recidivism among alcoholics, much as Alcoholics Anonymous and other self-help groups are seen as exerting an important influence today.

Within ten years, however, the movement collapsed as rapidly as it had arisen, its continuity broken by lack of systematic organization, its credibility plagued by frequent evidence of relapse among pledge signers, and its popular support reduced by criticism that it was irreligious and sensational.

ALCOHOLISM: A MORAL OR A MEDICAL DISORDER?

The rise of the Washingtonian Movement was paralleled by changes in both the conception of alcoholism and the perception of alcoholics. Displacement of the Puritan moral concept of the drunkard had begun in the late eighteenth century with Benjamin Rush's famous "Inquiry Into the Effects of Ardent Spirits." Rush's essay provided the first clinical description of the progressive nature of intemperance and, consistent with the prevailing use of distilled beverages, attributed a dominant etiological role to "ardent spirits." Later editions of the pamphlet included a graphic illustration of Rush's "Moral and Physical Thermometer," according to which different degrees of intemperance were directly proportional to the alcohol content of various popular beverages. When taken at meals in small quantities, even wine and small beer could promote "health and wealth." In contrast, various kinds of "ardent spirits," such as punch, rum, gin and brandy, were shown to be associated with specific "vices" (symptomatic behaviors such as "idleness" and "fighting"), a host of familiar physical "diseases" (e.g., tremors, polyneuropathy), and a variety of social consequences ("punishments" such as jail, hospital or poor house). In its final stages, intemperance was characterized by "drams of gin, brandy and rum" throughout the day, resulting in "epilepsy, Melancholy, palsy, apoplexy, Madness [and] Despair."

Fifty years after Rush's essay, Samuel Woodward, a prominent Massachusetts physician, outlined in a series of lectures and newspaper articles an even more sophisticated version of the disease concept of alcoholism. Like Rush, Woodward described alcoholism with a mixture of naturalistic and moralistic terms. Intemperance was seen as a "physical evil, a disease of the stomach and nervous system" characterized by tolerance ("after a time, the quantity must be increased"), psychological dependence ("a tormenting thirst, and insatiable craving") and physical withdrawal symptoms ("a sense of vacuity, faintness, and depression, which calls imperiously for a repetition of the stimulant upon which it depends"). Departing from previous views, Woodward attributes moral culpability to the external agent (alcohol) and to socially learned habits (spirits drinking), rather than to the behavior, disease or deviance resulting from dependence: "at that time *intemperance* only was considered as a *crime,* and a moderate use of ardent spirits was supposed to be salutary; while the truth is, the criminality consists in its *moderate* use; and the *intemperance* is *disease;* a man is no more to blame for intemperance, from this view of the subject, than for the gout, diseased liver, insanity, and delirium tremens — which the use of spiritous liquors also produces." Signaling an important shift in Temperance ideology, Woodward called for legal prohibition of all intoxicating beverages, arguing that it is the legal system that supports, if not encourages, excessive alcohol consumption, and that even fermented beverages induce "a love of something stronger."

The early involvement of prominent physicians like Rush and Woodward in Temperance activities provided an authoritative basis for a genre of literary propaganda collectively termed the "Temperance Tales." These appeared with increasing frequency and rhetorical fervor during the 1820 and 1830s. The genre reached its peak in the decade preceding the Civil War with the widespread dissemination of popular novels, magazine articles, pamphlets, plays, engravings and illustrations. In some cases the media presentations were synchronized, as in 1847 when George Cruikshank's influential series of engravings entitled *The Bottle* appeared simultaneously with a popular play of the same name. The story line depicted the tragic deterioration of James Latimer and family resulting from "too frequent use of the bottle." Chronic drunkenness is portrayed as the cause of family quarrels, economic privation, physical disease and antisocial behavior. The last scene, entitled "The Bottle Has Done Its Work," shows James "a hopeless maniac," interred in an institution, staring aimlessly at the bars of his padded cell.

The notion that social decline and behavior disorders were natural consequences of spirits consumption was even more vividly illustrated in Currier's famous engraving *The Drunkard's Progress*. First appearing in 1846, this tableau represented the progressive and apparently inexorable nature of intemperance as beginning with the first drink and ending with the drunkard's eventual suicide. In the popular literature of the period, autobiographical accounts and Temperance novels such as T. S. Arthur's *Ten Nights in a Bar Room* (1854) reiterated the infinite variety of social disorders and psychopathology engendered by even occasional use of spiritous liquors. The tremendous popularity of this literary genre suggests that the early disease conception of alcoholism was an ideological cornerstone of the Temperance Movement. This conception, which contains many elements of a negative stereotype, viewed intemperance as a progressive disease resulting more from the addictive nature of alcohol than from the moral weakness of the individual drinker.

Recent analyses of this stereotype have adopted a "sociology of knowledge" approach suggesting that the disease concept, as well as the Temperance Movement itself, emerged in response to social and structural changes taking place in American society. It is argued that rapid industrialization had created a need for a more disciplined work force at a time when the decline of traditional religion led the Protestant clergy to search for new ways to reassert its authority. Alcohol and alcoholism became convenient symbols of discontent with the changing social order, and the Temperance Movement became an effective vehicle for the emerging middle class to achieve social and political status.

The preoccupation of the American medical profession with intemperance, as both a social and a medical problem, can be inferred from a review of the period's medical journals. The *Boston Medical and Surgical Journal*, for example, printed over one hundred articles on alcohol-related topics between 1828 and 1860. These articles most often appeared in the form of news reports, theoretical debates and case histories, and were principally devoted to such topics as the pathology of alcohol, the advisability of alcoholic therapeutics (i.e., the medicinal value of alcohol), and the role of physicians in the Temperance Movement. Despite a general recognition that intemperance contributed to illness susceptibility, liver disease and mental disorder, there was little discussion of how intemperance was diagnosed and treated. One physician estimated that approximately 13 percent of his patients were "intemperate," and that their frequent illnesses contributed disproportionately to his fees. Of interest is the physician's definition of intemperate as "those who are sometimes drunken, or who are stupid by sottishness — whose bodily vigor is shaken, and whose mental faculties are clouded by frequent excitement of alcohol." In this rare account, behavioral and psychological symptoms predominate.

By far the most eloquent and detailed description of the medical response to intemperance is to be found in Woodward's *Essays on Asylums for Inebriates* (1838). Written at a time when Massachusetts led the country in establishing institutional treatment facilities for the mentally ill, it is not surprising that Woodward develops a similar model of "moral" treatment for the alcoholic. Woodward was superintendent at one of the first American insane asylums, at Worcester, and he alludes to his previous twelve years' experience at "two institutions in which numerous individuals, amounting to many hundreds, were confined, who by intemperance had become insane, or who had perpetrated crimes which rendered confinement necessary." This statement suggests that many of the more deviant problem drinkers were confined to the early insane asylums, in addition to those suffering from delirium tremens and the organic psychoses.

THE DECLINE OF THE TEMPERANCE MOVEMENT

By the middle 1800s, the first prohibition crusade had waned. The legal battles won by the teetotalers proved transient indeed. Public disenchantment culminated in repeal of many of the prohibition laws. Of the thirteen states with prohibition laws in 1855, only five remained in 1863. Most were found unconstitutional in judicial review. As the once unified Temperance Movement became splintered into factions ranging from teetotalers to moderate Temperance supporters, its cohesiveness and force diminished as well. The energy and passionate attention directed toward Temperance in the early 1800s was necessarily diverted to the dominant social issue of the time — slavery and the Civil War. As the impending social crisis neared and polarization of positions gained momentum, the problems associated with drunkenness seemed less significant. The nation was threatened with dissolution, and alcohol was nonpartisan and ubiquitous in the Civil War.

CHAPTER 4

Beyond the Civil War: Alcohol, Opium and Cocaine

The anguish of the American Civil War and its turbulent aftermath so disrupted the social fabric of this society that a full century was not long enough to heal all the wounds. The shattered union and divided families were reassembled within a political-legal structure, but the sense of shared obligations and reciprocal responsibilities of the Mayflower Compact had been tortuously distorted by all aspects of the war. Social unrest seems a mild euphemism for the economic and social chaos in the South, the emerging lawlessness of the Western frontier and the fragile structure of established order in the victorious North.

Reformers throughout history have alleged that unless draconian restrictions are imposed, alcohol will destroy society by so corrupting and debasing the morals of the inebriate that licentiousness, debauchery and anarchy will prevail. The Reverend Lyman Beecher and many after him saw in the inebriate a fearsome menace to the existing social order. Such fears thrive when signs of disorder and impermanence, poverty and squalor intrude on the serenity of established custom. Civil war had weakened the authority of the traditional ballasts of family and church. The social constraints on drunkenness once enforced by the respected tavernkeeper and general public sentiment were gone. A miscellany of "grog-shop" proprietors sold alcohol to whoever could pay. Moreover, the highly visible disabled veterans of the Civil War prompted fears of still another addiction, morphinism.

The survival of American society despite all the problems of the post–Civil War era seems presumptive evidence that the menace of alcohol was somewhat overstated. Nonetheless, a belief in the omnipotent evil of intoxicants is part of our traditional folklore. The clarion call of an imminent and overwhelming drug menace has been sounded rather frequently and, as we have seen, it usually appears justified by the particular economic and social conditions of the times. Yet, in the long view, unrelated agendas and political expediency often seem to prompt the alarm.

Prevailing fears about the social consequences of inebriety gradually became interwoven with fears about morphinism and, to a lesser extent, cocainism. The radical arm of the Temperance Movement, the Prohibitionists, had only paused during the Civil War. The Temperance Movement itself, a rational appeal for moderation, was in a terminal decline. The Prohibitionists gathered strength and advanced steadily towards their stated goal, the prohibition of all alcohol. These crusaders were by inclination receptive to the notion that morphine addiction and cocaine abuse were parallel social evils, but no coalition of specific antidrug factions emerged comparable to the unification of the Great Awakening and the American Temperance Movement of the first half of the nineteenth century.

We will trace patterns of alcohol and drug use through the late 1800s into the twentieth century in an effort to reconstruct the climate of experience and opinion that eventually led to passage of the Harrison Narcotics Act in 1914 and to national prohibition in 1919.

ALCOHOL, TEMPERANCE AND THE CIVIL WAR

The War between the States marks an important turning point in the nation's drinking history. As abolitionism gained favor during the 1850s, membership in Temperance societies declined. It was not until the 1870s that Temperance reform regained its vigor, and by that time the drinking habits of America had changed in many important respects. Not only did the fervor of the antislavery campaigns dilute the intensity of the Temperance Movement, the war itself created in its aftermath a whole new context for the development of troublesome social problems. Prominent among these were alcohol abuse and opiate addiction. The postwar era also brought new federal taxes on alcohol, the birth of the first organized opposition to Temperance, and the founding of the Prohibition Party.

One major change in American drinking patterns was a dramatic switch to beer. Between 1850 and 1860, annual beer consumption increased from 1.6 to 3.8 gallons per person. This transition was led by German and related Middle European immigrants who brought a taste for malt liquors and the skills necessary to brew them. Under the German brewmasters, beer and ale came to rank among the nation's favorite beverages. This new preference was aided by the fact that malt beverages were often exempted from the prohibition statutes imposed at the state and local levels. The beer boom occurred at a time of great legislative victories for Temperance reform and when per capita alcohol consumption had stabilized at the lowest levels in that century.

The most immediate effect of the war was to disrupt both production and consumption of alcoholic beverages. In addition to severe restrictions on commerce, the demand for alcohol was reduced by the recruitment of the nation's heaviest drinkers into the military. While this may have reduced per capita alcohol consumption to some extent, public intoxication by Union soldiers led to the imposition of antidrinking regulations in 1862.

At approximately the same time, the Lincoln administration levied a twenty-dollar license fee on liquor retailers. Taxes were also imposed on each barrel of malt beverages and distilled liquor produced. Not only did these taxes serve to legitimize the brewing and distilling industries, they also generated revenues that would be difficult to give up in exchange for prohibition. But these subtleties were not appreciated by the brewers and distillers who saw the need to organize in opposition to both the government and the Temperance forces. In the fall of 1862, a group dominated by German brewmasters organized the United States Brewers Association. Their goals were to lower the beer tax and support legislative candidates and newspaper publishers sympathetic to their interests. Their first goal was quickly achieved. In 1863, Congress moved to reduce the beer tax from one dollar to sixty cents a barrel. Not long after, the distillers were also rewarded with a reduction of the tax on spirits from two dollars to fifty cents.

Such victories for the liquor manufacturers helped to reawaken the dormant Temperance reformers. Mobilized by the Good Templars, a relatively new Temperance organization was founded in 1859 on principles borrowed from the Washingtonians; radical Temperance advocates assumed a new militancy at the close of the war. Unable to commit either the Republicans or the Democrats to a strong prohibition plank, Temperance reformers moved to create their own political party. The moving force behind this drive to politicize the Temperance Movement was James Black, a Pennsylvania Methodist, former Washingtonian and Good Templar. Black had organized the National Temperance Society and Publication House in 1865. Black's organization published monthly and produced more than a billion pages of Temperance propaganda during the next sixty years. Another founder of the Prohibition Party was Neal Dow, who had redeemed himself from financial scandal during the Civil War. Leading the Thirteenth Regiment of Maine Volunteers, "the most temperate and moral of all the regiments of the state," Dow had been captured by the enemy during a vigorous attack on a Confederate position. After the war, Dow joined with Black and other prominent temperance leaders to establish the National Prohibition Party in 1869. With victory over the abolitionists fresh in their minds, a national war of liberation was launched to free drunkards from the slavery of intemperance. From the very beginning, the Prohibition Party aligned itself with other reform causes such as the direct election of United States senators, full voting rights for blacks and unrestricted suffrage for women.

The renewed momentum of the Prohibition Movement occurred at a time when the nature of alcohol problems was also changing. The typical alcoholic was an unemployed man or a laborer who lived in a major industrial city. Alcoholics were more likely to be recent immigrants than native-born Americans. Most of the alcohol-related problems were concentrated in urban slums, which absorbed the successive waves of immigrants arriving from Europe. Public intoxication, delirium tremens and pathological behavior when drunk were the major reasons given for committing large numbers of these unfortunate men to a variety of public institutions. A significant proportion of these individuals were committed to prison, incarcerated in jails, admitted to psychiatric hospitals or forced to apply for public relief. While ethnic prejudice may explain why the Irish, English, Scots and Scandinavians exceeded native-born Americans in their proportional contribution to alcohol-problem statistics, this would not explain why Italians, Russians, Jews and blacks were underrepresented in proportion to their numbers. A more likely explanation relates to the social context in which each group lived and drank, as well as the culture and customs they brought from Europe (see Chapter 6).

OPIATES, COCAINE, TONICS AND ELIXIRS

The dislocation and disillusionment brought on by the Civil War and its aftermath served as a breeding ground for a new kind of substance abuse, destined one day to eclipse even alcoholism in its prominence as a public menace. Opiate addiction had been identified as a medical problem early in the nineteenth century, but it was not until after the Civil War that it took on the dimensions of a major public health problem.

Until the second decade of the twentieth century, there were no restrictions on the importation of opium. Synthetic morphine was developed in 1832 and rapidly became the medication of choice for the relief of pain. Morphine-based medications were also sold for the treatment of diarrhea, dysmenorrhea and some communicable diseases. Tincture of laudanum prepared from opium was readily available and said to be a frequent accompaniment of afternoon tea. Today, use of opiate analgesics (morphine, Dilaudid, Demerol) require a physician's prescription or illicit procurement. During the 1880s, opiates were available to everyone in various forms of patent medicines.

This was the era of the medicine show and unquestioning acceptance of the miraculous curative powers of a plethora of tonics, elixirs and patent medicines. These amazing potions usually contained alcohol and morphine or alcohol and cocaine or all three in combination. Self-medicators, then as now, tended to believe that "more" was better. Admonitions or recommendations for restraint in use usually were conspicuously absent from the patent medicine label and the pitchman's exhortations. Cocaine was a common additive in teas and general tonics and was available in wines, cordials, cheroots, cigarettes and inhalants. Until 1903, cocaine was an active ingredient in Coca-Cola.

Panaceas were the order of the day, and the purveyors of promises ranged from the unscrupulous medicine man to very respectable pharmaceutical manufacturers. Today, demonstration of the clinical efficacy of drugs is required by law. In 1885, Parke-Davis, a reputable pharmaceutical company, recommended cocaine preparations for gastric indigestion, cachexia, asthma, as an aphrodisiac, a local anesthetic, a stimulant and to combat the effects of alcohol and morphine. It is a curious chapter in the history of nineteenth-century medicine that cocaine was advocated for the treatment of opium addiction and the converse. Opiates were thought to be a potential pharmacotherapy for alcoholism, and heroin was introduced as a cure for morphinism.

Opiate analgesics were the primary drug of solace for the wounded of the Civil War. The introduction of the hypodermic needle and the replacement of opium with purified synthetic morphine were major advances in the alleviation of pain. The injection of morphine avoided many of the gastric side effects of oral opiate administration and resulted in a more rapid euphoretic effect. Opium and morphine were liberally administered to the wounded by physicians of both the North and the South. Following the war, disabled veterans were often victims of iatrogenic opiate addiction. But the Civil War was not the sole or even the most important factor in the spread of opiate addiction during the 1860s and 1870s.

Lax prescribing habits probably contributed to the spread of opiate addiction in the middle class even though more dilute opiate preparations were available in patent medicines. It has been estimated that as many as 10 percent of the physicians of the period were themselves addicted to morphine or to opium. Physicians were often unaware of the addiction potential of opiates for themselves or their patients.

During the nineteenth century, the majority of opiate addicts were women between the ages of twenty-five and forty-five. Female addicts included domestics, teachers, actresses, prostitutes, nurses, doctor's wives and even society ladies. Middle-class women were especially vulnerable to opiate addiction because they were more likely to call a physician for the treatment of "female complaints" like dysmenorrhea and other gynecological disorders. Insomnia, headaches, fatigue and anxiety were also often treated with opiates. Another reason for the increase in addiction among women is that opiates may have provided a more socially acceptable substitute for alcohol. With the growth of the male-dominated saloon, and the increasing involvement of women in the Temperance Movement, drinking by status-conscious women became a strict taboo. The use of opium or morphine, however, provided an acceptable substitute intoxicant for women who had free time, little male companionship and a legitimate medical excuse. Female prostitutes were an exception to this pattern. The high rate of addiction among prostitutes is more likely due to factors such as the contraceptive effects of opiates, and the apparent belief that narcotic intoxication provided a welcome escape from the frustrations of life. Opiate addiction was common among women until the early twentieth century, when milder and safer analgesics were introduced, new laws limited the availability of opiates in patent medicines, and physicians' prescribing procedures were brought under greater scrutiny.

Opium addiction also occurred among the Chinese immigrants and lower-class whites. Opium smoking was legal in America until 1909. The Chinese minority were viewed with suspicion by their Western neighbors. Discrimination became persecution, immigration quotas were established to stop the Chinese influx, and public opinion indicted opium in a paroxysm of antiminority sentiment. Most analysts believe that a climate of fear and prejudice were of crucial importance for the passage of sweeping drug control legislation. Yale historian David Musto describes the process this way:

> By 1914, prominent newspapers, physicians, pharmacists and Congressmen believed opiates and cocaine predisposed habitués towards insanity and crime. They were widely seen as substances associated with foreigners or alien subgorups. Cocaine raised the spectre of the wild Negro, opium the devious Chinese, morphine the tramps in the slums; it was feared that the use of all these drugs was spreading into the "higher classes."

But what of the Victorian lady who allegedly had been sipping laudanum since the mid-1850s? Perhaps a combination of inadequate medical reporting of opiate addiction cases and traditional chivalry protected her from public scrutiny. Perhaps, as in most emotionalized campaigns, "facts" were incidental and expendable. It is a surprising and somber postscript to the conspiratorial veil that enshrouded those nineteenth-century women that alcoholism among women today has only recently been "discovered" by the federal establishment responsible for investigating alcoholism problems (see Chapter 13).

CHAPTER 5

The Western Frontier and the Rise of the Saloon

W hile the Far West was not America's first frontier, nor its last, it certainly has been its most celebrated. And no scene epitomizes our image of the Old West more than the Rocky Mountain saloon, where men and women of the frontier gambled, fought, loved and drank. The steely six-gun, the faithful steed, and a shot glass of straight whiskey are symbols of the frontier spirit. How valid is the image that the West was won with a bottle and a six-gun? Did the miner and the cowboy differ from their Eastern counterparts primarily in the prodigious amounts of liquor they could hold?

·The Western frontier was a remarkable epoch in the history of American drinking patterns.·Abundant supplies of cheap whiskey and the proclivity of itinerant adventurers and drifters to abuse alcohol helped to give the "Wild West" its singular identity. The violence and lawlessness associated with drinking sprees seemed to confirm the worst fantasies of the most inventive Temperance reformers. The legendary saloon, Western counterpart of the grog shop, played a crucial central role in frontier social life until the final phases of the Western expansion. The American West of popular mythology developed from the experience of thousands of men and women who were drawn to California, Colorado, Arizona, New Mexico, Montana and Idaho in search of precious metals. The Western mining frontier lasted roughly half a century, beginning with the discovery of gold in California in 1848. Its history is told in the ledgers of towns such as Tombstone, Arizona, Leadville, Colorado, and Cripple Creek, Colorado. Following the end of the Civil War these towns attracted hordes of prospectors, miners, cardsharks and soldiers of fortune, all of whom demonstrated a strong taste for alcohol.

The rural mountain towns and seacoast villages established to serve the needs of the mining frontiers were also a temporary base for other Western immigrants, trappers, lumberjacks, and clipper ship merchant sailors. As the mining frontier began to assume an air of permanence, the Western territories also attracted cattle herders, who lived a seminomadic existence on the trails between Texas and Chicago. In the heyday of trail driving, roughly between 1865 and 1890, as many as forty thousand men worked as cowboys. Their rough demeanor and free and simple life-styles were glorified in popular Eastern magazines like *Harper's Monthly*. These Western folk heroes became role models for young men emigrating from Europe or the Eastern Seaboard.

As immigration to the Western territories accelerated, one institution came to epitomize the life-style of the frontier: the saloon. The term saloon is drived from the French *salon,* which was initially used in England and America to describe public meeting places or entertainment halls: By the end of the nineteenth century, its connotation had changed as the Colonial inns and taverns were replaced by a very different type of drinking establishment. Although the saloon was modeled initially after its Eastern counterpart, it quickly adapted to the social needs and drinking styles of the frontier. Indeed, the evolution of the Western saloon closely parallels the economic and social development of the Western town from frontier camp to rural community. As the central forum for Western drinking, its development helped to structure the very drinking style it was intended to serve.

The first saloons were typically rudimentary structures, often consisting of a tent or a shack with a row of barrels for a bar. Like the wood frame structures of the frontier camp itself, the saloon could be cheaply constructed with minimal investment. When decoration was added it often reflected a style best described as "Victorian macho," typified by paintings of naked women, pictures of boxer John L. Sullivan or illustrations of Custer's Last Stand. Such decorations mirrored the characteristics and aspirations of the saloon's patrons: male dominance, aggressive bravado, and the tendency to exploit both men and nature. The emerging image of the saloon was also reflected in the descriptive names that gave each a special identity. Names such as the Miner's Rest, the Pioneer, the Big Tent Saloon, or the Dew Drop Inn, often were associated with the most typical clientele, some characteristic of the structure or with the owner's sense of humor.

Records gleaned from license applications and census surveys reveal that the first Western saloonkeepers resembled their clientele in social background. They were predominantly single, Caucasian males, about thirty-five years of age. Approximately 50 percent were foreign born; Irish and Germans predominated among the recent immigrants. Saloonkeeping was frequently a temporary occupation attempted by unsuccessful miners and other manual laborers. The heavy demand for alcohol in the frontier camps promised a quick return on a small investment in a stock of cheap whiskey. Further, lack of regulation and the absence of license fees made it possible for almost anyone to open a saloon. There was a rapid turnover in saloonkeepers: few survived, much less prospered, for more than a few years. Competition between saloons was brisk, and the clientele were constantly changing. Nevertheless, the saloon proliferated on the Western frontier to such an extent that drinking establishments frequently outnumbered all other retail establishments in many Western towns. Throughout the 1880s, for example, Leadville, Colorado, boasted approximately one saloon for every one hundred people.

The Western saloon became the frontier's most versatile social institution. It served a variety of important needs that were critical to the growth and development of the West. Just as the New England tavern served as a forum for Colonial commerce and politics, so the Western saloon became a focal center of frontier society. In the absence of town halls and other public buildings, the saloon was often the seat of the county court, the site of the first post office, and host to public ceremonies. It was the meeting place where business deals were negotiated, and a recruitment center for laborers and volunteers. At the early stages of a town's development, it was the major clearinghouse for rumors, news and job-related information. Inasmuch as the saloon always contained a significant cross section of the town's population, its potential as a public meeting place was quickly recognized by churchless ministers in search of a congregation, frontier politicians in search of votes, and itinerant entertainers in search of an audience.

But the most important function of the saloon was to provide diversion, entertainment and refreshment. In addition to the standard games such as poker, billiards and checkers, the saloon frequently sponsored special activities like dances, lotteries, competitions, contests and prizefights. These entertainments were designed to attract large crowds and served as a focal activity around which drinking could be organized and prolonged, to the obvious delight of the saloonkeeper.

The convergence of large numbers of ambitious, unmarried male adventure seekers on the Western frontier created the kind of social disorder long celebrated in legend and film. Consistent with the image of the raucous and bawdy frontier settlement, historians of the period have noted that much of the brawling and killing was associated with drinking. Newspaper accounts suggest, for example, that one or both parties were drinking heavily in at least half of the fourteen murder incidents occurring in Leadville between March and October, 1880. In Silverton and Georgetown, Colorado, 60 percent of the misdemeanor arrests in the early years were for public intoxication. Among the most vexatious problems to the public authorities were barroom brawls, robbery, drunken assaults, the shooting of handguns in the air, rowdy behavior near saloons, and intoxicated patrons riding wildly through town. Another problem, frequently noted by the frontier physicians, was alcohol-related death from pneumonia and exposure. Cold winters and inhospitable terrain took a steady toll of the malnourished, chronically intoxicated "losers" who failed to find their fortunes in the West.

Thus by all accounts the explosive drinking style that emerged on the Western frontier conformed rather closely to the negative stereotype popularized in the "Temperance Tales" of the Prohibitionists. Four factors seem to have been paramount in the development of this drinking pattern. The first is the ready availability of cheap intoxicants. Because of the relative ease of transport, cheap whiskey was often the only beverage that could be hauled long distances over rugged terrain. For the same reasons that distillation proved to be so profitable in the East, teamsters and saloonkeepers in the West found that selling alcohol to the thirsty hordes of arriving immigrants was a sure way to make a profit. Frontier whiskey often consisted of raw alcohol mixed with water, and its wretched quality is reflected in such names as "extract of scorpions," "San Juan paralyzer" and "Taos Lightning." The latter was a wheat whiskey first distilled in New Mexico for fur traders and later adopted by the Southwestern miners.

By the early 1870s, beer became more popular than whiskey in the West as well as the East. Technological improvements like bottling and pasteurization led to the development of major regional brewers such as Schlitz, Pabst, Blatz and Budweiser. Many of the more successful Midwestern German brewers constructed mountain breweries in the larger and more established Western towns, which could then supply the local markets. Improved transportation also made possible the direct shipment of beer from the giant breweries of St. Louis and Milwaukee. By the 1880s most Westerners enjoyed a dependable and varied supply of local and regional beers.

A second determinant of the Western drinking style was the demand created by mobile men who were far from the moderating influences of family and community. In the mining camps especially, group drinking assumed ritualistic and symbolic functions as a facilitator of conviviality, solidarity and information exchange. Census reports reveal that approximately 50 percent of the early Western immigrants were American born, and the remainder were recent immigrants from Europe, Mexico and Asia. Of the Europeans, English, Irish and Germans accounted for the largest ethnic blocks. Men outnumberd women in the early days of most settlements by a factor of five to one. Because the Western frontier attracted a particular type of individual, these demographic statistics convey only a part of the picture of the Western immigrant. Men who were unemployed, unemployable, rootless and alienated were particularly attracted to the West, and their itinerant life-style placed them at risk for the development of drinking problems.

Yet another ingredient that contributed to the typical Western drinking pattern was the nature of work and leisure on the frontier. Work on the trail or in the mines was monotonous. Winters were long and imposed further isolation on those already separated from family and friends. Lack of entertainment and organized social activity increased the importance of communal drinking as a pastime, particularly following long periods of relative abstinence. Under these circumstances whiskey and a warm saloon became a welcome center of social life. Drinking for relief from drudgery and escape from worries was an accepted pattern. Thus, the cyclical contrast between long periods of hard, dangerous work and brief periods of intense recreation was common to the life-style of the cowboy and the miner, the northern lumberjack and the Far West trapper, the seaman whaler and the clipper ship merchant. To those whose work did not impose involuntary abstinence, these "blowouts" could assume a commonplace regularity. One Colorado editor observed that "It is a custom of the boys to 'irrigate' to a considerable extent, when they strike a bonanza. Some of them strike one every day."

Under these circumstances drinking became associated with almost all aspects of work and leisure in the West. Stage drivers, mule packers and wagon train guides often drank during their long journeys. Certain holidays, especially Christmas, Independence Day, and St. Patrick's Day, became occasions for drinking to intoxication for want of alternatives. The practice of treating, where one person buys a round of drinks with the expectation that all the other drinkers will eventually reciprocate, was especially conducive to heavy drinking.

A final feature that contributed to the development of the Western drinking style was the lack of social controls on the frontier. Not only were there no constraints on the marketing and production of alcoholic beverages, there were not enough peace officers to control legal offenders. In many towns, the saloon flourished even after the passage of statewide prohibition, because local legal authorities were not strong enough to confront the unruly mobs of drinkers or the strong economic interests that catered to them.

As the heterogeneous, transient nature of Western settlement gave way to a more organized management of economic and social development, other businesses and institutions gradually assumed the functions of the saloon. With the growth of public services and facilities, and the emergence of a more permanent work force, the image and the functions of the saloon began to change. Land and opportunity attracted men with families, and families had a far greater commitment to community development and long-range economic interests than their more adventuresome predecessors. Gradually, drinking establishments began to reflect the new social realities. Barrooms tended to become specialized around the needs of a particular clientele; some catered to political clubs, others to the town's business elite, still others to the predominant ethnic or occupational group. At the same time, the more successful saloons took on the appearance of permanence and refinement. Establishments like Tombstone's Crystal Palace, which graced the main streets of the more prosperous Western towns, were often more elaborately designed and furnished than the local church or the town hall. Undoubtedly the imported accoutrements, expensive fixtures, fine stemware and luxurious furniture added an aura of bourgeois respectability to drinking, as did the introduction of music and dancing. But the majority of saloons remained rather dingy enclaves serving a predominantly male clientele.

The arrival of a new type of settler, immigrants and native-born Americans less concerned with instant wealth did not directly threaten the existence of the saloon at first. However, the values and social outlook of these more stable settlers contrasted sharply with the sordid traditions of the hard-drinking frontier subculture. The clear association between drinking and disorder invited criticism from the new middle class and the older, more established, citizens. The inevitable result of midnight revelers shooting their pistols in the air or drunken brawls and homicides was increased concern with control of the liquor trade. These solid citizens were a ready source of sympathy and support for the Temperance organizations that developed in the Western territories and newly admitted states.

CHAPTER 6

Drinking and the Melting Pot

Westward expansion and the growth of the frontier drinking style were only one aspect of a much larger pattern of social change taking place in the United States during the latter part of the nineteenth century. During this period, massive immigration and rapid industrialization dramatically altered working conditions, family life and community living in the American towns and cities. While diversity and change had always been basic ingredients of the American experience, so too had conservative religious values and family-centered communities. What became evident after the Civil War was the disturbing realization that the fabric of American society had begun to unravel. Not only was lawlessness and licentiousness rampant on the frontier, social problems associated with the large industrial areas of the East and Midwest seemed to be increasing. When middle-class Americans sought an explanation for the disease and disorder they saw around them, Temperance advocates were only too ready to point out the apparent association between alcohol, the saloon and the problems of society.

INDUSTRIALIZATION AND THE URBAN SALOON

The trauma of the Civil War and Reconstruction was soon exacerbated by the disturbing changes brought on by industrial development, urbanization and immigration. The demand for a large, mobile work force for the railroads, mining, logging, seafaring industry and factories was answered with a liberal immigration policy that invited the resettlement of millions of discontented and adventurous Europeans to American cities. The rapid growth of factories, mills, railroads and related industries was associated with economic depressions, commercial panics and seasonal unemployment.

American cities swelled to accommodate factories and house the immigrants. Large numbers of disaffected and disaffiliated people became concentrated in segregated urban slums. In addition to the newly arriving immigrants, many of whom did not speak English, these areas were populated by unemployed and homeless vagrants, the urban poor and destitute elderly, pimps and prostitutes, organized criminals and petty thieves, and alcoholics and drug addicts. The development of slum environments and the Skid Row life-style on a mass scale was hastened by the demographic changes associated with the streetcar. Rapid transportation allowed middle-class and skilled workers to live a considerable distance from their places of employment, thereby opening areas close to the city center for the impoverished and the immigrants.

The fact that the slum population was disproportionately composed of unemployed or temporarily employed males, living on the streets or in crowded rooming houses, meant that there would be a ready market for masculine "diversions" such as drinking, gambling and prostitution. In many cases the slum areas had long been identified with crime and vice, but during the last part of the nineteenth century there was a rapid proliferation of saloons, brothels and gambling establishments. In most cities, the blatant illegality of these activities was testimony to the political corruption that pervaded every level of government.

Saloons became ubiquitous in the growing urban centers; their numbers doubled between 1880 and 1900. Saloons were concentrated in the slum areas where ethnic voting blocks and corrupt politicians could offer protection against the imposition of prohibition laws. Like its Western counterpart, the urban saloon emerged as the focal point of slum life, serving as a recreation center, poor man's club and employment agency. In ethnic neighborhoods, the saloon became the center of business transactions and of political activity, a place where deals were struck over a drink, or votes could be bought for a beer. Saloons often opened their backrooms and halls for meetings of social clubs, labor unions, lodges, and other organizations. Saloon halls also hosted christenings, weddings and dances, especially among immigrant groups. Other attractions included the infamous free lunch, which could be enjoyed for the cost of one or two glasses of beer. Because of the ubiquity and visibility of the urban saloon, many saw alcohol as a major factor in the problems of the slums.

In the early years of the Republic, the retail liquor trade was conducted in taverns, inns and groceries. Vendors were independent of one another and were never organized to promote the trade. During the last half of the nineteenth century, however, there was an important change in the saloon in both the structure of the industry and the conduct of the retail trade. In brief, the business of selling liquor replaced the business of selling food and hospitality. As billiards, pool tables, cards and free lunches were introduced, critics of the saloon saw it as an "evil," a breeding ground for crime and vice, a place that sheltered crooks and gamblers. It was further accused of harboring and encouraging the "white-slave" traffic (organized prostitution), being the birthplace of corrupt politics and a principal agent of poverty, wrecked homes, and insanity.

The growth of organized prostitution in the central city paralleled the growth of the saloon, and was undoubtedly related to the same set of causal factors. Once knowledge about syphilis increased at the turn of the century, there was virtual hysteria concerning the role of the saloon in the spread of the "social evil." By this time, the image of the saloon had become one with the image of the urban slum, arousing the concern of not only the Prohibitionists, but also the broader spectrum of opinion leaders such as newspaper editors, writers, churchmen and politicians. Even within the industry itself there was deep concern for the image of the saloon, as suggested by the following editorial in the New York *Wine and Spirits Gazette,* published on August 25, 1902:

> *The modern saloon has been getting worse instead of better. It has been dragged in the gutter; it has been made a cat's paw for other forms of vice; it has succumbed to the viciousness of gambling, and it has allowed itself to become allied with the social evil.*

What accounted for the gradual change in the character and function of the saloon? In part the transition can be attributed to the demographic changes that gave rise to the urban slums. The middle-class workers moved to the "streetcar suburbs," and as these areas progressively voted "dry" to eliminate the liquor traffic, the saloons concentrated in the core of the city were all that remained. Forced to serve a predominantly impoverished clientele, there is little wonder that the saloon became identified with the problems of the poor.

The evolution of the saloon was also affected by changes taking place in the liquor industry itself. In part this process was initiated by the Internal Revenue Act of 1862, which afforded the industry a certain amount of government sympathy because of the tax revenues it provided. Advances in the technology of growing, improvements in transportation, and the demand occasioned especially by the new waves of European immigrants, all hastened the trend toward greater concentration in the brewing and distilling industries. With the entry of large investors, competition became fierce along the large producers, and more systematic management was introduced. The brewers began to invest in the saloons, especially along the newly laid railroad routes and in the larger cities. As a result, the independent retailer virtually disappeared, replaced by the "tied house." The saloonkeeper became an agent of the brewer, who selected the site, provided the license, advanced the capital and held the mortgage. Approximately 75 percent of the country's saloons were owned or controlled at one time by the large liquor wholesalers.

To combat the influence of the antisaloon forces, industry associations levied assessments on their distillers, wholesalers and retailers, and raised enormous amounts that could be used to influence local and national elections. Newspapers were often subsidized in exchange for sympathetic reporting and editorializing. Politicians and police were often subverted by saloon money in exchange for their protection from legal harassment. "Tied house" saloonkeepers were under constant pressure to sell as much alcohol as possible. Spurred on by local competition, and constantly in debt to the brewers, many saloons stayed open twenty-four hours a day, seven days a week. Because the saloon was often the center of ethnic ward politics, elections, particularly those involving prohibition issues or prohibition candidates, were at times fraudulently decided by the saloonkeeper and his patrons.

ALCOHOL AND THE IMMIGRANTS

Closely related to the problems of urban disorder in general and the saloon in particular was the issue of foreign immigration. As immigration assumed a dominant role in the population expansion of the Eastern Seaboard cities, the drinking patterns of the new arrivals came under special scrutiny by the bluebloods and native-born Americans who had achieved middle-class status. What they saw tended to reinforce the pseudoscientific notions of racial inferiority in vogue during that period: uneducated, often unwashed foreigners living in squalor and constantly drunken and disorderly.

The available statistics tend to offer some support for the stereotype of the drunken foreigner, but like all stereotypes it is grossly inaccurate because it is applied to all cases. In fact, the drinking customs of different ethnic groups varied tremendously in terms of the beverages preferred, amounts consumed, occasions for drinking and patterns of consumption. While it is true that the foreign born were disproportionately represented in the statistical tabulations of alcohol-related problems, there were striking differences among the various ethnic groups.

In Boston and New York, the Irish consistently led all other nationalities in convictions for drunkenness and disorderly conduct, but the Scots and English were not far behind. When vital statistics were analyzed from the 1890 census data, Irish-born men and women were found to have the highest rate of mortality from alcoholism, far exceeding the rates for native-born Americans. The Irish also led in the category of liver disease, although here the rates are approximated by the Germans and the Italians. What is perhaps most striking about these statistics is the high rates of liver disease and alcoholism among the Irish females: rates that in many cases rival those of the English and Scottish males. The Scots and English were high on drunkenness indicators, while Italians, Russians, Poles, Germans and Scandinavians showed rates similar to native-born Americans.

Why were the Irish, English and Scottish immigrants overrepresented in the ranks of those who developed drinking problems? Regular alcohol consumption may have been a factor, but this was also a characteristic of the Italian, German and Jewish immigrants, who were ostensibly as temperate as native-born Americans. More important seems to be the common patterns of drinking which developed in England, Ireland and Scotland during the late eighteenth century and early nineteenth centuries. Each of these countries developed a large number of drinking customs related to the convivial usage of alcohol at marriage ceremonies, funerals, baptisms and during business transactions. Medicinal uses of alcohol in each country were widespread. Whiskey was taken as a remedy for disease, and as a palliative for the effects of the damp climate. Workers drank to produce stimulation and reduce fatigue. Throughout the British Isles the term "a hair of the dog that bit you" was used to describe drinking in the morning to cure a hangover. Of particular importance to an understanding of the hard-drinking habits of these ethnic groups are the social and occupational drinking habits that developed around the public house.

Throughout the British Isles, the pub emerged as the primary meeting place for recreational and occupational purposes. It served as a hiring hall for the trade unions. It hosted shop meetings, union meetings and political rallies. Military recruiting was even carried out on the premises. Many pubs featured food and entertainment. As one Englishman noted, as much of the history of England has been brought about in public houses as in the House of Commons. As in America, pubs proliferated in urban slums during the nineteenth century as an outgrowth of rapid industrialization. While the consequent alienation of the working class may have been a factor in the heavy drinking noted at this time in England, the various occupational rules which required workers to drink hard and frequently seem far more important. Far from being an occasion for dismissal, drinking on the job was encouraged through a variety of incentives, most of which were related to a system of occupational fines levied by members of a trade to finance communal drinking.

Drink fines were collected on becoming an apprentice, at the time of a marriage, and on the birth of a child. Fines were also paid in anticipation of union meetings, holidays and paydays. The result of these drink fines was to place enormous pressure on workers to drink with the work group. Hard drinking thus became an important means of achieving status within the work group, since the frequent indulgence in communal drinking was an indication of loyalty to the trade. Given the fact that these attitudes and customs were socialized during the teenage years, it is not surprising that hard drinking was a distinguishing characteristic of the young Irish, English and Scottish workingmen who immigrated to America. There were, nevertheless, some important differences in both the social background and the drinking habits of these three related nationalities, which may explain why the Irish exceeded all other groups in apparent drinking problems.

A major event in the historical development of Irish-American drinking customs was the Great Potato Famine of 1846–1851. Not only did the famine decimate the population of Ireland and precipitate the emigration of millions to America, it also changed the structure of Irish society in ways that contributed to future drinking problems. According to one interpretation of Irish drinking, drinking customs following the great famine were changed by three social factors. First, early marriage, along with the division of land among male sons, was discouraged in order to maintain the integrity of existing farms. The single-inheritance farm society led to a group of bachelors composed of landless farm laborers. Characterized by freedom from responsibility and devotion to leisure, the male bachelor group organized its life-style around the pub and hard drinking.

Although the Catholic Church was active in Temperance work, and even required a total abstinence pledge at the time of confirmation, there was considerable sympathy for the bachelor role, perhaps because drinking was considered a legitimate compensation for sexual denial. Condoned by the Church, and encouraged by social and economic circumstances, hard drinking fulfilled a variety of functions for the Irish bachelors, and they carried this custom with them to America. Hard drinking was valued as a symbol of manhood and masculinity, and may also have helped to compensate for the Irish man's lack of social status and political rights in a stagnant agrarian economy dominated by the hated English.

After immigration to America, hard drinking assumed new significance as a way to establish a separate cultural identity. The Irish ranked lowest among all ethnic groups in the proportion of family units among its emigrants. Thus, the Irish bachelor drinking group was not only preserved intact as a primary social unit, it was also invested with a new set of adaptive functions. In many respects hard drinking became a way of demonstrating one's Irishness, of providing a sense of ethnic identification that differentiated the Irish from other ethnic groups.

At first the Irish immigrants gravitated to the large Eastern cities like Boston, New York and Philadelphia. Their lack of industrial skills and financial resources forced them to settle with other ethnic minorities in the most dilapidated slum sections. They lived in cheap boardinghouses, tenements or flimsy one-room dwellings known as "shanties," all of which were characterized by overcrowding, poor sanitation, lack of privacy, little heat and poor ventilation. If grog shops and saloons were not already part of the buildings they occupied, the immigrants could readily find a large selection of drinking establishments, both licensed and unlicensed, within a short distance. The customary role of the saloon in immigrant social and economic life was expanded to become a medium of Irish political organization. If the leading politicians were not saloonkeepers themselves, the ward bosses often used the saloon to recruit voters over a round of free drinks.

Discrimination and prejudice may have placed special constraints on the adaptation of the Irish to America and probably contributed to their drinking problems. Along with the blacks and American Indians, the Irish were among America's most maligned groups. Despite the need for cheap labor, the Irish were often refused employment. When they were hired, they were more likely than other minorities to be employed as railroad workers, canal diggers, dockers, and in other low-status jobs, where rotgut whiskey was often provided as a substitute for full wages. Ironically, hard drinking and drunken fights were the reasons most often cited for refusing employment to the Irish. Regardless of the truth behind these beliefs and attitudes, by the end of the nineteenth century there was a pervasive stereotype, perpetuated by cartoons and newspaper articles, portraying the Irishman as a cheerful but slovenly drunkard.

Prejudice and discriminaton were undoubtedly implicated in the drinking problems of other minorities as well. The Reconstruction Period following the Civil War was associated with increased drinking among freed black males. Drinking had not been prevalent during slavery because of strict limitations placed on selling or serving alcohol to blacks. Nevertheless, it was not uncommon for slaves to be allowed to drink heavily on weekends and holidays, a custom that seems to have continued with greater intensity after the Civil War. Alcohol problems were most acute among black man, many of whom left their wives to enjoy the freedom permitted during the Reconstruction period.

These political and social freedoms were short-lived, however. By the 1870s blacks again faced severe restrictions on when, where, and with whom they could drink. Temperance advocates, who had provided so much moral support to the cause of abolition, unwittingly reinforced the prevailing racial theories by arguing that blacks should practice sobriety for the advancement of the "race." And, in many Southern states an argument advanced in support of prohibition was based on the fear of blacks, who were believed to be particularly violent and licentious when intoxicated. Economic factors also distinguished the black experience with alcohol from that of other ethnic groups. Because of their lack of financial resources, blacks were unable to engage in the large-scale manufacture and sale of alcoholic beverages. In spite of legal and financial obstacles, illegal alcohol production became widespread in the backwoods of the South. To the sharecroppers, tenant farmers, bayou fishermen, and Mississippi mudbank dwellers, the marketing of bootleg alcohol provided a welcome supplement to their meager incomes. It is reported that one theological student financed his seminary education by selling bootleg liquor to preachers and politicians.

TEMPERANCE RETURNS

By the latter part of the nineteenth century, despite the progressive increase in per capita beer consumption, drinking patterns in America had become as diversified and erratic as its restless and heterogeneous population. Both drinking and abstinence became closely identified with class, religion, and ethnicity. Middle- and upper-class Protestants saw total abstinence as a symbol of respectability and as a means to achieve health, success and happiness. Books on popular manners and advice constantly associated temperance with such virtues as industry and thrift. Among the socially conscious and educated, prohibition became the means through which America's social problems could be solved, its ethnic minorities assimilated, and its moral superiority demonstrated to all the world.

In face of the growing concern over urbanization, immigration and the pervasive social disorder Americans saw in almost every section of the country, alcoholic beverages and the urban saloon were singled out for particular attention. Reaction to the problems identified with alcohol took two basic forms: protest and reform. Disorganized and demoralized at the end of the Civil War, it took several decades for the forces of Temperance to regroup. In the process, an entirely new set of tactics, goals and organizational structure emerged.

Although women had always been active in Temperance programs, it was not until the postbellum period that they assumed an independent role in determining the nature and direction of the reform movement. At first isolated and spontaneous, their efforts gradually gained momentum as Temperance became more and more identified with the Women's Suffrage Movement. The first meager beginnings took place in 1873 and 1874 when the so-called Women's War against the saloon broke out spontaneously in many sections of the country. It began on Christmas Eve, 1873, in Hillsboro, Ohio, when a group of seventy women marched from a church meeting to a nearby saloon. After a reserved protest of song and prayer, they demanded that the saloonkeeper give up his business and that the patrons give up their drink. In the ensuing months, Committees of Visitation kept up the harassment there and at saloons in neighboring towns. News of their apparent success in closing several of these establishments (at least temporarily) spread rapidly, inspiring similar protests throughout the country. In the spring, antisaloon sentiment resulted in street protests in San Francisco and in other parts of California. In Portland, Oregon, women were hosed with cold water by one saloon proprietor and then arrested. The visitations continued, and eventually the saloonkeeper closed his business. In Cleveland, ten thousand women signed total abstinence pledges, and in Chicago, sixty-five thousand more petitioned for Sunday closing laws.

While some would interpret the women's crusade as an isolated instance of mass hysteria, several lines of evidence suggest the underlying motivation was more complicated and more political in nature. For example, the first uprising in Ohio occurred not long after an appearance there of Dr. Diocletian Lewis, a noted Chautauqua lecturer, fond of mentioning how his own mother had used prayer to close a saloon in order to save his father from drink. An advocate of homeopathic medicine, sex education and physical exercise for women, Lewis also preached the virtues of total abstinence to women's groups throughout the country.

Another indication of the significance of the women's crusade can be found in a careful study of the kind of women who participated. Seventy-five percent of those who protested in Hillsboro by signing petitions or conducting prayer vigils came from the highest social strata of the town. Uniformly white, Protestant and native born, they represented influential families that controlled most of the town's wealth. While their husbands may not have agreed with their tactics or their goals, they could not fail to notice a new sense of political consciousness emerging among women.

The women's crusade led directly to the formation of the Women's Christian Temperance Union (WCTU). This was the first mass movement of women in American history. Under the remarkable leadership of Frances Willard, the WCTU was to become one of the most influential organizations in the fight for Prohibition and Women's Suffrage. Frances Elizabeth Willard (1829–1898) was raised in a section of Wisconsin known as a hotbed of Methodist fundamentalism. Educated at Northwestern Female College, she taught at several schools and colleges before taking a faculty position at her alma mater at age twenty-three. In 1871, she became the first woman in the country to be named president of an institution of higher learning. Three years later she resigned her academic position, thereafter devoting herself entirely to Temperance and women's suffrage. Building on the Temperance leagues formed in many states during the waning years of the women's crusade, Miss Willard was instrumental in organizing these disparate groups into the National WCTU in 1874. Their emblem, a white ribbon, was chosen to symbolize purity. "For God, Home and Native Land," was their motto. Although their tactics were many and varied, one observer described their approach as "organized mother love."

Under Miss Willard's nineteen-year tenure as president, the WCTU adopted a broad-gauge, "do everything" policy. At the height of its power and influence, forty-five "reform departments" defined the organization's scope and ideological orientation. The Department of Christian Citizenship fought against gambling and race tracks. The Department for the Suppression of Sabbath Desecration worked to place legal constraints against the conduct of "nonessential" activities on Sunday, including public transportation, unessential labor, entertainment, and newspaper publishing. The Department of Unfermented Wine at the Lord's Table encouraged the clergy to confine their sacramental offerings to grape juice. Departments dealing with health, heredity and hygiene condemned patent medicines containing alcohol and narcotics, and worked for the suppression of tobacco use as well. A group of separate departments (e.g., Colored People, Foreigners, Railroad Employees, Indians) brought the Temperance message to those ethnic and occupational groups considered to be especially vulnerable to the effects of alcohol.

In addition to these forays into the territory of private morality and religion, the WCTU was heavily involved in humanitarian concerns, such as prison reform, improved employment codes, the eight-hour workday, world peace, international arbitration, the easing of racial tensions and adult education for native and foreign-born illiterates. Special attention was devoted to combating urban prostitution and the "white slave" trade. The Department of Social Purity demanded harsher penalties for those involved in organized prostitution, advocated raising the age of consent for intercourse to eighteen (in some states it had been as young as ten), and supported the establishment of rehabilitation centers for reformed prostitutes.

The issue of women's suffrage was especially important. The Franchise Department was organized to provide advice and support to suffrage groups. Under the theme of "home protection," both the WCTU and the Prohibition Party sought to identify liquor as a major factor in the social condition of women. It was argued that the law offered no protection to women whose husbands squandered their paychecks on alcohol, abused their families physically, or infected their wives with venereal disease. In a society where divorce stigmatized women more than men, where property rights favored the husband, and where restricted employment opportunities virtually prevented economic viability by divorced women, prohibition and equal suffrage were seen as two methods to protect the American home from the forces that threatened it. During the 1870s and 1880s the WCTU provided a more "respectable" alternative to existing women's rights organizations, which were often considered too radical by middle-class Victorian women.

Fearing that women's votes would be cast in favor of Prohibition, state suffrage amendments were actively opposed by a coalition of brewers, distillers, saloonkeepers and foreign-born voters. Although women had been actively working for the franchise since the Seneca Falls Convention of 1848, it was not until the late 1800s that the

Suffrage Movement gained momentum by joining forces with the Prohibition and Social Purity Movements.

Temperance education was perhaps the most active and influential activity of the WCTU. Created in 1880, the Department of Scientific Temperance Instruction devoted itself to the advocacy of health education in schools and colleges. A major theme reiterated in much of their graphic propaganda, and reluctantly supported by well-known scientists, was that alcohol was a poison and that total abstinence was an absolute prerequisite for good hygiene. While much of their material was simplistic and unscientific by contemporary standards, it is interesting to note that the WCTU was one of the first organizations to draw attention to the birth defects now believed to result from excessive drinking by pregnant women.

To spread these views to the youth of America, the WCTU conducted an intensive and relentless campaign directed at legislators, publishers, educators and teachers. As a result of their pressure, few publishers could market a book on hygiene without WCTU endorsement. These efforts were so successful that by 1900 most states had enacted laws mandating Temperance education in the schools. Such laws often required that teachers affirm each year that Temperance instruction had been given. While the effects of this propaganda are impossible to evaluate, it is testimony to the ingenuity of the WCTU that by the time the national Prohibition amendment came before the American electorate, almost all of the nation's voters had been exposed as children to the propaganda for total abstinence.

At the end of the nineteenth century, Frances Willard was regarded as one of the most celebrated and influential women of the day. Her gospel politics, though applied vigorously in every state in the union, never achieved the grand objective she had envisioned: a national political party molded out of a harmonious coalition of reform organizations. At various times the WCTU established working relations with such groups as the Prohibition Party, the Suffragists, the socialists' Knights of Labor, the antimonopolist agrarian Grangers, and the Populists. Although Miss Willard managed to bring all of these groups together at a mammoth "People's Convention" in 1892, the hoped-for merger was never achieved.

Whereas Miss Willard epitomized a kind of quiet diplomacy that sought to implement political solutions to solve the nation's social problems, there was a radical element within the WCTU that adopted a more direct approach to the abolition of the saloon. Among the most famous (or infamous) crusaders of this persuasion was Carry Nation.

Born in Kentucky in 1846, Carry Amelia Nation spent her childhood in Missouri and Texas. Her mother had intermittent psychotic delusions of being Queen Victoria, and Carry herself was given to convulsions and hallucinations. Within two years her first marriage ended in divorce, followed by the death of her former husband from alcohol-related causes. Her second marriage to David Nation, an editor and self-styled minister, brought her to Kansas where she was elected county president of the WCTU in 1888. In a later reminiscence on the significance of her married name, she wrote that: "My right name is Carry A. Nation — carry a nation for temperance. . . . After I entered temperance work I then became convinced that my father had christened me 'Carry' for a purpose. It was to carry on the work of temperance."

Her great campaign against the illegal saloons of Kansas began at the turn of the century. It was directed at the ostensible failure of public authorities to enforce the state prohibition amendment. In 1880 Kansas had enacted statewide prohibition under the leadership of John St. John, the then successful Prohibition Party candidate for governor. Throughout this period Kansas was one of the most violent and disorderly sections of the country, plagued by bank robberies, cattle wars, farmer protests, rapid population growth and economic dislocations. By 1890 Prohibition was a dead letter in Kansas, with hundreds of saloons operating openly in all parts of the state. Her most famous "hatchetation" campaign brought her across the country in 1900 by way of Topeka, Des Moines, Chicago, St. Louis, Cincinnati, Atlantic City, Philadelphia and New York. Armed with a hatchet in one hand and a Bible in the other, Mrs. Nation rallied her followers with chants of "Smash, smash, for Jesus' sake, smash." Marshaling crowds of women, students and other onlookers, she would attack the saloons and demolish windows, mirrors, paintings, glassware and bottles. Although her assaults resulted in her forceable removal and arrest on numerous occasions, the publicity she attracted was often sufficient to bring reluctant officials and police to close down the illegal saloons.

When Carry Nation arrived in New York she went directly to the saloon of the great pugilist John L. Sullivan. Feigning illness, Sullivan wisely decided to avoid what could well have been the most difficult match of his career. Along the way she inspired women to follow her example, and in many areas the hatchets were applied with equal ferocity. Her notoriety as a lecturer, crusader and, in her later years, as a side-show curiosity, enabled her to attract large audiences, raise funds for charitable causes, and draw attention to Temperance reform and law enforcement. While most historians have focused on the propriety of her tactics and the fanaticism of her cause, it is important not to lose sight of the social disorder her campaigns were directed at, and the receptive audience she encountered during her campaigns.

Led by the WCTU and the Prohibition Party, a second wave of Prohibition legislation was enacted throughout the country in the 1880s and early 1890s. Whereas the first wave of the 1850s had unsuccessfully tried to enforce Prohibition in thirteen states by legislative statutes, the second wave focused on constitutional amendments, which would be presumably more difficult to repeal. Unfortunately for the Prohibitionists, amendments were also more difficult to pass. While five states (Kansas, Maine, Rhode Island, South Dakota and North Dakota) adopted constitutional Prohibition, twelve other states rejected such amendments between 1887 and 1890. An alternative tactic was to pass local option laws in fifteen states, many of which were required to place the wet/dry issue before the voters in each annual election.

By 1892, this wave began to subside as it became apparent that the Prohibition Party would never achieve status as a serious threat to the Republicans and Democrats. Although they never polled more than 270,000 votes in a national election, the Prohibitionists' influence was felt for a time at the state level, and strong presidential candidates like John St. John could at times sway enough votes to spoil a national election for one of the major parties.

CHAPTER 7

The New Century and Prohibition

As the second great wave of Prohibition sentiment began to fade during the "Gay Nineties," a new phenomenon emerged to complement the waning influence of the Prohibition Party. What the Prohibitionists lacked in youth, dynamism, ruthlessness and bureaucratic efficiency was amply provided by their kindred spirits in the newly formed Antisaloon League of America. The league grew out of a movement in Ohio to persuade state legislators to adopt a local option statute. The success of this campaign, led by the Reverend H. H. Russell, led to the first convention of the Antisaloon League in 1895. While their ultimate and avowed purpose was to abolish the personal use of alcoholic beverages, they wisely sought to focus public attention on the abuses of the saloon. In the process they effectively molded a coalition of Prohibitionists, evangelical Christians, feminists, political activists and social reformers into one of the most influential pressure groups in American political history. More than any other group, the Antisaloon League was responsible for passing the Eighteenth Amendment to the Constitution.

The unprecedented success of the league can be traced to a number of innovations in the tactics of Temperance reform and to the inability of the saloon interests to marshal an adequate defense. Borrowing from the blueprint of the defunct American Temperance Society, Russell used the Protestant congregation as the league's basic organizational unit. The pulpits were a platform for soliciting donations, recruiting volunteers and disseminating propaganda; the league managed to enlist the support of approximately sixty thousand churches.

The second ingredient in the league's recipe for Prohibition was money. Large sums were raised from such millionaire notables as Andrew Carnegie, Pierre du Pont, Henry Ford, John D. Rockefeller, John Wanamaker and Samuel Kresge, but the bulk of the league's approximately two-million-dollar-a-year budget came from individual donations of less than a hundred dollars.

Yet another factor in the league's success was bureaucratic organization. Strict control was exercised from the central leadership in Ohio, which coordinated its national campaigns and assisted its state leagues with propaganda, finances and expertise. A paid, full-time staff of young, energetic, well-educated clergymen and professionals was responsible for the daily business of the league. Their work consisted of mapping strategy, mobilizing volunteers, recruiting support through paid or volunteer speakers, disseminating massive amounts of literature, studying candidates' voting records, compiling voting lists, working to support or defeat political candidates, lobbying legislators and drafting and introducing legislation, circulating petitions and organizing letter-writing and telegram campaigns. At the height of their influence, the league directed the world's largest prohibition propaganda mill. They were capable of fielding a veritable army of fifty thousand trained speakers nationwide to communicate their message. By the time national Prohibition was introduced in Congress in 1916, the league had developed the tactics of pressure politics to a fine art.

Wheelerism and the Booze Busters:

A final factor in the league's winning ways was its leadership. In contrast to the leadership of the Prohibition Party, the league's staff were on average ten years younger. While in many respects they were as provincial as the Prohibitionists, the league's penchant for rational, bureaucratic political techniques reflected the modernism of its professional organizers. This style was epitomized to such an extent by the indefatigable Wayne B. Wheeler that the league's critics frequently substituted the word "Wheelerism" for pressure politics. Once described by a "wet" editor as a man who "works with the zeal of a Savonarola and the craft of Machiavelli," Wheeler devoted his adult life to the destruction of the saloon. As a protégé of the Reverend Russell, he studied law, rose through the ranks, and became general counsel of the national league in 1916. At the height of his power Wheeler humbled congressmen, dictated to presidents, intimidated the major political parties, drafted the Prohibition Amendment, and wrote a major part of the Volstead Act that enforced it.

Another member of the league's hierarchy was James Cannon, Methodist bishop and senator from Virginia. Rabidly anti-Catholic, he described the Roman Church as "the mother of ignorance, superstition, intolerance and sin." In his opposition to dancing, drinking, the theater and other occasions of immorality, Cannon represented a strain of self-righteous bigotry and moral intolerance that surfaced from time to time in the WCTU, the Prohibition Party and the Antisaloon League itself.

Another of the league's exemplars was William "Pussyfoot" Johnson, whose nickname was earned by the stealth and cunning he demonstrated in stalking illegal saloonkeepers and corrupt brewers. Traveling around the country as a Temperance correspondent Johnson wrote exposés on drinking in colleges, voting frauds, bootlegging in the South, and illegal tippling by Washington state's legislators. In 1906 Teddy Roosevelt appointed him, at the recommendation of the league, a special officer to enforce the ban on liquor sales in the Indian territory of Oklahoma. Deputizing a band of gunslingers in Muskogee, Johnson spent three years as that territory's chief "Booze Buster." He later extended his operations to the Indian territories of Minnesota, Idaho, Montana, New Mexico and the Dakotas. Johnson and his men were responsible for shooting twenty-five outlaws, capturing 5,473 criminals (of whom more than 80 percent were convicted), and destroying hundreds of illegal liquor caches and frontier saloons. Later, he carried on his prohibition work as director of the World League against Alcoholism and as managing editor of the Antisaloon League's American Issue publications.

In one of those publications he is listed, along with a distinguished group of fifty-three other "Temperance Heroes and Martyrs," for an injury suffered during a Temperance debate in London. The debate ended prematurely when some college students began chanting:

> *Pussyfoot, Pussyfoot, why are we here?*
> *We've come to prevent you from stopping our beer.*
> *Pussyfoot, Pussyfoot, there'll be a big riot.*
> *We drink in pubs, you on the quiet.*

In the riot that ensued, Johnson lost an eye from a well-directed stone. Examples of other Temperance heroes, some of whom made the "supreme sacrifice," include the Reverend John A. Wilson, struck in the head with a ten-pound weight while conducting an antisaloon campaign in Delaware in 1874; John M. Pinkney, U.S. congressman from Texas, who was shot and killed while participating in a mass-meeting of Prohibitionists in 1890; Osborne Congleton, thrown into San Francisco Bay in 1890 while delivering a Temperance lecture; Isaac Cowen, a Prohibition candidate for Congress who was beaten by a drunken mob in Cleveland in 1892; Daniel E. Garry, who had his home dynamited in 1909 because of his leadership role in the Zanesville, Ohio, Civic League; and the Reverend R. E. McClure, an advocate of liquor-law enforcement whose breast pocket Bible stopped a bullet in 1913.

The De-Salooning of America:

Initially the Antisaloon League built broad support by concentrating on local and state legislation, which addressed the most widely recognized abuses associated with the saloon. These included laws against gambling and prostitution, restrictions on Sunday alcohol sales and the serving of minors, as well as the creation of dry zones around schools, colleges and military posts. In many states "high license" laws were supported to drive out marginal operators who could not afford the costly license fees.

Another part of the league's strategy was the support of various democratic reforms in the political process, particularly reforms that would facilitate the passage of anti-liquor legislation. For example, once a state had enacted the initiative and referendum amendments to the state constitution, party bosses could no longer block the direct introduction of antisaloon legislation to popular vote. For the same reason the league advocated laws substituting primary elections for the traditional nominating process controlled by party leaders. Now the league could support candidates whom they could trust to work for prohibition. Suspecting the potential of women in the electoral process, the league actively supported the Equal Rights Movement. This tactic seems to have worked to their advantage. Of the twelve states in which women could vote before 1920, ten had adopted laws prohibiting drinking by popular vote.

The ultimate tactic in the league's single-minded march toward national Prohibition was to dry up the country piecemeal through local option laws and statewide prohibition. In order to grant local towns and cities the power to ban saloons, it was first necessary to pass enabling laws in each state. Typically, petitions would be circulated through the churches, meetings and rallies would be organized, and volunteers would conduct door-to-door campaigns before the critical election. By 1900, thirty-seven states had enacted such laws, often through the newly acquired referendum mechanism. This permitted the Prohibition forces gradually to isolate the wet areas. In general, rural, Protestant and native-born voters tended to support local restrictions on the sale of alcohol, although it was not unusual for industrial cities to vote dry from time to time, as did Worcester, Massachusetts, and Birmingham, Alabama, in 1908.

Once the league had consolidated large areas of local prohibition within a state, an attempt would be made to eliminate the recalcitrant wet areas by state laws or amendments. The last great wave of prohibition sentiment began in 1907 when Georgia and Oklahoma enacted statewide prohibition. By 1916, twenty-three states had adopted prohibition, seventeen of these through referenda. Indicative of the popular support given to the league's campaigns, in 1914 the voters in Virginia, Oregon, Washington, Colorado and Arizona enacted prohibition laws after their state legislators had refused to pass similar statutes. As Bishop Cannon noted in a rare moment of "dry" humor, even the wets in Virginia were willing to support prohibition as long as they did not have to give up drinking.

Although a majority of the population was wholly or partially deprived of legal liquor by the beginning of the First World War, many states permitted beer and wine for personal use, and many others failed to enforce the laws they had enacted. Since interstate commerce was regulated by the federal government, liquor dealers in wet states conducted a lucrative business by exporting liquor to dry states. Even when interstate transportation of liquor was outlawed by the Webb-Kenyon Bill in 1913, enforcement was neither vigorous nor systematic. In general, voters supported prohibition in

Washington, Oklahoma and the South, while other areas, particularly around the large urban population centers, were more resistant.

THE MARCH TOWARD PROHIBITION

In 1913, the Antisaloon League declared itself in favor of national prohibition by amendment to the federal Constitution. To demonstrate their resolve to Congress, a parade led by children was organized on December 10 in Washington. Marching down Pennsylvania Avenue to the tune of "Onward, Christian Soldiers," four thousand members of the Antisaloon league and the WCTU converged on the capital where they were met by Congressman Richard Hobson and Senator Morris Sheppard. There the legislators were handed copies of a proposed Eighteenth Amendment to the Constitution, which was prefaced by the following paragraph:

> *Whereas, exact scientific research has demonstrated that alcohol is a narcotic poison, destructive and degenerating to the human organism, and that its distribution as a beverage or contained in foods, lays a staggering economic burden upon the shoulders of the people, lowers to an appalling degree the average standard of character of our citizenship, thereby undermining the public morals and the foundation of free institutions, produces widespread crime, pauperism and insanity, inflicts disease and untimely death upon hundreds of thousands of citizens, and blights with degeneracy their children unborn, threatening the future integrity and the very life of the nation; Therefore be it resolved: . . . that the following amendment of the Constitution . . . become valid as part of the Constitution. . . .*
>
> *Sec. 1. The sale, manufacture for sale, transportation for sale, importation for sale, and exportation for sale, of intoxicating liquors for beverage purposes in the United States and all territory subject to the jurisdiction thereof are forever prohibited. . . .*

The Hobson-Sheppard resolution, reported out of committee in 1914, was actively opposed by the U.S. Brewers Association and the National Wholesale Liquor Dealers' Association, whose maladroit tactics may have contributed to the eventual passage of the Eighteenth Amendment as much as the Prohibitionists' zeal. While the Antisaloon League and other prohibition organizations were sympathetic to equal rights issues, the brewers antagonized women's groups by their active opposition to women's suffrage. The brewers and distillers generated further hostility by their involvement in questionable election practices in Texas and Pennsylvania. What emerged was an image of the liquor industry as more concerned with their own profits than with social problems associated with alcohol. In the absence of any obvious attempts to reform the abuses of the saloon, public opinion was clearly on the side of the Prohibitionists, even while many of their supporters had no desire to ban the use of alcoholic beverages for personal use.

Because 1914 was an election year, the Antisaloon League mobilized its considerable resources in a massive effort to elect a majority of Prohibitionists to the next Congress. Operating through the thousands of Protestant denominations affiliated with the league, the campaign began with an avalanche of letters, telegrams and petitions. At the same time, the league supported speakers and sought every opportunity to speak in favor of the amendment in villages, towns and cities throughout the country. Their attack on the saloon was buttressed with a massive outpouring of literature designed to portray the liquor industry as the "drunkard-making business." Finally, intensive lobbying was conducted to persuade candidates that their position on prohibition could mean the difference between election and defeat.

Despite an increase in dry seats gained in both houses of Congress, the Hobson-Sheppard resolution fell short of the two-thirds majority required before it could be sent to the states for ratification. Following Wayne Wheeler's advice, the resolution was not resubmitted again until after the 1916 elections, giving the league additional time to consolidate further its gains on the state and local levels. By that time, the First World War was in its second year, and the impending involvement of U.S. forces threatened to overwhelm the prohibition issue. As President Wilson prepared to ask Congress for a declaration of war, the Prohibition forces seized the moment to include dry legislation in the President's war program, under the guise of food conservation. Noting that liquor production used up large quantities of grain, sugar and other foodstuffs, the Prohibitionists asserted that drinking was unpatriotic. To reinforce their point, the league argued that "Liquor is a menace to patriotism because it puts beer before country." Thanks to the beer industry, their argument was given some credence by evidence that the brewers actively supported the antiprohibitionist national German-American Alliance. At the instigation of Wayne Wheeler, whose rallying cry was "Kaiserism abroad and booze at home must go," the alliance's charter was revoked and its leadership disbanded.

Although the liquor forces could not prevent prohibition of distilled beverages, they did secure a compromise that allowed production of beer and wine. The final version of the Food Control Bill took effect on September 8, 1917. Realizing that the Prohibition Amendment now could be pushed as a wartime measure, the drys recommended several changes in the original resolution before resubmitting it to the Congress. Most notably, the words "for sale" were omitted as qualifiers of manufacture, transportation, importation and exportation, thereby prohibiting the private, as well as the public, use of alcohol. In addition, the word "purchase" also was excluded, thus absolving the drinker from legal responsibility when not involved directly in the manufacture and sale of alcohol.

THE SUMMIT SCALED

By December 1917, the modified amendment had passed handily in both the House and the Senate. To take effect as the Eighteenth Amendment, however, ratification by thirty-six states was required within the next seven years. Given the fate of numerous other proposed constitutional amendments, the wets felt confident that the Prohibition Amendment would languish and eventually die as the war effort dominated the national agenda. But as subsequent events proved, this was a serious miscalculation. Given the Antisaloon League's skill in exploiting the propaganda value of the war effort, and their organizational experience in dealing with state legislatures, the amendment progressed through the states in record time to become law on January 16, 1919. To add insult to injury, further wartime restrictions were included in the Agricultural Appropriations Bill of November 1918. Ironically, both laws were passed within a short time after the Armistice had been signed in Paris.

The Volstead Act:

In order to provide for enforcement of both the Eighteenth Amendment and wartime prohibition (which by now applied by extension to the period of demobilization), it was necessary to draft the National Prohibition Act. Although this law was drafted by Wayne Wheeler, it took its popular name from the man responsible for introducing it into the House, Representative Andrew Volstead. Contrary to his later reputation, Volstead was not an ardent Prohibitionist. In fact, he actually opposed prohibition candidates in several elections. He did believe, however, that "law has regulated morality since the Ten Commandments," and because prohibition was law in many of the townships under his jurisdiction, he considered it a moral imperative to prosecute violators.

The Volstead Act was designed to "prohibit intoxicating beverages"; to regulate the "manufacture, production, use and sale of high proof spirits for other than beverage purposes"; to "insure an ample supply of alcohol, and promote its use in scientific research and in the development of fuel, dye and other lawful purposes." The act was intended to combine selected aspects of state prohibition laws into a single statute. The result was not a simple legal statement, but an eclectic assortment of often contradictory codes, totaling seventy-two sections in all.

President Wilson took issue with Wheeler and Volstead and their supporters and vetoed the Volstead Act on the grounds that the war emergency had long passed, and it had been used as a pretense for constitutional prohibition. In his explanation of the veto, Wilson stated prophetically that:

> In all matters having to do with personal habits and customs of large numbers of our people, we must be certain that the established processes of legal change are followed. In no other way can the salutary objects sought to be accomplished by great reforms of this character be made satisfactory and permanent.

Unconvinced by these arguments, Congress easily overrode the President's veto and the bill became law. The Antisaloon League had triumphed, and an estimated 170,000 saloons were destroyed.

PROHIBITION IN PERSPECTIVE

The enactment of the Volstead Act culminated one hundred years of Prohibition agitation aimed at the legal control of alcoholic beverages. Only by viewing national prohibition in the broadest histori-

cal context can an explanation be found for why Prohibition succeeded, and how it ultimately failed. Historical explanations of Prohibition have been as varied and divided as the long debate over
legal control itself, reflecting as much about the spirit of the times
and the individual historian's biases as they do about the evidence
available from historical sources.

For a period after repeal, historians saw Prohibition as a pragmatic reform in the great tradition of other amendments to the
Constitution, such as the abolition of slavery, extension of voting
rights to women, the direct election of U.S. senators, the institution
of a progressive income tax and the electoral reforms permitting the
initiative recall and referendum. Samuel Eliot Morison and Henry
Steele Commager in their popular textbook *The Growth of the American Republic* (1930) described prohibition and women's suffrage as
companion reforms that called on the state to "do whatever might
be necessary to promote the health, happiness, and morality of its
citizens."

Interpretations focusing on the motives and psychology of the
Prohibitionists gained precedence during the 1950s and 1960s. From
these interpretations have emerged an image of the Prohibitionists
as a mixed group of austere Puritans, health cranks, commercial
evangelists, designing capitalists and bigoted racists who perpetrated a hypocritical deception on the American people. Andrew
Sinclair, for example, in his book *Prohibition: The Era of Excess* (1962),
saw Prohibition as an attempt by latter day Puritans "to extend
their own repression to all society." According to this interpretation,
the Prohibitionists succeeded in spreading this "rural-evangelical"
virus by masking their religious zeal in "aggressive prurience." One
of the most influential interpretations, Joseph Gusfield's *Symbolic
Crusade* (1963), did not relegate Prohibition to the category of mass
hysteria, but nevertheless focused more on the motives of the reformers than on the drinking of the reformed. Gusfield's explanation
relies on the assumption that abstinence was a symbol used by middle-class Americans to recoup the status they had lost as guardians
of the American dream.

In recent years, historians and social scientists have approached
Prohibition with a new set of theories and analytic methods. These
newer interpretations have brought perspective and depth to the
paradox of Prohibition, focusing less on the motives of the reformers
and more on the social conditions that gave rise to reform. By sifting

through the historical records of those who supported Prohibition and what they attempted to accomplish, many previous assumptions about the nature of Prohibition have been called into question, not because they were erroneous, but rather because they were incomplete. Thus, serious objections can be raised to the argument that Prohibition was a fanatical movement, although fanatics certainly took part in it; that Prohibition was a consequence of war hysteria, although the war certainly accelerated the prevailing social trend; or that Prohibition was foisted on America by a radical minority, even though the majority never expected America to become "bone dry."

The first objection to these traditional historical interpretations can be found in Krout's classic on the *Origins of Prohibition* (1925). To Krout, Prohibition was not the result of temporary conditions, but the final expression of a fundamental process that had been more than a century in the making. The abolition of alcohol became the final objective of the Prohibition Movement, but its ultimate goal had always been the prevention of alcohol problems. Its supporters were drawn from every walk of life, in various times and places and a multitude of motives inspired those who labored in its cause. For many, abstinence embodied not only a symbol of moral superiority, but also the key to a more perfect social order, and more pragmatically a "cure" for disease, poverty and alcoholism.

There is little doubt that the Antisaloon League constituted a radical minority obsessed with the goal of abolishing alcohol entirely from American society. However, without the support of millions of Americans, in every state in the union, their pressure politics and propaganda would have achieved little. In only four years, between 1916 and 1919, prohibitionists won twenty-five out of thirty-five state referenda on the constitutional amendment, a striking testimony to Prohibition's popular support. Impressive as these victories were, they did not necessarily mean that people were opposed to drinking *per se*, since many interpreted the issue in terms of the abolition of the saloon. Further, careful analysis of the voting trends indicates that a majority of Americans may not have supported Prohibition before their legislatures ratified the Eighteenth Amendment. In general, the larger the state, the less likely it was to approve Prohibition by referendum, although prohibition sentiment was strong in rural as well as urban states.

The twin influences of class and ethnicity appeared even more powerful than ruralism in garnering support for prohibition. Voters in the dry areas were middle-class, Protestant, native born and socially mobile. Those in the predominantly wet areas tended to be working-class ethnic minorities and immigrants. In both rural areas and cities, it was the middle class that consistently voted for alcohol controls. To these Americans, Prohibition was both a conservative and a humanitarian reform. It was conservative in the sense the middle class believed that through the control of alcohol, it could protect home and family from the massive social disorder, social change and confusion of values they perceived all around them. If the Prohibition crusade symbolized anything to these people, it was the abolition of the saloon, and with it the rampant social disorder they believed saloons were responsible for. As seen from the perspective of 1930, Prohibition had tried to cleanse America of the demoralizing and corrupting influence of alcohol and restore her lost innocence. Prohibitionists envisioned an "unsullied generation which would regard drinking as a moral perversion and the purveyor as a felon."

The perennial seekers of simple solutions to complex problems did not appreciate that foreign immigration, westward migration, industrial expansion and urban growth were as much a part of this process as the saloon. Nor was the saloon the only enemy of the prohibition advocates. Poverty, tuberculosis, unsanitary living conditions, prostitution, and inhuman working conditions were related targets for the Prohibitionists. While few would argue that the progressive movement, which Temperance supporters were frequently allied with, was not humanitarian, what seems most inconsistent with those ideals was the Prohibitionists' insistence that all others adhere to their standards of morality.

The fact that religious leaders were in the forefront of the prohibition movement has contributed to the impression that this was a moral crusade against the irreligious or unholy masses. Undoubtedly many Protestant evangelicals used prohibition as a means to spread salvation, but it seems that the majority of Prohibition supporters were searching for an alternative to the Church's waning moral authority. If the Church could not exert sufficient authority, the alternative was to demand that state governments assume the role of moral stewardship.

In an era when science rivaled religion as the key to progress and salvation, the arguments advanced by the Prohibitionists were decisive in molding antialcohol sentiment. Prohibitionist propaganda was a curious mixture of scientific fact and fancy, which relied on the American medical profession and the emerging social science disciplines as unwitting accomplices. Medical studies showing that alcohol was a depressant drug totally lacking in nutritional value were used to support the Prohibitionists' claim that alcohol reduced industrial efficiency. Evidence of the toxic effects of chronic drinking was used to convince abstainers and schoolchildren that alcohol was a poison. Life insurance data showing that heavy drinkers died younger, and medical reports that deformed children were born to parents who drank, were presented as further justification for Prohibition. Finally, the growing interest in "statistical science" gave credibility to the argument that the mere association of alcohol with crime, disease, poverty and insanity was proof that drinking "caused" these conditions. At a time when public health workers were using government intervention to fight dysentery, tuberculosis and other communicable diseases, many found the Prohibitionists' proposed solution to social problems to be within the proper sphere of government.

Perhaps the most convincing evidence that Prohibition was more than a mindless aberration of rural American evangelicals is the remarkable parallelism between prohibition movements throughout the industrialized world. During the first quarter of the twentieth century almost every European nation joined North America in viewing alcohol control as an issue of urgent social concern. This concern was translated into prohibition legislation in Austria, Belgium, England, Finland, Iceland, Norway and Russia. While most of these nations lifted these restrictions after the First World War, Finland and the United States maintained prohibition for some time after the Armistice. The major characteristics that distinguished these two nations were political pluralism and social heterogeneity.

From the perspective of today, it seems there is still much to be learned from Prohibition as an experiment, however misguided, in social reform. So little attention has been focused on the ways people drank, and the problems associated with alcohol abuse, that it is premature to reject the Prohibitionists' arguments as completely distorted fabrications. Between 1900 and 1914 per capita alcohol consumption rose to the highest levels since the 1830s. Given the legal restrictions on alcohol availability in most areas of the country outside the major urban areas, the greatest part of the nation's drinking was heavily concentrated in a small proportion of the male population: to most Americans, particularly those living close to the urban saloons, the stereotype portrayed in the "Temperance Tales" was consistent with what they saw. A simplistic dichotomy between the evils of alcohol and the merits of abstemiousness seemed confirmed in common experience. Those who drank, drank excessively, and caused problems for themselves and others. Those who did not drink, on the other hand, appeared largely free of problems. In a society where heavy drinking and abstinence seemed to be the only visible alternatives, moderation was not considered achievable. In countries like France, Spain and Italy, where heavy drinking was masked by the majority's example of moderation, temperance and prohibition movements were virtually unknown, not because there were no alcohol problems, but rather because moderation was the norm rather than the exception.

CRIME AND THE CALL FOR REPEAL

Given the preceding account of why Prohibition succeeded, how then are we to explain the definitive rejection of the Eighteenth Amendment only thirteen years after its passage? In many respects the repeal movement bears a striking resemblance to the movement that culminated in Prohibition: both succeeded during periods of national emergency; both were spearheaded by powerful, well-organized, political pressure groups; each was seen as a solution to the social problems of the day; and both emerged from a complex matrix of social and economic changes. However, the obvious major difference between Prohibition and repeal movements was the change in public attitudes toward alcohol beverage control. Prohibition was consistent with the attitudes of those supporting social reform, but once the saloon had been destroyed, repeal of Prohibition became consistent with a host of additional reform attitudes, including prevention of crime, corruption and disrespect for the law.

Ironically, the seeds of repeal were sown by the compromises that the Antisaloon League made to secure passage of the Volstead Act and the Eighteenth Amendment. From the very beginning there was ambiguity about the definition of "intoxicating liquors," which the Eighteenth Amendment prohibited. Until the courts fixed the alcohol content of "intoxicating liquors" at five percent, many believed that beer and wine would be permitted. In fact, the minimal alcohol content provision provided a loophole that was exploited by the brewers, since they continued production of "near-beer," which ostensibly contained less than five percent alcohol. The lack of federal inspection soon made it possible to produce large quantities of beer at the normal 3 to 5 percent range.

Another source of ambiguity in the law was the provision concerning enforcement; this responsibility was assigned to both the states and the federal government. Although some states eventually passed "Baby Volstead" acts to authorize enforcement at the state level, most were content to rely on the federal authorities. The enforcement division of the federal Prohibition Bureau was originally staffed with fifteen hundred agents. The division grew to as many as three thousand, with additional help from Customs agents, the Coast Guard and the Immigration Service, but this number proved no match for the challenge posed by millions of ingenious smugglers, home brewers and commercial distillers.

Enforcers and Evaders:

Illegal smuggling, at first amateurish and disorganized, grew into a major enterprise involving an elaborate network of land, sea and even air transportation. Beer was brought across the Canadian and Mexican borders by rail, truck, passenger cars, speedboats and, according to one account, by airplanes practicing midair "refueling" during clandestine rendezvous over the northern border. Rum and Scotch whiskey were supplied to the Atlantic Coast from the West Indies at first by independent entrepreneurs, and later by organized crime rings. The rum trade, which had been a lucrative industry for British Nassau and other Caribbean islands in Colonial times, was revived with the full acquiescence of government authorities, who realized the enormous profits involved in charging import duties on liquor transshipped from the British Isles. Though huge amounts of contraband were seized each year, in 1926 Treasury Secretary Andrews estimated that only 5 percent of the actual liquor smuggled was being seized.

Home Brews:

Another source of illegal supply was home brewing. In a veritable renaissance of domestic spirits production reminiscent of Colonial times, many relearned and improved techniques common in seventeenth-century America. Beer had maintained its favored status, and enormous sums were expended for home-brewing supplies like malt, hops, yeast, bottles, crown caps and other paraphernalia. Perhaps as many as a hundred thousand malt and hop shops were in operation by the time of repeal. Beer brewing became a family pastime in many ethnic neighborhoods, and the custom inspired the following poem:

> *Mother's in the kitchen*
> *Washing out the jugs;*
> *Sister's in the pantry*
> *Bottling the suds;*
> *Father's in the cellar*
> *Mixing up the hops;*
> *Johnny's on the front porch*
> *Watching for the cops.*

Because Section 29 of the Volstead Act permitted the manufacture and possession of fruit juices for home consumption, the home winemaking industry also enjoyed enormous growth. In addition to the winemaking supplies that were marketed in every major city, California's grape growers increased their acreage about 700 percent during the first five years of Prohibition. Fresh grapes, grape concentrate, frozen "bricks of Bacchus," and grape juice were readily available to consumers. Grape juice was often labeled with the warning: "CAUTION: WILL FERMENT AND TURN INTO WINE." Customers could select grape concentrates packaged as particular wines, blocks of Port, or Rhine wine, etc., and dissolve these in water. Warning labels advised against storing the solution for twenty days because it would then turn to wine.

With appropriate technical guidance and the purchase of a cheap portable still, many Americans took up home distillation as well. Once alcohol had been produced from boiling mash generated from crushed fruits, grain, sugar beets or potato peelings, various types of liquor could be simulated by means of the proper additives. A popular favorite was "bathtub gin," produced by combining the distillate with water, glycerine and juniper oil.

The diversion of commercial liquor, ostensibly produced for medicinal, religious or industrial purposes, was yet another source of concern for the harried Prohibition agents. Spurious prescriptions for alcohol were written by thousands of accommodating physicians. The production of sacramental wines was soon diverted from the sacred to the secular market. The output of industrial alcohols provided the greatest source of illegal diversion. These highly toxic wood alcohols found their way into much of the available bootleg liquor. When denatured industrial alcohol was not sufficiently diluted, or was consumed in large quantities, the result was paralysis, blindness and death. In 1927, almost twelve thousand deaths were attributed to alcohol poisonings, many of these among the urban poor who could not afford imported liquors. In 1930, U.S. public health officials estimated that fifteen thousand persons were afflicted with "jake foot," a debilitating paralysis of the hands and feet brought on by drinking denatured alcohol flavored with ginger root. These casualties were particularly numerous in the South among both poor blacks and whites.

The Speakeasy:

Given the abundant supply of both domestic and imported alcohol, there was no shortage of retail outlets to bring it to the waiting consumer, who, thanks to the provisions of the Volstead Act, was under no legal obligation to refrain from drinking. Although liquor was relatively difficult to obtain in rural areas, particularly those that had been dry even before Prohibition, there were no similar impediments in most urban areas. One Prohibition agent conducted an informal study of the time required to purchase liquor in the cities he visited. A New Orleans taxi driver purchased a drink for the agent in thirty-five seconds, garnering first place for that city. The runners-up were Pittsburgh at eleven minutes, Atlanta at seventeen minutes, Chicago at twenty-one minutes and Washington at one hour. Despite the inconvenience a stranger might encounter in obtaining a drink in the nation's capital, there was apparently little effort involved for the resident lawmakers. Bootleggers were known to roam the halls of Congress peddling their wares with impunity, and

President Harding served liquor to his Ohio cronies during their weekly poker games at the White House.

New York City was perhaps the easiest place to drink during Prohibition, as suggested by the results of a *New York Telegram* investigation conducted in Manhattan. The newspaper's reporters were able to purchase liquor in drugstores, delicatessens, confectionaries, soda fountains, cigar stores, barbershops, hotels, paint stores, malt shops, groceries, athletic clubs, taverns, tearooms, spaghetti houses, political clubs (Republican and Democratic), laundries, and newspapermen's associations.

By all accounts the speakeasy was the most popular place to buy and enjoy alcohol. Many speakeasies resembled the urban saloon in size and appearance, but the clientele were often quite different from the saloon patrons of old. In Manhattan, for example, many private clubs catered to the rich, the famous and the affluent, in part because these were the only people able to afford the exorbitant prices charged for imported alcohol or quality domestic products. During the Prohibition period, the price of liquor by the drink increased by 300 percent to 600 percent. However, by excluding the urban poor from the speakeasy, and by limiting the availability of alcohol through retail outlets, Prohibition actually succeeded in reducing the average consumption of working-class men and women. This factor, and the necessity for sober comportment in public settings, contributed to a dramatic decline in a large number of alcohol-related problems. After the imposition of war restrictions on alcohol in 1917, there was a marked reduction in arrests for drunkenness, hospital admissions for alcoholic psychosis, as well as mortality from alcoholism and liver cirrhosis. This decline continued until about 1922, when a reverse trend began. By the 1930s many of these indicators were approaching their prewar levels, providing ample justification for the wets' contention that Prohibition was not working.

These trends closely parallel the changes in per capita consumption during the 1920s, which in turn reflected the improved marketing of illegal alcohol. Accompanying the temporary suppression of average alcohol consumption and its rapid resurgence was a major shift in who was drinking and what was being drunk. The best available evidence suggests that the business, professional and salaried workers were consuming disproportionate amounts of alcohol, primarily in the form of distilled liquor. Thus it appears that Prohibition succeeded in abolishing the saloon and sobering up the working class at the expense of introducing hard liquor to a whole new generation of young, upwardly mobile middle-class Americans.

This change in the drinking public and the nation's drinking patterns contributed to the erosion in public confidence in Prohibition after 1930. Once the stereotyped images of the saloon and the drunkard had faded from public consciousness, a new image of alcohol emerged, which drew on its associations with the "Roaring Twenties." This view, soon to be refined with great success by the image-makers of Hollywood and Madison Avenue, portrayed drinking as fashionable, defiant, trend-setting, sophisticated and convivial; a perfect complement to the new leisure-class life-style made possible by the automobile, the cinema, the radio and the phonograph.

Scandals and Confusion:

The glamorous image of drinking was further enhanced by the various problems related to the enforcement of the Volstead Act. A major problem dating from the inception of the act was the exclusion of the Prohibition Bureau from Civil Service requirements. Intended as a compromise to win wet votes, this loophole permitted congressmen to use the bureau for political patronage. Not only did dry agents justifiably earn their reputation as being generally incompetent, they also seemed to be more susceptible to political corruption than other parts of the government. Payoffs and bribes were commonplace in and out of the Prohibition Bureau, particularly as organized crime became entrenched in the liquor business. Major scandals appeared with increasing frequency, and in 1928 virtually the entire political establishment of Pittsburgh was indicted for conspiring with criminals to monopolize the city's liquor traffic. In New York, racketeer "Big" Bill Dwyer built an empire of nightclubs, breweries, gambling casinos, racetracks and organized sports on the profits of his East Coast rum-running operation, the success of which depended on the collusion of four Coast Guard ships which he counted in his employ. In Chicago, the infamous Al Capone made the successful transition from saloon to speakeasy by building on his experience with organized prostitution.

In spite of these insurmountable obstacles to effective enforcement, the Prohibition Bureau continued to conduct its affairs under the watchful eye of the Antisaloon League. Between 1920 and 1930 its agents arrested more than 500,000 offenders, and seized more than 1,600,000 distilleries, 9,000,000 gallons of spirits and one billion gallons of malt liquor. In the process, Prohibition agents, by conservative estimate, shot and killed 137 persons, prompting some newspapers to condemn the bureau's "license to kill." Nor was the image of the Prohibition Bureau improved by their actions against an illegal distillery in northern California. Discovering a fifty gallon still secreted within a hollowed-out redwood tree, the dutiful agents posted the "premises" with the following notice: "CLOSED FOR ONE YEAR FOR VIOLATION OF THE NATIONAL PROHIBITION ACT."

Given the absurdities of closing down redwood trees while at the same time countenancing the most blatant violations of the Volstead Act, it is surprising that the enforcement of Prohibition met with so little public disfavor until 1930. Indeed, the liquor question was a major issue in the presidential election of 1928, when Al Smith was defeated in part because he advocated a change in the definition of "intoxicating liquors" to exclude beer and wine. With the continued endorsement of dry referenda throughout the nation, and the election of a strong majority of "dry" legislators to Congress, Prohibition seemed to be well entrenched in the American legal structure.

The fragile nature of national Prohibition as an experiment in social reform became apparent following the crucial events that occurred between the 1928 and 1932 presidential elections. The accumulating evidence that Prohibition was failing to prohibit prompted the drys to demand more stringent enforcement of the Volstead Act. Republican Herbert Hoover was not elected entirely by the drys, but nevertheless was committed to defending the Eighteenth Amendment. This commitment was expressed in his now famous and often misquoted statement that "Our country has deliberately undertaken a great social and economic experiment, noble in motive and far-reaching in purpose. It must be worked out constructively." By this he meant that honest enforcement and the elimination of abuses would be a primary concern of his administration.

To this end, Hoover endorsed the Jones "Five and Ten Law,"

which increased the penalties for bootlegging to five years in prison and/or a $10,000 fine, and, for the first time, threatened purchasers with felony convictions if they failed to inform on their sources. The law was castigated by the Hearst newspapers, which accused the Prohibitionists of creating a new category of political prisoner. Supported by the Antisaloon League, the Jones law was another example of the radical changes taking place in that organization since the death of Wayne Wheeler in 1927. With Bishop James Cannon now in charge, the league's still considerable propaganda machine was used to spread Cannon's own distinctive brand of nativist and anti-Catholic bigotry, and Governor Smith was one of the primary targets. When accusations of adultery and misappropriation of campaign funds were brought against Bishop Cannon in 1930, the league shared in his tarnished image.

Repeal Attained:

As the Antisaloon League's support began to erode, the movement for repeal, active since the early 1920s, gained momentum. At the forefront of the movement was the Association Against the Prohibition Amendment (AAPA), an organization supported by several dozen millionaires who were convinced that repeal would reduce their tax burden and, after the stock market crash, hasten economic recovery by putting people to work. Actively supported by the brewers and distillers, the AAPA skillfully manipulated public opinion in a propaganda campaign reminiscent of the Antisaloon League. They argued that Prohibition actually caused an increase in drinking, eroded public confidence in government, corrupted public officials, interfered with individual rights, and threatened to bring on a Bolshevist revolution. In a flurry of rhetoric worthy of the century-long debate, AAPA described Prohibition as "wrong in principle, has been equally disastrous in consequences in the hypocrisy, the corruption, the tragic loss of life and the appalling increase of crime which have attended the abortive attempt to enforce it; in the checking of the steady growth of temperance which had preceded it; in the shocking effect it has upon the youth of the nation; in the impairment of constitutional guarantees of individual rights; in the weakening of the sense of solidarity between the citizens and the government which is the only sure basis of a country's strength."

Clearly the AAPA and its supporters offered a new vision of the proper role of government in the management of social problems. They challenged government interference in local decision-making and individual initiative. But perhaps the most telling argument of all was the AAPA's economic analysis of the real cost of Prohibition. It was estimated that eleven billion dollars in tax revenues from alcohol had been lost between 1920 and 1931 and projected that these lost revenues would have "balanced the budget" by 1933. Then as now, balancing the budget was seen by many as the ultimate justification for political action. Economic improvement and civil liberty proved an irresistible combination in Depression-racked America.

By 1932, the AAPA claimed 550,000 members — a formidable coalition of small contributors and prominent millionaires. Joining forces with other pro-repeal organizations, the AAPA lobbied successfully with a "wet" plank in the Democratic party platform. Roosevelt promised repeal, later denounced Prohibition as a damnable affliction, and legalized beer sales within days of his inauguration. The Democratic landslide brought a wet majority to Congress that immediately called for Prohibition repeal. Once the amendment resolution passed the Congress, it was sent to specially elected state constitutonal conventions for ratification. Delegates who pledged to support repeal received overwhelming majorities in the popular elections held throughout the country in 1933. The long thirst was almost over.

CHAPTER 8

The Legacy of the Radical
Temperance Movement

Prohibition was repealed on December 5, 1933, with the adoption of the Twenty-first Amendment to the Constitution. The amendment contained two short sentences.

Section 1: The eighteenth article of amendment to the Constitution of the United States is hereby repealed.

Section 2: The transportation or importation into any State, Territory, or Possession of the United States for delivery or use therein of intoxicating liquors, in violation of the laws thereof, is hereby prohibited.

Both sections were crucial to the subsequent development of drinking in America. While the first made possible the importation, manufacture and sale of alcohol, the second delegated to the states all authority for regulation. One reason why the federal government relinquished its authority to the states is undoubtedly ideological. Repeal occurred just before the New Deal was to revolutionize thinking about the role of government in local affairs, and it reflected the prevailing philosophy that the individual states were the best authorities on how to control the manufacture, use and availability of alcoholic beverages. Given the ethnic, religious and social diversity of the United States, the government decision not to impose any guidelines on the states was probably wise. Nevertheless, it is significant that the federal government did retain the authority to tax alcohol that it had assumed during Prohibition. Inasmuch as the revenue potential of alcohol was a primary argument for repeal, it is conceivable that the amendment would not have passed without the tax provisions.

The basic mechanism by which the federal government sought to control alcohol's manufacture and distribution at the federal level was the Federal Alcohol Administration Act passed in 1935. Responsibility for the administration of this and related laws was invested in what has become the Bureau of Alcohol, Tobacco and Firearms in the Department of the Treasury. The bureau's functions have come to include the issuing of "basic permits" to importers, manufacturers, wholesalers, and warehousers to conduct business; tax collection; interdiction of illicit alcohol; regulation of labeling and advertising of alcoholic beverages.

A curious example of the government's role in labeling and advertising of alcohol can be found in a regulation that prohibits malt beverages and wine under 14 percent alcohol to advertise or to list their alcohol content (unless required by state law). Implemented in 1935, this regulation was apparently designed to correct advertising abuses that followed repeal. Some manufacturers and distributors, taking advantage of the pro-liquor psychology that followed repeal, began to use such terms as "strong," "extra strength," "high test," "high-proof," and "pre-war strength," in their labeling and advertising.

Another example of the government's exercise of its regulatory power lies in its differential taxation of beer, wine and distilled spirits. From the very beginning of repeal, spirits were taxed at a much higher rate than fermented beverages, reflecting Congress's belief that the latter were beverages of moderation. These examples show that the federal government did not withdraw completely from alcoholic beverage control after repeal; considerations of public health, consumer protection and tax collection guided its policies.

The states used their authority to control alcoholic beverages to develop a variety of different systems to govern manufacture, distribution and sale. Thirty states decided to allow legal availability of alcohol, while eighteen states continued Prohibition. It was not until 1966 that the last state went wet. Almost two-thirds of the states adopted some form of local option, permitting citizens in a designated area to vote for or against local Prohibition. Thus, despite the repeal of national Prohibition in 1933, 38 percent of the population still lived in dry areas where alcohol remained illegal.

In states that permitted the sale of alcoholic beverages, two kinds of distribution systems emerged: the control, or state monopoly system, and the license system. Since repeal, thirty-four states have chosen to regulate availability by the license system, whereby the state grants manufacturers, wholesalers and retailers the privilege of conducting business. Sixteen states adopted the control system, in which ownership and operation of wholesale and retail operations are controlled by the state government.

The enabling legislation adopted in most states following repeal often discussed the purposes for which the specific alcohol-control regulations were being adopted. Typically a variety of socially desirable ends were mentioned, including health, morals, protection, safety and welfare, as well as the promotion of temperance and the prohibition of the saloon. The provision of an orderly, regulated, gangster-free alcoholic beverage market was considered a prime goal of the new laws. The relative attention given to economic, health and moral issues shifted over time. The concern for public morality was initially reflected in laws restricting alcohol sales to certain times and locations (e.g., away from churches) as well as restrictions on advertising. Over the years, however, health and moral concerns have diminished to the point that regulations are primarily seen in terms of tax revenues or industry concerns. Federal taxes on alcohol remained stable after 1951, whereas state taxes tended to increase to meet revenue needs, rather than as an instrument for control.

As the Depression Era was brought to a close by the Second World War, Americans had experienced almost ten years of legalized drinking to form an opinion on the necessity for wartime restrictions. In the First World War, drinking was strictly controlled and labeled unpatriotic and wasteful of food resources. The Second World War prompted no dramatic increase in sentiment for liquor control. The contrast is striking, and it suggests that there is no necessary relation between wartime conditions and alcohol control.

Despite the recommendations of Temperance and religious leaders, there was little support in Congress or in the military for restrictions on manufacture or limitations on availability of alcohol. Although scarcity of raw materials and the absence of a large segment of the drinking age population reduced both supply and demand, beer was readily available at the military camps. What accounted for this dramatic change in opinion concerning the need for alcohol control during national emergencies? One factor may have been the relative sobriety of the drinking population during the Depression years. From repeal until the outbreak of war, beer consumption had remained well below the pre-Prohibition high attained in 1914. The Depression economy, high prices, and the lingering effects of Prohibition undoubtedly affected the demand for alcohol during this period. Another contributing factor may have been the restrictions of immigration imposed during the 1920s. As the previous waves of hard-drinking immigrants became older and more assimilated, they may have abandoned their harmful drinking habits.

Whatever the reason, drinking problems were far less visible on the eve of the Second World War, and there were relatively few temperance advocates proselytizing for a return to Prohibition. Discredited by the negative image of the Volstead years, an image promoted with great skill by the brewery industry, the "drys" could no longer command the kind of support they previously enjoyed. Lawmakers were reluctant to tamper with what had become a valued revenue source. In the military, antiprohibition sentiment was strong, particularly in view of the threat that Prohibition would again be imposed while servicemen were away.

Disenchantment with the temperance cause and the lack of tangible evidence that restrictions were needed thus seem to account for the different approaches to alcohol control assumed during the two world wars. What is common to these different approaches, however, is the fact that both were consistent with the social trends prevailing on the eve of hostilities. As America emerged from the Second World War, it appeared that the long national conflict over the temperance issue had finally been put to rest.

COULD IT HAPPEN AGAIN?

Has support for Prohibition disappeared in America today, or does it lay dormant, awaiting the right issue, the new popular coalition? In 1981, historians Aaron and Musto concluded that

> *Prohibition will certainly never return. Above and beyond the mechanical problems of enforcement, it failed originally because it created no stabilizing vested interests. No reform movement can survive unless it is rooted in*

new institutions. But while extreme forms of controlling consumption of alcohol are utterly lacking in feasibility, there is the chance that state policy may once again assume a more interventionist role. The boundaries between personal and governmental responsibility constantly shift. Although a mass movement to curb drinking will never re-emerge, one can conceive that new, extensive regulation of the liquor industry might be integrated into a paradigm of environmental safeguards and corporate responsibility.

Aaron and Musto go on to qualify this conclusion, pointing out that "The conditions are present for a revival of widespread interest in the problem of alcohol. No one can predict if such a resurgence of popular concern will in fact develop, but the record of the past suggests that movements once thought safely interred do not always remain in their graves."

Contemporary Drinking Patterns:

As we have seen, the prevailing pattern of drinking in the early 1800s and again at the end of that century influenced public opinion and helped to justify the evolution of the Temperance Movement into Prohibition. Following the Second World War and throughout the decades of the fifties, sixties and seventies, American drinking behavior has remained relatively stable. But a number of subtle and important changes in both economic diversification and consolidation marked the beginning of a new era of alcohol production as well as alcohol consumption by the American public.

Wine has traditionally been associated with moderate drinking — it is usually drunk with meals, and this has generally been applauded as a sensible drinking pattern. Rapid advances in technology plus economic incentives have stimulated a great increase in viniculture and wine production, first in California and subsequently in many other regions of the United States. Wine consumption has increased while consumption of distilled spirits has shown moderate fluctuation and most recently some decline. The American wine industry once focused upon production of large quantities of good-quality standardized white and red wines, which were usually disdained by connoisseurs. Recently, the wine industry has expanded into the production of *varietal* wines, many of extraordinary excellence, which now challenge the status of the most prestigious French imports. The growing number of independent wine producers has been paralleled by the acquisition of quality-oriented vintners by major conglomerates. For example, the Coca-Cola Corporation owns and operates the Sterling vineyards in the Napa Valley and the Paul Masson Winery is currently owned and operated by Joseph E. Seagram and Sons, a major producer of distilled spirits.

At the same time technologic advances in brewing, plus fierce competition among brewers, have resulted in a major consolidation within the brewing industry; fewer facilities now produce larger quantities of beer.

A Faint Echo of Temperance:

If the Temperance Movement has not disappeared entirely, it has gained a new identity and ideology. New, at least, for the 1980s but in some ways rather similar to the early arguments that supported temperance reform. In the 1980s, the effect of alcohol use and abuse on health and the attendant economic costs to the American nation have commanded increasing attention. Well-publicized fears about the impact of alcohol upon the health and well-being of Americans often exaggerate justifiable concerns and ascribe blame to alcohol without consideration of other contributing factors (see chapters 13, 16, and 17). For example, the enormous toll of alcohol-related automobile fatalities, particularly among the youth, is undeniable. But are Americans and particularly adolescents drinking more or does the access of a relatively affluent population to automobiles more accurately explain this tragedy? The possible adverse effect of alcohol on prenatal development was noted in the Bible and has been rediscovered by medicine (and reformers) several times over the past 250 years. The most recent incarnation in the early 1970s was labeled the "fetal alcohol syndrome." The debate still rages over whether alcohol alone or a combination of malnutrition, poor prenatal care and abuse of other drugs as well as alcohol leads to the pattern of malformations and delayed development associated with the syndrome.

The creation of federal agencies such as the National Institute on Alcoholism and Alcohol Abuse has undoubtedly contributed a great deal to advancing knowledge about alcoholism and its consequences. But growth of the bureaucracy, which has facilitated support of science, has also engendered bureaucratic incentive for convincing the people and members of Congress (who appropriate funds) of the perils and dangers of contemporary alcohol problems. Reports of achievements in science command less attention than pseudoscience stories or gothic horror tales about alcohol, somewhat reminiscent of the "Temperance Tales." This mixed anthology often stimulates a desire for quick answers and solutions. Although it is widely recognized that alcohol is a necessary but not sufficient condition for the emergence of human alcohol problems, the latter-day temperance enthusiast may be tempted once again to target alcohol availability as the villain and the issue of primary concern.

In contemporary America, both the tactics and the tone of temperance sentiment have changed appreciably from the 1800s. Inebriety, licentiousness, moral depravity and sin have all but vanished from the extant vocabulary. The new contender for the status of moral purity would seem to be health (although ill-health has not yet achieved equivalence with religious fundamentalists' conceptions of sin). Today, rallying cries once structured in terms of social order, home and basic decency are now framed in terms of health promotion and disease prevention. By any logic, health is an unequivocally laudable goal, to be sought relentlessly for the good of the individual and society. But if "health," like its predecessor goals, becomes a slogan to justify political action that limits individual choice, then like most slogans promising simple solutions to complex problems, "health" will become unhealthy for this society.

The vigilant may have noted that in September 1979, Senator Thurmond of South Carolina proposed an amendment to the authorizing legislation for the National Institute on Alcohol Abuse and Alcoholism which would require a warning label on any alcoholic beverage containing more than 24 percent alcohol. The proposed warning: "Caution: Consumption of alcoholic beverages may be hazardous to your health, may be habit forming, and may cause serious birth defects when consumed during pregnancy." Thurmond explained his reasons for introducing this amendment as follows:

> *A health warning label that I am proposing will be an effective method of keeping the public constantly aware of the dangers of alcohol and encouraging responsible decisions with respect to alcohol consumption. If this warning label deters one potential alcoholic from taking his first drink, if it encourages one expectant mother to refrain from or substantially moderate her drinking habits, or if it makes a casual drinker who*

drives stop and think before having "one for the road," then this legislation will be effective and worthwhile.

Senator Thurmond said that the purpose of the warning was not "to regulate people's lives, but simply to provide the consumer with information." Federal officials agreed that the government has an obligation to inform the public about scientifically established health hazards and denied that warning labels were intended for regulation or control. The warning label debate continues with accelerating acrimony — its eventual fate remains to be determined.

Richard Bonnie, a legal scholar at the University of Virginia, has observed that persuasion through regulation of information is one of four major modes of legal regulation of drug use employed by the government. In 1981, Bonnie commented:

> *Most of the federal government's "Prevention" program (its implementation of "discouragement" policies) concerning tobacco and alcohol use has been the transmission of government-sponsored messages advising the general population, or specific target populations, of the risks (or health dangers) of smoking and excessive or irresponsible drinking. Traditionally, the government has rooted these efforts in the ideology of free choice — the government aims mainly to facilitate informed and prudent personal choices. However, recent regulatory and legislative initiatives strongly suggest that the government is likely to speak more often, and more forcefully, in coming years. In what may be a watershed policy development in behavioral health, the declared government policy appears to be shifting away from an "informed choice" model and toward the public health model — reducing aggregate consumption of alcohol and discouraging its use by target groups and discouraging cigarette smoking by everyone. I predict, then, a transition from information to persuasion — or, as the industry might say, to propaganda.*

Another form of legal suasion is regulation of the conditions under which alcohol is available. Long-term plans described by officials of the Federal Alcoholism Agency in 1980 included re-examination of alcohol beverage controls. The agency expressed particular interest in "examining and analyzing local and state laws and regulations that might influence drinking patterns. Such studies might examine changes in drinking age, pricing, and/or taxation policies, zoning issues, distribution and consumption policies. Other areas of interest are the impact of beverage alcohol advertising, labeling and other trade practices."

EPILOGUE OR PROLOGUE?

An interweaving of the pleasures of drinking and the problems of alcoholism have occurred since the beginning of recorded history, and this society is no exception. Penalties for disruptive public drunkenness and concerns about inebriety began in Colonial times and continue in modified form to the present day. With the importation of opiates and cocaine, society became gradually aware of the human propensity for the abuse of other drugs as well as alcohol. Legal restrictions on the availability of these drugs followed in the wake of the Temperance Movement. By 1870, the Temperance reformers had included cigarettes in their campaign, and between 1895 and 1914, fourteen states had passed cigarette prohibition laws. This early antismoking movement presaged a wave of sentiment that has gathered momentum since the late 1950s and federal mandate of package warning labels in 1967. The "marijuana menace" and cocaine, as well as cigarettes, have tended to divert attention from alcohol of late. However, the strategies used to combat use of these substances suggests the tenacious persistence of Prohibitionist sentiment in America.

The time-honored aphorism that "history repeats itself" has become a tired cliché, but it has a ring of authenticity. It is highly likely that America has not seen the end of efforts to solve alcohol-related problems through moral and political suasion, embellished with inflammatory rhetoric. Lyman Beecher and H. H. Russell may well preach again from a secular pulpit. The crucial prohibition coalition and probable leadership are anyone's guess, but a number of ingredients that were important in past movements exist today. A new fundamentalist religious revival is now in process, and it is too early to know if it will ever reach the scale of the "Great Awakening." Suffrage was won, but the constitutional amendment to guarantee equal rights has languished. Some economists call the massive unemployment and decline in productivity afflicting America in the 1980s a depression instead of a recession — to the dismay of political spokesmen responsible for molding public opinion. How disruptive this "depression" will be to our existing social order cannot be predicted. The masterful organizing ability and shrewd insight of a Wayne Wheeler is not unique today — a cadre of talented political analysts and media specialists guide and package the electoral process. As America jogs, health-bent, toward the next century, what fears and hopes may beguile it into another regulatory excess? And will the best-educated and best-informed people in America's history critically examine each issue or accept conglomerate political packages? And will our government expand its paternalism to promote health and discourage disease in a way that eventually compromises individual choice?

In 1859, John Stuart Mill expressed his belief about the proper role of representative government as follows:

> *The only purpose for which power can be rightfully exercised over any member of a civilized community, against his will, is to prevent harm to others. His own good, either physical or moral is not sufficient warrant. He cannot rightfully be compelled to do or forbear because it will be better for him to do so, because it will make him happier, because in the opinion of others, to do so would be wise, or even right. These are good reasons for remonstrating with him, or reasoning with him, or entreating him, but not for compelling him, or visiting him with any evil in case he do otherwise. To justify that, the conduct from which it is desired to deter him must be calculated to produce evil to some one else. The only part of the conduct of any one, for which he is amenable to society, is that which concerns others. In the part which merely concerns himself, his independence is, of right, absolute. Over himself, over his own body and mind, the individual is sovereign.*

In 1978 a contemporary legal scholar, Richard J. Bonnie, examined the evolving federal policies designed to discourage alcohol, drug and tobacco use and commented:

> *Any regulatory approach whose ultimate aim is to orchestrate changes in mass behavior implies a significant sacrifice in human freedom. Over the long term, systematic implementation of a policy of discouragement — through law and other more subtle devices which behavioral technologists might conceive — might well change the normative climate in society, not only by changing attitudes towards substance consumption, but also toward human liberty itself.*

Bonnie concludes that "Lifestyle modification may be a legitimate approach for improving the health of the citizenry and for re-

ducing health care costs, but it may also be a prescription for op-
pression. Those who make policy should take care to recognize the
difference . . . and the risk" (1978, p. 215).

And so should we! The complexity of government and the pre-
emptive demands of daily life encourage acquiescence to policies
with obvious "good" intentions. There is no safe or simple way to
insure the fragile balance envisioned by the framers of the Constitution will prevail. Only future historians can judge how well this de-
mocracy has kept its covenant with America's past.

CHAPTER 9

The Industry and the Regulators

The major types of beverage alcohol consumed today are distilled spirits, wine and beer, and each is produced by a distinct industry. Each industry has its own trade association, as well as technical, industrial and marketing experts. At one time these industries were largely independent and in open competition for the alcohol consumer market. Today, there is considerable overlap in ownership between the major producers. For example, some major wineries have been acquired by distilled spirits corporations. Whatever the corporate structure, all beverage alcohol producers share one important common factor — all are supervised by governmental regulatory agencies.

Regulation of alcohol availability was an integral part of American colonial governments, as described earlier. As government became more complex so did the regulatory structure.

Today, the production and sale of beverage alcohol is regulated by the federal government, the states, the counties and the municipalities. Consistency has never been a prominent part of the regulatory process; alcohol control laws are no exception. The mosaic of often conflicting laws within a state as well as between states could baffle the most diligent legal scholar. Each town and village may enact laws by popular vote which limit or ban the sale of beverage alcohol within their jurisdiction. States may elect to control the sale of beverage alcohol through private or state monopoly retail stores. States as well as the federal government collect millions of dollars in excise taxes on beverage alcohol, and these potential revenues have led to the establishment of specific state and federal agencies to monitor production, distribution and sales of wine, beer and distilled spirits.

It is the revenue imperative that has nurtured and sustained these regulatory agencies across the centuries. The very term "regulation" suggests that in its absence something unregulated and perhaps excessive might occur. Consider license — the term used for formal, legal authorization, i.e., a permit. But license is also freedom to deviate (e.g. poetic license); excessive, undisciplined freedom; and licentious comes from the same Latin root. These implicit meanings, so much a part of the basic structure of our language, have mitigated for an acceptance of regulation and some vague, seldom-articulated suspicion that what is regulated deserves to be.

The beverage alcohol industry has often been criticized, even vilified, as the "cause" of alcoholism. National Prohibition represented the peak expression of such sentiments but the issue remains alive even today. However, it is important to recognize that the industry has not been indifferent to the problem of alcohol abuse. Responsible members of the distilled spirits, wine and beer industries acknowledge that alcoholism is a public health problem which requires constructive solutions from all segments of society. Each industry recognizes that abuse of its products damages its reputation and impairs its economic stability.

The remainder of this chapter describes the production and consumption pattern for each of the major industries as well as the revenues generated. The evolution and impact of government regulations is described in the last section.

THE DISTILLED SPIRITS INDUSTRY

Data compiled by the Distilled Spirits Council of the United States, a major trade organization representing the distilled spirits industry, indicate that eight hundred thousand full-time (or equivalent) employees are involved in the production of distilled spirits and other alcoholic beverages. In 1982, the distilled spirits industry alone generated federal, state, and local tax revenues of 6.7 billion dollars. These taxes represented 57.5 percent of the $11.7 billion tax revenue recovery from production and sales of all alcoholic beverages.

Production

The location of distilleries in the United States is relatively concentrated, as shown in Figure 1. The largest number of distilleries are in California but these primarily produce brandy, a grape fermentation product. Kentucky is the largest producer of distilled grain spirits in the United States, and Indiana and Tennessee rank high in production.

The amount of whiskey, brandy, and rum distilled in the United States, from the repeal of Prohibition through 1982, are shown in Table 1.

Distillation of whiskey reached its peak in 1935 at 184,865 tax gallons. More recent trends in distilled spirits production are shown in Figure 2.

Consumption

The total amount of distilled spirits produced during any calendar year is not actually consumed during that period. Therefore, pro-

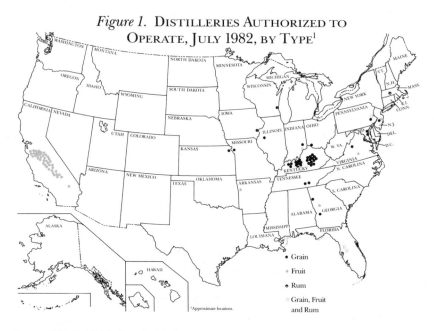

Figure 1. Distilleries Authorized to Operate, July 1982, by Type[1]

• Grain

* Fruit

• Rum

◦ Grain, Fruit and Rum

[1]Approximate locations.

Courtesy of Distilled Spirits Council of the United States, Inc.

Table 1. DISTILLATION OF WHISKEY, BRANDY, AND RUM IN THE
UNITED STATES, 1934–1982
(000 Tax Gallons)

Calendar Year	Whiskey	Brandy	Rum[a]
1934	107,901	12,475	2,172
1935	184,865	18,921	2,997
1940	111,699	25,631	2,384
1945	101,627	28,996	2,766
1950	174,817	10,795	1,878
1955	120,542	7,006	1,928
1960	148,915	9,054	1,704
1961	145,601	6,870	1,793
1962	112,952	8,816	1,775
1963	104,858	10,213	2,227
1964	112,871	11,013	2,163
1965	126,878	15,881	2,426
1966	128,506	18,549	2,523
1967	153,780	15,227	992
1968	178,049	16,705	1,187
1969	169,875	14,329	903
1970	146,360	15,272	1,108
1971	119,377	13,556	1,011
1972	116,561	12,540	1,781
1973	108,393	16,606	1,162
1974	75,151	22,242	1,574
1975	59,637	20,529	795
1976	79,119	19,287	1,365
1977	80,597	18,504	1,175
1978	79,151	18,588	1,528
1979	101,265	18,552	2,206
1980	84,312	21,926	1,938
1981	96,680	24,235	1,879
1982[b]	90,956	15,329	1,192

[a] Excludes distillation in Puerto Rico and the Virgin Islands.
[b] Preliminary.
Source: Bureau of Alcohol, Tobacco, and Firearms, U.S. Treasury Department.

Courtesy of Distilled Spirits Council of the United States, Inc.

Table 2. APPARENT CONSUMPTION OF DISTILLED SPIRITS IN THE UNITED STATES, LICENSE AND CONTROL STATES, 1934–1982

	License States[a]			Control States		
Calendar Year	Wine Gallons (000)	Per Capita (Resident)	Percentage of Total Consumption	Wine Gallons (000)	Per Capita (Resident)	Percentage of Total Consumption
1934	45,892	0.78	79.1	12,073	0.41	20.9
1935	65,219	0.91	72.7	24,452	0.65	27.3
1940	101,736	1.22	70.2	43,256	1.03	29.8
1945	142,220	1.68	74.8	47,911	1.15	25.2
1950	138,966	1.40	73.1	51,053	1.07	26.9
1955	145,422	1.34	72.9	54,148	1.05	27.1
1960	175,573	1.44	74.8	59,141	1.09	25.2
1961	182,707	1.46	75.7	58,742	1.06	24.3
1962	192,693	1.52	76.0	61,008	1.09	24.0
1963	196,780	1.51	76.0	62,200	1.11	24.0
1964	209,721	1.59	76.0	66,141	1.16	24.0
1965	223,889	1.66	76.1	70,355	1.23	23.9
1966	233,071	1.71	75.4	75,845[b]	1.27	24.6
1967	243,902	1.78	75.3	80,094	1.32	24.7
1968	261,514	1.89	75.5	84,730	1.39	24.5
1969	272,534	1.95	75.3	89,240	1.46	24.7
1970	279,534	1.98	75.4	91,039	1.47	24.6
1971	288,368	2.00	75.4	94,156	1.51	24.6
1972	296,281	2.03	75.2	97,555	1.55	24.8
1973	306,727	2.08	75.4	100,332	1.58	24.6
1974	314,444	2.11	75.4	102,841	1.61	24.6
1975	318,685	2.12	75.3	104,735	1.62	24.7
1976	320,086	2.11	75.1	106,307	1.63	24.9
1977	323,813	2.11	75.0	107,740	1.64	25.0
1978	332,925	2.14	75.1	110,536	1.67	24.9
1979	335,418	2.12	75.0	112,078	1.68	25.0
1980	337,631	2.11	75.1	111,786	1.66	24.9
1981	339,177	2.10	75.5	110,276	1.63	24.5
1982	330,677	2.02	75.6	106,982	1.57	24.4

NOTE: Because of rounding, detail may not add to total. Excludes Hawaii, 1934–1964.
[a] Includes District of Columbia.
[b] Includes Mississippi gallonage from July through December 1966.

Source: Distilled Spirits Council of the United States, Inc.; Bureau of the Census, U.S. Department of Commerce.

Where available, license state consumption based on taxed gallonage reported by state tax authorities. In other states, estimated from state treasury tax collections by dividing revenues collected by tax rate per gallon. In Alaska, Georgia, Hawaii, Missouri, South Carolina, and South Dakota, consumption represents sales by distillers to whole-

Total	All States Allowing Sale of Distilled Spirits				
Wine Gallons (000)	Number of States[a]	Resident Population (000)	Per Capita (Resident)	Population 18 Years and Over (000)	Per Capita (Adult)
57,965	28	88,707	0.65	—	—
89,670	41	109,042	0.82	—	—
144,992	46	125,662	1.15	—	—
190,131	46	126,338	1.50	—	—
190,020	47	146,823	1.29	—	—
199,571	47	159,959	1.25	—	—
234,715	49	176,512	1.33	—	—
241,449	49	180,082	1.34	—	—
253,701	49	182,879	1.39	—	—
258,979	49	185,687	1.39	—	—
275,862	49	188,365	1.46	—	—
294,244	50	191,496	1.54	—	—
308,917	51	195,857	1.58	—	—
323,996	51	197,863	1.64	—	—
346,244	51	199,312	1.74	—	—
361,774	51	201,306	1.80	—	—
370,574	51	203,302	1.82	133,218	2.78
382,523	51	206,172	1.86	136,629	2.80
393,836	51	208,654	1.89	139,527	2.82
407,059	51	210,761	1.93	142,293	2.86
417,285	51	212,779	1.96	145,101	2.88
423,420	51	214,943	1.97	148,102	2.86
426,394	51	217,074	1.96	151,146	2.82
431,553	51	219,308	1.97	154,190	2.80
443,461	51	221,665	2.00	157,241	2.82
447,497	51	224,068	2.00	160,373	2.79
449,417	51	227,156	1.98	162,761	2.76
449,453	51	229,307	1.96	166,147	2.70
437,659	51	231,534	1.89	168,769	2.59

salers and in Wisconsin sales by wholesalers to retailers. Hawaiian estimates are based on semiannual tax data.

Control state consumption is actual retail sales in state liquor stores as reported to the National Alcohol Beverage Control Association, except in Mississippi and Wyoming where statistics represent wholesale sales by state liquor control boards to private retailers.

Note: Actual consumption takes place *after* tax collections and wholesale shipments are made. Also, stock changes at the retail level and in the home not included in the estimates.

Courtesy of Distilled Spirits Council of the United States, Inc.

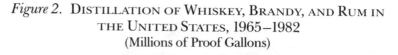

Figure 2. DISTILLATION OF WHISKEY, BRANDY, AND RUM IN
THE UNITED STATES, 1965–1982
(Millions of Proof Gallons)

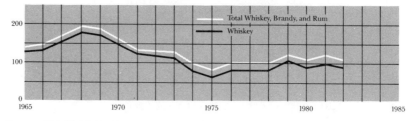

Courtesy of Distilled Spirits Council of the United States, Inc.

duction does not necessarily predict actual consumption. Variable amounts of time may elapse between production of distilled spirits, processing and aging, and subsequent bottling before consumption. One of the best ways to assess apparent consumption is to take into account existing inventories of distilled spirits at the beginning of a calendar year, depletion of those inventories at the end of a calendar year as well as the total production during the calendar year.

Table 2 shows apparent whiskey use from 1934 through 1982. The peak in apparent whiskey use occurred during 1974 (2.88 gallons per adult), followed by a small but steady decline through 1982. At the time of the preparation of this material, no final data were available for 1983.

Seasonal fluctuation in the production of distilled spirits is caused, in part, by seasonal variations in the availability of grains and fruits from which beverage alcohol is produced. But there is also seasonal variability in the bottling of distilled spirits, as shown in Figure 3. The largest amount is bottled during the month of October and the lowest during the month of February. More distilled spirits are consumed in the United States during the Christmas holiday season, a reaffirmation of the holiday accolade of good cheer to all.

It is very difficult to determine the exact consumption of distilled spirits over short periods of time. Therefore, estimates of apparent consumption represent the closest approximation to actual use. The Distilled Spirits Council of the United States points out "consumption estimates are computed from monthly state tax collections at wholesale level with the following exceptions: Estimates for Alaska,

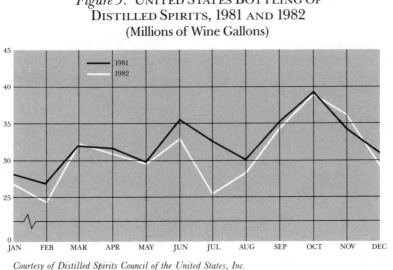

Figure 3. UNITED STATES BOTTLING OF
DISTILLED SPIRITS, 1981 AND 1982
(Millions of Wine Gallons)

Courtesy of Distilled Spirits Council of the United States, Inc.

Georgia, South Carolina and South Dakota are based on shipments of distillers to wholesalers. Hawaiian estimates are based on semi-annual tax data. Wisconsin estimates are based on wholesale ship-ments. Actual consumption (as opposed to apparent consumption) takes place after tax collection and wholesale shipments are made. Moreover, stock exchanges at the retail level and the home have a bearing on actual consumption but such stock changes are not in-cluded in the estimates. Irregularity in compiling tax receipts on which consumption data in most states are calculated can also cause distortions. While distortions may occur in relatively short periods, over a period of time the data generally reflect actual consumption levels in the United States as a whole."

The Distilled Spirits Council of the United States also stresses that "consumption in some states is affected by special considera-tions such as sales across state borders resulting from lower cross-bor-der prices, which raise apparent consumption in some states while decreasing consumption in other states. Similarly consumption in some states is inflated by sales to tourists."

Given these limitations in arriving at actual consumption, it is still possible to compute apparent consumption of distilled spirits in the United States over annual intervals. Figure 4 shows the apparent consumption of distilled spirits in the United States from 1965 to

Figure 4. APPARENT CONSUMPTION OF DISTILLED SPIRITS IN
THE UNITED STATES, 1965–1982
(Millions of Wine Gallons)

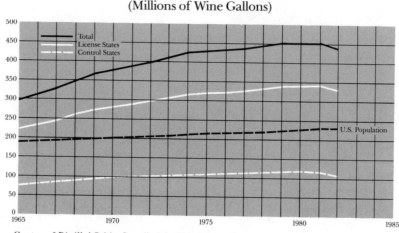

Courtesy of Distilled Spirits Council of the United States, Inc.

1982. The apparent consumption in both licensed and controlled states as well as the rate of increase in population of the U.S. from 1965 through 1982 is also shown in Figure 4. Licensed states are those states in which distilled spirits and other forms of beverage alcohol are sold in privately owned retail outlets, in comparison to controlled states, which regulate sales of beverage alcohol through state-owned stores.

The apparent consumption in licensed states is greater than in controlled states because there are more licensed states and many of these states have significantly larger populations than the controlled states. As can be seen in Figure 4, total consumption as well as consumption in licensed states and controlled states increased from 1965 to 1979. Thereafter consumption appears to have leveled off and was relatively stable until 1982, when a small decline occurred. The increase in consumption of distilled spirits from 1965 through 1979 could not be explained solely by the increase in U.S. population since the proportional increase in total consumption exceeded the increase in U.S. population growth.

Beverage Choice

Americans seem to enjoy changes in life-style, and changes in American drinking habits are reflected by significant differences in

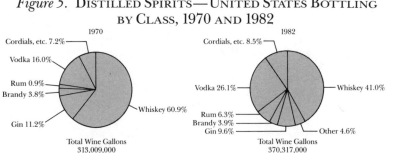

Figure 5. DISTILLED SPIRITS—UNITED STATES BOTTLING
BY CLASS, 1970 AND 1982

Courtesy of Distilled Spirits Council of the United States, Inc.

types of distilled spirits bottled over a twelve-year period. During
1970, whiskey accounted for 60.9 percent of bottled distilled spirits
but by 1982 whiskey had dropped to 41 percent (Figure 5). During
the same twelve-year period, bottling of vodka increased from 16
percent to 26.1 percent, rum from 0.9 to 6.3 percent, cordials from
7.2 to 8.5 percent. Bottling of gin decreased from 11.2 to 9.6 percent.

There are also differences in the types of whiskey consumed by
Americans over the decade of the seventies (Figure 6). In 1970,
blended whiskey accounted for 38.8 percent of bottling but dropped
to 21.6 percent in 1982. Imported bulk whiskey bottling more than
doubled over twelve years from 16.8 to 35 percent.

Imports and Exports

Figure 7 shows the U.S. imports of distilled spirits in 1972 and 1982.
Whiskey imports decreased from over 87 million wine gallons* in
1982 to approximately 76½ million wine gallons during 1982. However, imports of gin, rum, brandy, cordials and tequila all increased.

In 1982, 10.1 million proof gallons of distilled spirits were exported from the U.S., an increase of 11.4 percent above 1981. Whiskey was the major type of distilled spirits exported from the U.S.
Whiskey exports in 1982 increased by 3.3 percent over exports in

---

*A wine gallon (bulk or liquid gallon) — measures liquid volume, regardless of alco-
hol content. The standard U.S. wine gallon contains 231 cubic inches at 60 °F.

A proof gallon is the standard U.S. gallon of 231 cubic inches at 60 °F. containing 50
percent by volume of ethyl alcohol (100 ° proof).

A tax gallon is the unit of distilled spirits on which the federal excise tax is levied. The
federal excise tax on all spirits is $10.50 per 100 ° proof gallon so that a tax gallon is
equivalent to a proof gallon.

The above definitions are taken from the *Annual Statistical Review* 1982.

Figure 6. WHISKEY—UNITED STATES BOTTLING
BY TYPE, 1970 AND 1982

Courtesy of Distilled Spirits Council of the United States, Inc.

1981. Other spirits such as vodka and cordials increased by 25.1 percent over 1981. The major importers of American distilled spirits were Canada, the Federal Republic of Germany, Australia, Norway, Japan, Sweden, and the United Kingdom. Although these exports represented sizable volumes, they were small in comparison to imports from foreign countries.

Although United States imports of distilled spirits decreased by 8.6 percent in 1982, over 31.4 percent of the total U.S. supply of distilled spirits was imported in 1982 — 106.3 million gallons in all.

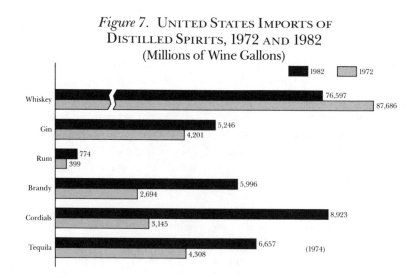

Figure 7. UNITED STATES IMPORTS OF
DISTILLED SPIRITS, 1972 AND 1982
(Millions of Wine Gallons)

Courtesy of Distilled Spirits Council of the United States, Inc.

Whiskey accounted for 72.1 percent of these imports. Cordials, tequila, brandy and gin were also imported in large quantities. The major exporters of distilled spirits to the United States were Canada, the United Kingdom, Mexico and France.

Tax Revenues

The sale of distilled spirits provides substantial tax revenues for federal, local and state governments. Figure 8 shows the cost components of an average 750 ml bottle of spirits sold during 1982. The average retail price was $6.74. Federal and local taxes accounted for over 43.8 percent of the retail cost. Federal taxes alone accounted for a greater proportion of the cost than either retail costs, wholesale costs, or distillers' or importers' cost.

Public revenues from distilled spirits production and sale in the U.S. since Repeal are shown in Figure 9. During 1979, 1980, and 1981, a staggering $7 billion per year was collected, and in 1982, $6.7 billion (a decrease of 3.5 percent) was collected. Since Repeal,

Figure 8. COST COMPONENTS OF AN AVERAGE 750 ML BOTTLE
OF SPIRITS SOLD AT RETAIL, 1982

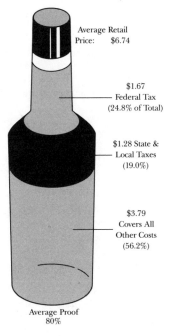

Average Retail
Price: $6.74

$1.67
Federal Tax
(24.8% of Total)

$1.28 State &
Local Taxes
(19.0%)

$3.79
Covers All
Other Costs
(56.2%)

Average Proof
80%

Courtesy of Distilled Spirits Council of the United States, Inc.

Table 3. INTERNAL REVENUE COLLECTIONS BY PRINCIPAL SOURCES.
Fiscal Years 1950 Through 1981 (in Thousands of Dollars)
(Income and Profits Taxes)

Fiscal year ended June 30	Total internal revenue collections	Alcohol taxes	Total	Corporation income and profits taxes
1950	38,957,132	2,219,202	28,007,202	10,854,351
1951	50,445,686	2,546,808	37,384,879	14,387,569
1952	65,009,586	2,549,120	50,741,017	21,466,910
1953	69,686,535	2,780,925	54,130,732	21,594,515
1954	69,919,991	2,783,012	54,360,014	21,546,014
1955	66,288,692	2,742,840	49,914,826	18,264,720
1956	75,112,649	2,920,574	56,636,164	21,298,522
1957	80,171,971	2,973,195	60,560,425	21,580,653
1958	79,978,476	2,946,461	59,101,874	20,533,316
1959	79,797,973	3,002,096	58,826,254	18,091,509
1960	91,774,803	3,193,714	67,125,126	22,179,414
1961	94,401,086	3,212,801	67,917,941	21,764,940
1962	99,440,839	3,341,282	71,945,305	21,295,711
1963	105,925,395	3,441,656	75,323,714	22,336,134
1964	112,260,257	3,577,499	78,891,218	24,300,863
1965	114,434,634	3,772,634	79,792,016	26,131,334
1966	128,879,961	3,814,378	92,131,794	30,834,243
1967	148,374,815	4,075,723	104,288,420	34,917,825
1968	153,636,838	4,287,237	108,148,565	29,896,520
1969	187,919,560	4,555,560	135,778,052	38,337,646
1970	195,772,096	4,746,382	138,688,588	35,036,983
1971	191,647,198	4,800,482	131,072,374	30,319,953
1972	209,855,737	5,110,001	143,804,732	34,925,546
1973	237,787,204	5,149,513	164,157,315	39,045,309
1974	268,952,254	5,258,477	184,648,094	41,744,444
1975	293,822,726	5,350,858	202,146,097	45,746,660
1976	302,519,792	5,399,055	205,751,753	46,782,956
TQ 1976	75,462,780	1,305,841	49,567,484	8,808,905
1977	358,139,417	5,406,633	246,805,067	60,049,804
1978	399,776,389	5,612,715	278,438,289	65,380,145
1979	460,412,185	5,647,924	322,993,733	71,447,876
1980	519,375,273	5,704,768	359,927,392	72,379,610
1981	606,799,103	5,688,413	406,583,302	73,733,156

Individual income taxes	Employ-ment taxes	Estate and gift taxes	Tobacco taxes	Manufac-turer's excise taxes	All other taxes
17,153,308	2,644,575	706,227	1,328,464	1,836,053	2,214,951
22,997,309	3,627,479	729,730	1,380,396	2,383,677	2,392,719
29,274,107	4,464,264	833,147	1,565,162	2,348,943	2,507,933
32,536,217	4,718,403	891,284	1,654,911	2,862,788	2,647,492
32,813,691	5,107,623	935,121	1,580,229	2,689,133	2,464,859
31,650,106	6,219,665	936,267	1,571,213	2,885,016	2,018,866
35,337,642	7,295,784	1,171,237	1,613,497	3,456,013	2,019,380
39,029,772	7,580,522	1,377,999	1,674,050	3,761,925	2,243,856
38,568,559	8,644,386	1,410,925	1,734,021	3,974,135	2,166,675
40,734,744	8,853,744	1,352,982	1,806,816	3,958,789	1,997,292
44,945,711	11,158,589	1,626,348	1,931,504	4,735,129	2,004,394
46,153,001	12,502,451	1,916,392	1,991,117	4,896,802	1,963,582
50,649,594	12,708,171	2,035,187	2,025,736	5,120,340	2,264,817
52,987,581	15,004,486	2,187,457	2,079,237	5,610,309	2,278,536
54,590,354	17,002,504	2,416,303	2,052,545	6,020,543	2,299,645
53,660,683	17,104,306	2,745,532	2,148,594	6,418,145	2,453,406
61,297,552	20,256,133	3,093,922	2,073,956	5,613,869	1,895,909
69,370,595	26,958,241	3,014,406	2,079,869	5,478,347	2,479,809
78,252,045	28,085,898	3,081,979	2,122,277	5,713,973	2,196,909
97,440,406	33,068,657	3,530,065	2,137,585	6,501,146	2,348,495
103,651,585	37,449,188	3,680,076	2,094,212	6,683,061	2,380,609
100,752,421	39,918,690	3,784,283	2,206,585	6,684,799	3,179,985
108,879,186	43,714,001	5,489,969	2,207,273	5,728,657	3,801,104
125,112,006	52,081,709	4,975,862	2,276,951	5,395,750	3,750,104
142,903,650	62,093,632	5,100,675	2,437,005	5,742,154	3,572,217
156,399,437	70,140,809	4,688,079	2,315,090	5,515,511	3,665,182
158,968,797	74,202,853	5,307,466	2,487,894	5,486,106	3,855,998
39,758,579	19,892,041	1,485,247	622,821	1,543,339	1,046,007
186,755,263	86,076,316	7,425,325	2,389,501	6,068,682	3,958,893
213,058,144	97,291,653	5,381,499	2,450,913	6,555,681	4,045,639
251,545,857	112,849,874	5,519,074	2,495,517	7,057,612	3,848,450
287,547,782	128,330,479	6,498,381	2,446,416	6,487,421	9,980,416
332,850,146	152,885,816	6,910,386	2,583,857	6,089,000	26,058,329

Courtesy of United States Brewers Association, Inc.

Figure 9. FEDERAL, STATE, AND LOCAL REVENUE PER GALLON
OF DISTILLED SPIRITS SINCE REPEAL
(Dollars per Gallon)

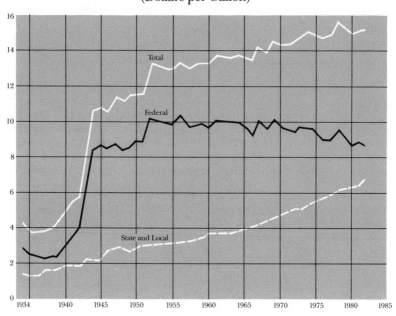

Courtesy of Distilled Spirits Council of the United States, Inc.

over $167 billion in taxes have been paid to federal, state, and local
governments from production and sale of distilled spirits.

Table 3 shows internal revenue collections by principal sources for
the fiscal years 1950 through 1981. Total federal internal revenue
tax collections amounted to almost $607 billion. Alcohol taxes ac-
counted for almost $6 billion, an amount over twice that collected
from tobacco taxes, and approximated collections from state and
gift taxes and manufacturers excise taxes. These data leave no doubt
that alcohol taxes are a major source of income for the United States
government.

THE BEER INDUSTRY

The popularity of beer in America is strongly rooted in history. A
1767 survey estimated that each person in Massachusetts consumed
a keg (up to five gallons) of the beverage daily. The United States

Brewers Association, the major trade association for the industry, founded in 1862, is the oldest trade organization to be continuously incorporated in America.

The production of beer has been an important American enterprise since colonial times. Currently, beer is consumed in two out of three American homes, and 50 percent of the nation's male and female population are beer drinkers. Consumption demands in America require an annual production of more than 64 *billion* twelve-ounce bottles or cans. Moreover, the popularity of beer has continued to grow during the decade of the seventies; the brewing industry reported an average growth rate of over 3 percent annually. In 1981, the sale of beer and ale amounted to 176,675,741 barrels. The United States Brewers Association points out that in 1863, brewers sold 1,765,827 barrels: "that means that per capita consumption in 1863 was 1.7 gallons, in 1981 it was 24.6 gallons."

The United States is the world's largest producer of beer. Table 4

Table 4. BEER PRODUCTION (IN BARRELS) THROUGHOUT THE WORLD

West Germany	79,870,000
Russia*	59,653,000
England	52,410,00
Japan	39,610,000
Brazil	25,140,000
Mexico	24,987,000
East Germany	20,452,000
Czechoslovakia	20,396,000
France	18,622,000
Canada	17,908,000
Spain	17,831,000
Australia	17,189,000
Netherlands	14,180,000
Belgium	12,783,000
Colombia	10,226,000
Yugoslavia	10,226,000
Venezuela	10,226,000
Romania*	9,800,000
Poland	9,630,000
Denmark	8,630,000

* Estimate   *Courtesy of United States Brewers Association, Inc.*

Figure 10. BREWERIES AUTHORIZED TO OPERATE—JULY 1, 1981
By Census Regions and Geographic Divisions of the United States

Courtesy of United States Brewers Association, Inc.

compares United States production of over 193 million barrels with
the amount of beer produced by other nations. West Germany produced less than one-half of the amount of beer brewed in the United
States. Estimates indicate that Russia ranks third in world beer production.

Figure 10 shows the location of breweries in the United States.
The combined industry produces products whose retail value exceeds $30 billion and the industry employs approximately 43,000
persons.

According to the United States Brewers Association, here is what
the industry buys and produces.

Agricultural commodities, the output of more than 4 million acres
of farm land, worth over $1.1 billion, are used each year by the
brewing industry. These include:

Five billion pounds of malt from 147,500,000 bushels of choice
malting barley — value, $738,000,000.

Other select grains, chiefly corn, wheat and rice — value, $236,-
000,000.

Hops — value to the grower, $66,350,000.

Some 87.5 percent of all beer sold is packaged in cans or bottles. In one year, the brewing industry buys:

At least 29.5 billion steel cans and aluminum cans. Unknown prior to 1935, beer cans now account for 25 percent of the can industry's total production.

A total of 16,300,000,000 bottles, which in returnable and nonreturnable form amounts to 40 percent of package sales.

A sum of $470,000,000 for interest, rentals, repairs, maintenance, etc.

The industry's annual bill for containers — cans, bottles, kegs — and related packaging materials purchased from other American industries is close to $4.5 billion.

Supplies and services of numerous kinds, providing jobs for thousands of people, are also required in brewing and distributing malt beverages. Annual outlays for these include:

Fuel, power and water $350,000,000
Wholesale payroll $1,500,000,000
Brewery equipment and improvements $700,000,000

Much of the estimated $700 million or more the brewing industry expends a year for plant modernization goes into processing equipment to assure top-quality beer and ale.

Choice-grade grains, pure water, exact temperatures at various stages of brewing, precise timing of operations, close control of yeast action, clinic-clean apparatus and a dust-free atmosphere are all vital to the production of malt beverages.

Although about 90 percent of the brewing industry's output consists of "bottom-fermented" lager beer, there are other forms of malt beverages including malt liquors and packaged draught beer. Ale, produced by "top fermentation" at high temperatures, is usually paler in color and differs slightly in flavor from lager beer.

Porter and stout are also malt beverages. They are darker in color and sweeter or "maltier" than ale. Bock beer is a special springtime brew made like lager beer but of somewhat deeper color and more pronounced flavor.

Yeast causes fermentation. It grows and increases its own mass during the fermentation process. The excess, an important by-product, is sold as brewer's yeast. A rich source of B-complex vitamins,

Table 5. STATE PER CAPITA CONSUMPTION BY RANK 1979–1981
(in gallons)

1981 Rank	State	Per Capita Consumption	1980 Rank	Per Capita Consumption	1979 Rank	Per Capita Consumption
1	Nevada	37.1	1	36.9	1	37.0
2	Wisconsin	34.3	2	34.3	3	33.4
3	New Hampshire	33.4	3	34.1	2	33.8
4	Montana	32.8	4	31.5	4	32.1
5	Wyoming	32.0	5	31.2	5	31.5
6	Texas	31.9	6	30.1	7	29.2
7	Arizona	31.3	7	30.0	6	29.4
8	Hawaii	30.2	9	28.6	13	26.4
9	District of Columbia	29.1	16	26.2	12	26.5
10	New Mexico	28.8	12	27.3	9	27.4
11	Colorado	28.6	10	27.8	11	26.7
12	Alaska	28.4	18	25.8	23	24.4
13	Florida	28.3	8	29.4	8	28.5
14	Nebraska	27.7	11	27.6	10	26.8
15	North Dakota	27.0	14*	26.5	15	25.7
16	Delaware	26.9	20	25.5	29*	23.6
17	Massachusetts	26.7	13	26.9	14	25.8
18	Vermont	26.2	22	25.2	16	25.6
19	California	25.7	19	25.6	18	25.3
20	Pennsylvania	25.6	17	26.0	19*	25.1
21	Idaho	25.5	21	25.4	19*	25.1
22	Illinois	25.2	24	25.0	21*	24.8
23	Louisiana	25.1	31	23.4	33	22.5
24	Iowa	24.9	25	24.7	21*	24.8
25*	Maryland	24.8	22	25.2	26	24.1
25*	Missouri	24.8	26	24.6	25	24.2
27	Rhode Island	24.7	14*	26.5	17	25.5
28	Minnesota	24.5	28	24.4	28	23.8
29	Washington	24.4	27	24.5	24	24.3
30	Oregon	24.2	30	24.1	27	23.9
31	Ohio	23.6	29	24.3	31	23.5
32	Michigan	23.2	32	23.2	29*	23.6
33	Maine	23.1	33	23.1	32	22.8
34	South Dakota	22.7	34	22.5	36	21.3
35	New York	22.5	35	22.3	34	21.9
36	Indiana	22.4	36*	21.9	38*	20.9
37	New Jersey	22.2	36*	21.9	38*	20.9
38	Virginia	21.9	38	21.8	35	21.4
39	Kansas	21.8	39	21.7	37	21.2
40	South Carolina	21.6	40	21.1	41	20.2
41	Oklahoma	20.8	41	19.8	44	18.9
42*	Georgia	20.1	42	19.5	43	19.1
42*	Mississippi	20.1	46	19.0	46	18.5

1981 Rank	State	Per Capita Consumption	1980 Rank	Per Capita Consumption	1979 Rank	Per Capita Consumption
42*	Tennessee	20.1	44	19.3	42	19.3
45	Kentucky	19.6	47	18.5	47	18.4
46	North Carolina	19.4	43	19.4	45	18.6
47	Connecticut	19.0	45	19.1	40	20.4
48*	Arkansas	18.4	48*	17.5	50	16.1
48*	West Virginia	18.4	48*	17.5	49	16.3
50	Alabama	17.5	50	16.4	48	16.7
51	Utah	15.6	51	15.0	51	14.7
	U.S. Average:	24.6		24.3		23.8

* tied

Courtesy of United States Brewers Association, Inc.

brewer's yeast is of great benefit as a human nutrient and as a supplement to livestock feedstuffs and pet foods.

Animal and poultry diets are made more nutritious by the addition of enriched grains, another brewing by-product. Of high protein content, these are the particles or barley, corn, rice and similar cereal grains which remain, after the starch-sugars have been extracted in brewing.

Table 5 shows per capita consumption of beer ranked from highest to lowest for the fifty states and the District of Columbia. During 1979 through 1981, Nevada led the nation in consumption, with approximately thirty-seven gallons per capita. In contrast the neighboring state of Utah ranked lowest with approximately fifteen gallons per capita consumption. The differences in beer consumption between the two states illustrate the influence that laws enacted by local jurisdictions can have upon sale and consumption of beverage alcohol. Because many of the residents of the state of Utah are members of the Church of Latter Day Saints (Mormon), their religious beliefs militate against beverage alcohol consumption and this point of view has become incorporated through the political process into local law.

Exports and Imports

The United States exports of malt beverages are shown in Table 6. During 1981 the United States exported over nine million cases of beer containing twenty-four twelve-ounce containers. This figure

Table 6. UNITED STATES EXPORTS OF MALT BEVERAGES

Quantities 24/12 Oz. Cases — All Types of Containers — Calendar Years

(Exclusive of Shipments to Armed Forces Overseas)

Country of Destination	1971	1972	1973	1974**	1975	1976	1977	1978	1979	1980	1981
Canada	28,735	7,508	17,700	37,859	97,159	261,525	583,760	6,002,883	964,176	7,727,465	1,789,970
Mexico	—	—	416	5,092	—	86,940	29,857	—	94,093	164,572	204,160
Other North America	—	—	—	—	—	—	—	—	—	—	—
Total North America	28,735	7,508	18,116	42,951	97,159	348,465	613,617	6,002,883	1,058,269	7,892,037	1,994,130
Nicaragua	3,114	932	2,497	4,026	—	18,982	15,893	—	—	—	—
Costa Rica	1,252	1,309	392	—	—	—	—	—	—	—	—
Panama Canal Zone	—	—	—	—	—	—	—	—	—	—	—
Republic of Panama	4,200	4,978	3,986	10,896	—	—	72,139	62,421	39,822	26,000	12,013
Other Central America	21,245	22,146	7,588	—	—	21,399	25,582	—	—	—	—
Total Central America	29,811	29,365	14,463	14,922	—	40,381	113,614	62,421	39,822	26,000	12,013
Bermuda	16,416	13,075	36,117	71,758	177,529	248,721	215,722	205,414	238,913	204,038	192,543
Bahamas	106,893	209,727	172,461	158,035	193,637	177,775	219,537	222,477	354,247	395,485	372,088
Cuba	—	—	—	—	—	—	—	—	—	—	—
Haiti	3,946	6,415	11,706	11,452	45,133	—	—	—	—	—	—
Dominican Republic	2,629	9,475	1,628	14,648	—	—	—	—	—	—	—
Curacao (N.W.I.)	—	—	—	—	—	—	—	196,851	283,640	220,164	264,527
Other Caribbean	111,243	204,551	234,810	137,000	93,903	147,788	171,936	—	136,522	170,972	360,981
Total Caribbean	241,127	443,243	456,722	392,893	510,202	574,284	607,195	624,742	1,013,322	990,659	1,190,139
Surinam	180	470	235	—	—	25,219	29,754	45,860	37,642	18,547	22,665
Brazil	—	—	—	—	—	—	—	—	—	—	—
Venezuela	417	7,362	14,383	34,424	30,796	81,998	143,426	226,236	447,612	546,461	498,269
Other South America	5,186	—	—	—	—	—	—	—	—	—	—
Total South America	5,783	7,832	14,618	34,424	30,796	107,217	173,180	272,097	485,254	565,008	520,934

Country											
France	—	—	—	—	—	—	—	—	—	—	—
Portugal	346	2,520	3,725	5,129	—	—	—	—	—	—	—
Italy	—	—	252	—	—	—	—	—	—	—	—
Azores	—	—	—	—	—	—	—	—	—	—	—
Iceland	4,837	9,399	—	—	—	—	—	—	—	—	—
Other Europe	—	—	33,755	18,287	—	31,806	60,919	91,442	73,518	139,507	147,772
Total Europe	5,183	11,925	37,732	23,416	—	31,806	60,919	91,442	73,518	139,507	147,772
Syria and Lebanon	16,743	19,362	34,200	46,433	50,006	—	42,978	53,438	88,992	55,839	76,411
Saudi Arabia	—	—	504	—	—	—	—	—	65,097	1,220	—
Palestine and Transjordan	—	—	—	—	—	—	—	23,429	—	30,454	111,560
India, Pakistan and Ceylon	3,448	1,784	4,255	—	—	—	—	—	—	—	—
Thailand	—	—	—	—	—	—	—	—	—	—	—
Philippines	59,919	98,910	171,976	149,956	222,516	441,674	751,876	1,210,566	1,514,285	2,199,048	1,672,126
Hong Kong	361	26,290	16,535	20,892	63,540	93,316	198,702	289,109	471,927	813,534	1,112,380
Japan	—	—	—	—	—	—	—	—	—	—	—
British Malaya	34,238	19,399	26,771	439,358	—	18,612	362,082	1,918,468	3,179,133	1,720,206	800,321
Other Asia	—	—	—	—	—	—	—	—	—	—	—
Total Asia	114,709	167,331	254,241	656,639	336,162	553,602	1,355,638	3,495,001	5,319,434	4,820,301	3,772,798
Egypt	373	—	784	—	—	31,966	—	19,451	3,318	2,818	6,745
Liberia	—	—	4,333	8,262	—	—	—	—	—	—	—
Union of South Africa	—	—	—	—	—	—	—	—	—	—	—
Tangier	—	—	—	—	—	—	—	—	—	—	—
Portuguese Guinea and Angola	—	—	11,276	21,892	—	—	—	—	—	—	—
Other Africa	7,714	10,425	12,908	5,794	1,174,538	1,492,175	624,301	324,019	—	59,318	1,982
Total Africa	8,087	10,425	29,301	35,948	1,174,538	1,524,141	624,301	343,469	3,318	62,136	8,727
Australia and Other Oceania	80,714	210,726	290,159	345,142	256,358	276,136	433,860	302,666	313,347	385,823	574,989
Total	514,149	888,355	1,115,352	1,590,901	2,598,147	2,772,404	5,036,327	11,487,522	8,735,732	15,633,424	9,051,474

Other countries (breakdowns not available) — 1974 — 44,566 cases; 1975 — 192,932 cases; 1976 — 216,212 cases; 1977 — 154,163 cases; 1978 — 292,801 cases; 1979 — 429,488 cases; 1980 — 751,953 cases; 1981 — 829,972 cases.

Courtesy of United States Brewers Association, Inc.

Figure 11. 1981 PROPORTION OF U. S. MALT BEVERAGE
EXPORTS BY CONTINENTS OF DESTINATION

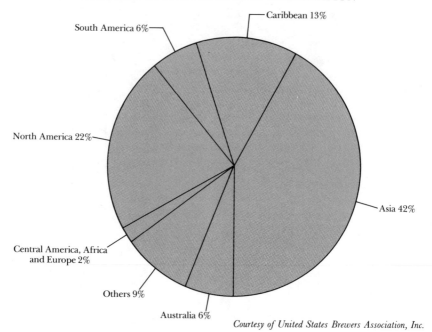

Courtesy of United States Brewers Association, Inc.

represented a decrease from 1980 when exports exceeded over 15½ million cases.

Figure 11 shows the 1981 proportion of U.S. malt beverage exports by continents of destination. Over 42 percent of all exports went to Asian countries.

Imports of malt beverages are shown in Table 7. Imports during 1981 exceeded exports by over sixty-two million cases. The Netherlands shipped almost twenty-nine million cases to the United States, 20 million more than we exported to the entire world. Our neighbors in Canada, Newfoundland, and Laborador shipped over twenty-three million cases of beer to the United States during 1981.

Taxes

Federal internal revenue collections from the malt beverages industry from 1901 through 1981 are shown in Table 8. Over $1.5 billion in excise and other taxes were collected by the U.S. government as a consequence of the sale of beer during 1981 alone.

THE WINE INDUSTRY

When the early Viking explorers described newly discovered America as "Vineland," they were probably responding to the profusion of growth of wild berries but not grapes. Neither Native Americans nor early European settlers in the New World developed a distinct preference for wine over forms of beverage alcohol. Some authorities attribute this to the lack of success in viticulture during our nation's eary history. While numerous attempts were made to cultivate grape vines imported from Europe these were decimated by insects and plant disease unique to the American environment. But as agriculture science evolved so did viticulture, and by the nineteenth century production of quality wine in the United States was well under way.

Enactment of Prohibition was in many ways more devastating to the wine industry than to the beer or distilled spirits industries. Most wine production before enactment of Prohibition, and to some extent today, is a relatively small scale family business. When the Eighteenth Amendment prohibited production and sale of wine the agricultural base for the industry, the vineyards, was destroyed. Restoration of grape growing and wine production came slowly following repeal of Prohibition. However, during the past decade there has been a steady increase in wine production and an even more extraordinary increment in the quality of wine which is being produced in the United States.

Wine Production

Figure 12 shows the volume of wine entering distribution channels in the United States from 1972 through 1982. It is clear that most wine consumed in America is produced in California. Over a ten-year period there has been a steady growth in the California wine industry amounting to over seventy million gallons. During the same period however, growth in foreign wine distribution has been proportionately greater.

Table 9 presents the statistical highlights of wine production and distribution for the years 1981 and 1982 (these data were the latest available at the time of preparation of this chapter).

Table 7. UNITED STATES IMPORTS OF MALT BEVERAGES
Quantities 24/12 Oz. Cases — All Types of Containers — Calendar Years

Country of Origin	1972	1973	1974	1975
Canada, Newf. and Lab.	1,879,698	2,357,363	3,968,034	5,672,339
Mexico	686,037	916,328	1,401,515	1,522,834
Other North America ...	—	—	—	—
Total North America .	2,565,735	3,273,691	5,369,549	7,195,173
Central America	—	3,200	3,648	2,115
Caribbean	27,556	38,985	21,034	8,500
South America	3,100	9,269	11,709	6,900
United Kingdom	552,293	765,852	712,958	928,027
Eire	336,511	248,956	361,347	283,779
Denmark	318,162	240,951	258,024	271,814
Netherlands	4,075,350	5,483,094	6,755,330	8,797,096
Czechoslovakia	32,415	40,250	79,059	79,337
Germany	3,780,510	4,242,450	4,219,275	3,793,003
France	**	**	**	**
Sweden	5,996	7,799	15,904	9,489
Norway	443,088	431,907	451,179	367,426
Other Europe	108,026	137,709	172,846	211,921
Total Europe	9,652,351	11,631,473	13,062,313	14,741,892
Japan	189,631	246,165	245,973	269,344
Philippines	333,879	349,293	228,728	284,146
Other Asia	7,833	31,838	13,426	27,667
Total Asia	531,343	627,296	488,127	581,157
Africa	—	12,340	1,260	3,936
Australia & Other Oceania	30,917	34,923	171,036	590,727
Totals	12,811,002	15,598,672	19,092,285	23,130,400

* Subject to revision   1976 Other Countries 447,851 cases   1977 Other Countries 80,898 cases   1979 Other Countries 2,833 cases   1981 Other Countries 15,904 cases.
** Included in Other Europe

1976	1977*	1978*	1979	1980	1981
8,141,973	10,461,669	13,722,710	17,930,352	21,108,964	23,046,574
2,442,873	2,418,577	2,819,915	4,041,913	4,554,911	4,822,976
10,584,846	12,880,246	16,542,625	21,972,265	25,663,875	27,869,550
—	1,850	6,999	2,558	32,851	—
—	33,776	38,045	23,393	44,585	49,375
—	16,992	16,574	28,437	30,402	28,428
825,426	613,731	1,026,059	1,260,688	1,372,061	1,447,298
491,945	665,125	823,859	858,330	896,578	985,923
319,913	202,272	317,204	276,000	348,511	266,944
12,956,638	14,093,025	21,116,122	27,404,579	24,674,531	28,991,494
—	80,242	163,427	81,933	127,299	131,772
5,394,146 **	4,538,920 **	5,328,471 **	6,315,111 **	6,574,588	8,101,485
				436,657	708,539
—	11,082	8,206	—	—	47,771
360,159	286,234	292,882	266,460	219,070	208,508
26,236	290,697	348,687	563,941	390,330	471,597
20,374,463	20,781,328	29,424,917	37,027,042	35,039,625	41,361,331
301,272	316,661	410,813	520,854	554,648	704,574
298,163	369,665	390,268	390,167	296,207	515,778
—	50,606	67,321	149,973	257,563	440,740
599,435	736,932	868,402	1,060,994	1,108,418	1,661,092
—	1,690	573	—	1,667	—
859,877	625,416	779,965	1,095,651	1,006,963	946,871
32,866,472	35,078,230	47,678,100	61,213,173	62,928,386	71,932,551

Courtesy of United States Brewers Association, Inc.

Table 8. INTERNAL REVENUE PAID TO U.S. GOVERNMENT BY THE
DISTILLING AND BREWING INDUSTRIES SINCE 1901

Fiscal Year Ending June 30th	Distilled Spirits And Wine	Malt Beverages	Total Revenue
1901	$116,027,980.00	$ 75,689,908.00	$191,698,888.00
1902	121,138,013.00	71,988,902.00	193,126,915.00
1903	131,953,472.00	47,547,856.00	179,501,328.00
1904	135,810,015.00	49,083,459.00	184,893,474.00
1905	135,958,513.12	50,360,533.18	186,319,046.30
1906	143,394,055.12	55,641,858.56	199,035,913.68
1907	156,336,901.89	59,567,818.18	215,904,720.07
1908	140,158,807.15	59,807,616.81	199,966,423.96
1909	134,868,034.12	57,456,411.42	192,324,445.43
1910	148,029,311.54	60,572,288.54	208,601,600.08
1911	155,279,858.25	64,367,777.65	219,647,635.90
1912	156,391,487.77	63,268,770.51	219,660,258.28
1913	163,379,842.54	66,266,989.60	230,146,332.14
1914	159,098,177.31	67,081,512.45	226,179,689.76
1915	144,619,699.37	79,328,946.72	223,948,646.09
1916	158,682,439.53	88,771,103.99	247,453,543.52
1917	192,111,318.81	91,897,193.81	284,008,512.62
1918	317,553,687.33	126,285,857.65	443,839,544.98
1919	365,211,252.26	117,839,602.21	483,050,854.47

PROHIBITION ERA

1934	89,951,747.45	168,959,585.17	258,911,322.62
1935	195,457,893.33	215,563,879.02	411,021,772.35
1936	256,337,600.14	149,126,436.96	505,464,037.10
1937	312,660,986.64	281,584,099.63	594,245,086.27
1938	294,786,143.24	273,912,458.29	567,978,601.53
1939	324,458,706.36	263,340,994.32	587,799,700.68
1940	356,476,968.79	267,776,187.32	624,253,156.11
1941	497,349,942.18	322,706,236.15	820,056,178.33
1942	678,844,626.47	369,672,080.09	1,048,516,706.56
1943	961,036,260.17	462,610,196.27	1,423,646,456.44
1944	1,051,607,390.98	567,167,764.95	1,618,775,155.93
1945	1,667,064,026.33	642,801,763.74	2,309,865,790.07
1946	1,872,215,494.24	653,949,191.43	2,526,264,685.67
1947	1,809,681,946.79	665,081,495.63	2,474,763,442.42

Source: U.S. Treasury Department Internal Revenue Service Alcohol Tax Unit.
Note: Figures include special or occupational taxes

Fiscal Year Ending June 30th	Distilled Spirits And Wine	Malt Beverages	Total Revenue
1948	$1,554,207,444.04	$701,119,310.13	$2,255,326,754.17
1949	1,519,803,979.41	690,803,188.60	2,210,607,168.10
1950	1,547,007,213.04	672,149,871.81	2,219,202,084.85
1951	1,877,335,742.75	669,471,182.09	2,546,807,924.84
1952	1,816,367,790.00	732,751,899.00	2,549,119,689.00
1953	2,012,244,108.00	768,681,070.00	2,780,925,178.00
1954	2,008,112,000.00	774,900,000.00	2,783,012,000.00
1955	2,000,055,000.00	742,784,000.00	2,742,840,000.00
1956	2,149,993,000.00	770,581,000.00	2,920,574,000.00
1957	2,207,630,000.00	765,565,000.00	2,973,195,000.00
1958	2,183,801,000.00	762,660,000.00	2,946,461,000.00
1959	2,229,591,000.00	772,505,000.00	3,002,096,000.00
1960	2,392,793,000.00	800,921,000.00	3,193,714,000.00
1961	2,412,549,000.00	800,252,000.00	3,212,801,000.00
1962	2,523,252,000.00	818,030,000.00	3,341,282,000.00
1963	2,610,801,000.00	830,855,000.00	3,441,656,000.00
1964	2,685,563,000.00	891,936,000.00	3,577,499,000.00
1965	2,862,315,000.00	910,319,000.00	3,772,634,000.00
1966	2,922,350,000.00	892,028,000.00	3,814,378,000.00
1967	3,130,708,000.00	945,015,000.00	4,075,723,000.00
1968	3,324,175,000.00	963,062,000.00	4,287,237,000.00
1969	3,548,212,742.00	1,007,347,737.00	4,555,560,479.00
1970	3,664,874,647.00	1,081,507,252.00	4,746,381,899.00
1971	3,692,760,060.00	1,107,722,039.00	4,800,482,099.00
1972	3,942,138,000.00	1,167,863,000.00	5,110,001,000.00
1973	3,946,540,000.00	1,202,973,000.00	5,149,513,000.00
1974	4,092,487,000.00	1,265,990,000.00	5,358,477,000.00
1975	4,042,275,000.00	1,308,583,000.00	5,350,858,000.00
1976	4,056,625,000.00	1,342,430,000.00	5,399,055,000.00
1977	4,008,136,000.00	1,398,497,000.00	5,406,633,000.00
1978	4,189,101,000.00	1,423,614,000.00	5,612,715,000.00
1979	4,143,323,000.00	1,504,601,000.00	5,647,924,000.00
1980	4,156,915,000.00	1,547,853,000.00	5,704,768,000.00
1981	4,082,085,000.00	1,606,328,000.00	5,688,413,000.00

Courtesy of United States Brewers Association, Inc.

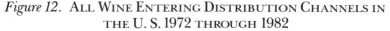

Figure 12. ALL WINE ENTERING DISTRIBUTION CHANNELS IN
THE U. S. 1972 THROUGH 1982

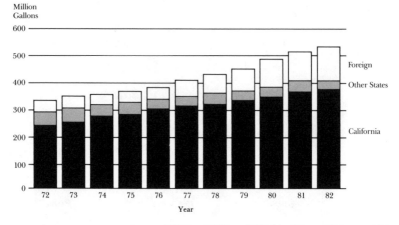

Courtesy of the Economic Research Department of Wine Institute

Wine Consumption

Figure 13 shows all ages per capita quantity of commercially produced wine entering distribution channels in the U.S. by states during 1982. The District of Columbia and Nevada showed the largest
per capita figures. It is likely that significant wine consumption by
nonresident tourists or visitors may account for this phenomena.
California, which produces the largest quantity of domestic wine,
also is the largest consumer. The U.S. per capita average is 2.21 gallons annually.

Table 10 presents an interesting comparison of the estimated per
capita consumption of specific beverages in the U.S. from 1978
through 1982. A small but steady decline in consumption of distilled
spirits was paralleled by a small increment in the consumption of
wine and beer. Coffee consumption has also declined somewhat, as
has tea and milk. In contrast, consumption of soft drinks in the U.S.
has risen from 35.36 gallons annually to almost 40 gallons annually
over a five-year period.

Table 9. STATISTICAL HIGHLIGHTS

Item	Units	Calendar Year 1981	Calendar Year 1982	Percent Change
Utilized Grape Production in California	1000 Tons	3993	5358	34.2
Grape Crush in California	1000 Tons	2416	3116	29.0
Standard Wine Removals[1]				
In California	1000 Gallons	421330	513238[2]	21.8
In Other States	1000 Gallons	41294	41650[2]	0.9
Total	1000 Gallons	462625	554888[2]	19.9
Wine Inventories on December 31				
In California	1000 Gallons	569366	661212	16.1
In Other States	1000 Gallons	46570	47093	1.1
Total	1000 Gallons	615936	708306	15.0
Bonded Winery Premises in California on July 1	Premises	540	591	9.4
Shipments of California Wine				
To California Markets	1000 Gallons	98214	99623	1.4
To Other Markets	1000 Gallons	259985	258911	−0.4
Total	1000 Gallons	358199	358534	0.1
Wine Entering Distribution Channels in the U.S., by Types of Wine				
Table	1000 Gallons	386696	396867	2.6
Dessert	1000 Gallons	42287	39150	−7.4
Sparkling	1000 Gallons	34319	37410	9.0
Vermouth	1000 Gallons	8007	7625	−4.8
Other Special Natural	1000 Gallons	34375	31342	−8.8
Total	1000 Gallons	505684	512394	1.3
Wine Entering Distribution Channels in the U.S., by Areas Where Produced				
California Produced	1000 Gallons	347077	349692	0.8
Other States Produced	1000 Gallons	43894	40613	−7.5
Foreign	1000 Gallons	114713	122089	6.4
Total	1000 Gallons	505684	512394	1.3

Table 9. CONTINUED

Item	Units	Calendar Year 1981	Calendar Year 1982	Percent Change
U.S. Per Capita Wine Consumption				
Adult Per Capita	Gallons	3.30	3.28	−0.6
All Ages Per Capita	Gallons	2.20	2.21	0.5
Five Leading Wine Consuming States				
California	1000 Gallons	108791	109921	1.0
New York	1000 Gallons	52833	52845	0.0
Florida	1000 Gallons	25174	26642	5.8
Illinois	1000 Gallons	24910	25062	0.6
New Jersey	1000 Gallons	23484	24383	3.8
Total	1000 Gallons	235192	238853	1.6

[1] Removals from fermenters. Crop year basis.
[2] Estimated
Courtesy of the Economic Research Department of Wine Institute

Wine Imports

Table 11 shows foreign wine shipments into the U.S. by country of origin from 1978 through 1982. Italy ranks first in shipments of wine to the United States; France and West Germany rank second and third respectively. The increase in imported wine consumption in the United States during the past decade has been truly extraordinary. Whether this primarily reflects an increasing sophistication of the American consumer, lower costs of wine from foreign countries as the American dollar has gained strength or both cannot be clearly ascertained. In the long run, this phenomenon bodes well for American wine producers since there is growing world recognition that America's best wines are among the finest produced anywhere.

The Regulators

Many years before passage of the Eighteenth Amendment, which brought national Prohibition to America in 1920, state laws that regulated the manufacture, distribution and sale of beverage alcohol affected over half of the American population. Enactment of na-

Figure 13. ALL AGES PER CAPITA QUANTITY OF COMMERCIALLY PRODUCED WINE ENTERING DISTRIBUTION CHANNELS IN THE UNITED STATES, BY STATES, 1982

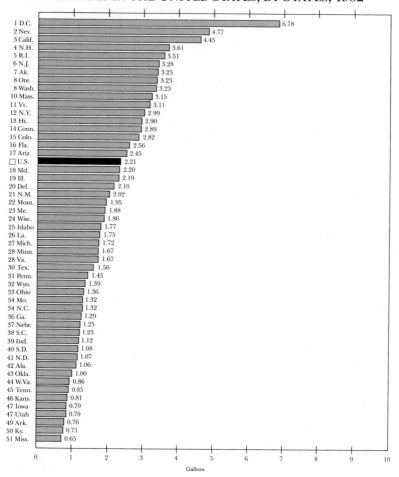

State	Gallons
1 D.C.	6.78
2 Nev.	4.77
3 Calif.	4.45
4 N.H.	3.61
5 R.I.	3.51
6 N.J.	3.28
7 Ak.	3.25
8 Ore.	3.23
8 Wash.	3.23
10 Mass.	3.15
11 Vt.	3.11
12 N.Y.	2.99
13 Hi.	2.90
14 Conn.	2.89
15 Colo.	2.82
16 Fla.	2.56
17 Ariz.	2.45
☐ U.S.	2.21
18 Md.	2.20
19 Ill.	2.19
20 Del.	2.10
21 N.M.	2.02
22 Mont.	1.95
23 Me.	1.88
24 Wisc.	1.86
25 Idaho	1.77
26 La.	1.75
27 Mich.	1.72
28 Minn.	1.67
28 Va.	1.67
30 Tex.	1.56
31 Penn.	1.45
32 Wyo.	1.39
33 Ohio	1.36
34 Mo.	1.32
34 N.C.	1.32
36 Ga.	1.29
37 Nebr.	1.25
38 S.C.	1.23
39 Ind.	1.12
40 S.D.	1.08
41 N.D.	1.07
42 Ala.	1.06
43 Okla.	1.00
44 W.Va.	0.86
45 Tenn.	0.85
46 Kans.	0.81
47 Iowa	0.79
47 Utah	0.79
49 Ark.	0.76
50 Ky.	0.71
51 Miss.	0.65

Gallons

tional prohibition would have been impossible without an existing mandate for "local option" afforded to towns, villages, counties, cities and states for the regulation of beverage alcohol production and sales. Following repeal of Prohibition by the Twenty-first Amendment in 1933 "local option" choices were again instituted by voters in thirty-nine states. Jurisdictions by popular vote could decide to become "wet" and allow the sale of legal beverage alcohol, or

Table 10. Estimated per capita consumption of specified beverages in the United States 1978 through 1982[1]

	1978	1979	1980	1981	1982[2]
Wine (Gallons)[3, 4]	3.01	3.01	3.20	3.30	3.28
Brandy (Gallons)[3]	0.124	0.127	0.124	0.145	0.134
Distilled Spirits (Gallons)[3, 5]	3.07	3.04	3.00	2.93	2.80
Beer (Gallons)[3]	35.5	36.2	36.8	36.9	36.2
Coffee (Pounds)[3, 6]	16.2	17.2	15.7	15.4	15.0
Tea (Pounds)[3, 7]	1.23	1.07	1.21	1.20	1.19
Fluid Milk (Gallons)[8, 9]	28.9	28.4	28.1	27.6	[10]
Soft Drinks (Gallons)[8, 11]	35.36	36.67	37.90	38.85	39.51

[1] Figures for 1980 based on United States population as of April 1. Other years based on U.S. population as of July 1. Civilian population data are used in determining per capita consumption for coffee and tea. Resident population data are used in determining per capita consumption for other beverages. Resident population includes military services personnel stationed in the U.S. but excludes U.S. civilians and military services personnel living abroad.

[2] Preliminary.

[3] Based on population of age 21 years and older.

[4] See Table 15 for per capita wine consumption based on population of all ages.

[5] Includes brandy.

[6] Green-bean equivalent of instant and regular.

[7] Leaf equivalent of instant and regular.

[8] Based on population of all ages.

[9] Fluid whole and low-fat milks; includes quantities used in flavored drinks but excludes reconstituted condensed, evaporated or dried milk.

[10] Not available.

[11] Data are incomplete because sales by all classes of soft drink manufacturers are not included. Data are on a consistent basis from year to year and are satisfactory for use in indicating trends in total soft drink sales.

Sources: Prepared by Economic Research Department, Wine Institute, from reports of Bureau of Alcohol, Tobacco and Firearms, U.S. Treasury Department; Bureau of the Census, U.S. Department of Commerce; Economics and Statistics Service, U.S. Department of Agriculture; Distilled Spirits Council of the United States; U.S. Brewers Association, Inc. and National Soft Drink Association.

Courtesy of the Economic Research Department of Wine Institute

remain "dry" and prohibit the sale of distilled spirits, beer, wine or all three beverages.

Most states decided to become "licensed states" where private individuals could secure licenses to carry out all phases of wholesale and retail sales of beverage alcohol. In 1983, thirty-two states and the District of Columbia operated on this basis. Licenses for private sales of beverage alcohol are also available in Wyoming and Mississippi but wholesale transactions are handled by state-owned warehouses.

In 1983, there were eighteen "controlled states," in which a state government monopoly was the primary source for sales of distilled spirits. These states do not regulate sales of beverage alcohols which are consumed on the premises of restaurants, cocktail lounges, taverns and bars. In addition, most controlled states permit the sale of beer and wine in private outlets in addition to the state-owned stores. The controlled states are Alabama, Idaho, Iowa, Maine, Michigan, Mississippi, Montana, New Hampshire, North Carolina, Ohio, Oregon, Pennsylvania, Utah, Vermont, Virginia, Washington, West Virginia, Wyoming, and Montgomery county, Maryland.

THE STATES

The following summary prepared by the United States Brewers Association (in the *Brewers Almanac*, 1982), presents a brief picture of local option throughout the nation, on a state-by-state basis:

ALABAMA — The Alabama Beverage Control Law, enacted in 1937, provided for simultaneous county local option elections on March 10, 1937, resulting in forty-three counties voting to remain "dry" and twenty-four counties voting to come under the provisions of the law. The law provided that after six months had elapsed, any county could hold another election; and thereafter elections could be held every two years on petition. Since 1937, there have been numerous county local option elections. To date, forty Alabama counties are "wet" with the remaining twenty-seven counties being "dry." The population residing in "dry" counties totaled 967,761 or 24.7% of the state population. Prior to 1971 total prohibition existed in the dry counties. The 1971 Legislature passed a law which allows the possession of one case of Alabama tax-paid beer in any of the dry areas.

ALASKA — Whenever 35% of the voters at the last general election petition the council of a city or 35% of the registered voters within an established village petition the Lieutenant Governor on the question of the sale of intoxicating liquors, the

Table 11. FOREIGN TABLE WINE SHIPMENTS INTO THE UNITED
STATES, BY COUNTRIES OF ORIGIN 1978 THROUGH 1982[1,2]

Country	*1000 Gallons*				
	1978	*1979*	*1980*	*1981*	*1982*
Italy	39422	43188	54259	59860	63024
France	13964	12879	11381	15054	18050
West Germany	13781	11684	11659	13034	13198
Portugal	6269	5935	5683	5366	5053
Spain	1680	1724	1634	1464	1501
Greece	693	586	573	571	627
Yugoslavia	90	88	312	350	562
Romania	220	199	431	289	353
Bulgaria	0	0	268	184	294
Chile	98	188	222	272	287
Hungary	129	146	233	306	286
Israel	406	396	349	311	258
Japan	211	163	211	291	257
Netherlands	186	228	62	157	206
Austria	119	136	220	238	195
Argentina	248	220	177	193	189
Australia	57	47	101	151	125
Algeria	107	81	17	51	122
South Africa	7	19	14	35	113
United Kingdom	92	94	59	71	83
Belgium	113	79	23	49	66
East Germany	3	10	5	11	38
New Zealand	4	6	7	19	28
Switzerland	44	32	29	17	27
Brazil	23	55	49	29	23
Mexico	22	35	22	13	23
Canada	119	48	36	82	19
Cyprus	5	8	15	15	19
China	1	4	11	1	16
Morocco	6	2	9	0	12
Other Countries	107	52	79	78	39
Total	78226	78332	88150	98562	105093

[1] Import for consumption. Includes grape still wine not over 14 percent alcohol and un-
flavored fruit wine from Japan.
[2] Amount less than 500 gallons and percentages less than 0.05 percent are rounded to
zero.
[3] Average annual compounded growth from 1977 to 1982.
[4] Undefined.

Sources: Prepared by Economic Research Department, Wine Institute, from reports of
Bureau of the Census, U.S. Department of Commerce.

Percent Change 1981–82	Average Annual Percent[3]	Percent of Total				
		1978	1979	1980	1981	1982
5.3	20.8	50.4	55.1	61.6	60.7	60.0
19.9	9.3	17.9	16.4	12.9	15.3	17.2
1.3	4.9	17.6	14.9	13.2	13.2	12.6
−5.8	−2.9	8.0	7.6	6.4	5.4	4.8
2.5	1.5	2.1	2.2	1.9	1.5	1.4
9.8	2.6	0.9	0.7	0.7	0.6	0.6
60.6	47.8	0.1	0.1	0.4	0.4	0.5
22.1	29.5[4]	0.3	0.3	0.5	0.3	0.3
59.8		0.0	0.0	0.3	0.2	0.3
5.5	35.0	0.1	0.2	0.3	0.3	0.3
−6.5	42.9	0.2	0.2	0.3	0.3	0.3
−17.0	−8.0	0.5	0.5	0.4	0.3	0.2
−11.7	0.3	0.3	0.2	0.2	0.3	0.2
31.2	14.9	0.2	0.3	0.1	0.2	0.2
−18.1	28.8	0.2	0.2	0.2	0.2	0.2
−2.1	3.6	0.3	0.3	0.2	0.2	0.2
−17.2	26.9	0.1	0.1	0.1	0.2	0.1
139.2	8.0	0.1	0.1	0.0	0.1	0.1
222.9	56.6	0.0	0.0	0.0	0.0	0.1
16.9	13.5	0.1	0.1	0.1	0.1	0.1
34.7	1.6	0.1	0.1	0.0	0.0	0.1
245.5	66.2	0.0	0.0	0.0	0.0	0.0
47.4	47.6	0.0	0.0	0.0	0.0	0.0
58.8	14.0	0.1	0.0	0.0	0.0	0.0
−20.7	−30.4	0.0	0.1	0.1	0.0	0.0
76.9	7.5	0.0	0.0	0.0	0.0	0.0
−76.8	−14.3	0.2	0.1	0.0	0.1	0.0
26.7	36.6[4]	0.0	0.0	0.0	0.0	0.0
1500.0[4]		0.0	0.0	0.0	0.0	0.0
	43.1	0.0	0.0	0.0	0.0	0.0
−50.0	−15.8	0.1	0.1	0.1	0.1	0.0
6.6	13.4	100.0	100.0	100.0	100.0	100.0

Courtesy of the Economic Research Department of Wine Institute

voters may by a majority vote adopt or reject the proposition:

(1) Shall the sale of alcoholic beverages be prohibited?

(2) Shall the sale of alcoholic beverages be prohibited unless sold under a community liquor license?

(3) Shall the sale and importation of alcoholic beverages be prohibited?

(4) Shall the sale of alcoholic beverages be prohibited except by ... (listing the types of licenses which premises would be exempted from the prohibition).

If the vote is to restrict or prohibit the effective date is January 1, of the year following the election. If the vote is in an established village it is effective not only within the village but the unincorporated area within five miles of the village.

While the number of dry communities has increased in recent years the number of their residents is less than 10,000 or 2% of the state population.

ARIZONA — There are no local option provisions in the Arizona law.

ARKANSAS — Beer sales became legal August 24, 1933. Before January 1, 1943, two counties voted to prohibit beer. Effective that date Initiated Act No. 1 provided for local option elections by petition on manufacture and sale of "intoxicating liquor" defined as any beverage containing more than ½ of 1 per cent of alcohol by weight. Under a law which became effective Janu-

ary 28, 1955, these local option elections are limited to regular November biennial election dates (even-numbered years.) Of Arkansas' seventy-five counties, the sale is prohibited in forty-three as well as four cities, three towns, seventy-five townships and six precincts of twenty other counties. The dry areas have a population of 846,965 or 36.9% of the total state population of 2,296,000.

CALIFORNIA — There are no local option provisions in the California law.

COLORADO — There are no local option provisions governing the sale of 3.2 beer. Beer containing more than 3.2 per cent of alcohol by weight is classed as "malt liquor," subject to the local option provisions of the Liquor Code, and its sale is prohibited in seven towns with a population of 3,562 or .12 per cent of the state population of 2,965,000. An election may not be held in the same place more than once in any four-year period.

CONNECTICUT — The sale of beer became legal on April 20, 1933, the law permitting the sale and manufacture of all types of alcohol beverages in every town in Connecticut until a contrary preference has been indicated by town ordinance. Three propositions were provided to be voted upon, after petition, at the annual town meeting: (1) permits for all alcohol beverages; (2) permits for beer; (3) no permit. At the end of 1981 two towns in Connecticut were dry for all alcohol beverages, including beer. The total population of these dry towns is 2,800 or .09 per cent of the state population of 3,134,000. In addition, there are three towns that allow beer only. A 1965 law eliminated the question on beer only and substituted question on whether sale of alcoholic beverages by one or more classes of permits specific in petition shall be allowed.

DELAWARE — Article XIII of the Delaware Constitution of 1897 as amended deals with local options. Section One permits the General Assembly from time to time, to submit the question of licensing of the sale or manufacture of alcoholic liquors, including "vinous or malt liquors," to the voters. The state is divided into four areas for the purposes of such an election (Wilmington, the remainder of New Castle County, Kent County and Sussex County), and a majority vote would determine licensing in the area until resubmission of the question.

By a further provision of Section One, the question must be put to the electors in any of these four districts if a majority of each house of the legislature of that particular area request this from the General Assembly.

The only statutory sections dealing with these local option provisions, 4 Del. C. Sec. 102,103, merely permit such elections and removes an area from the effect of the Chapter's general liquor regulations if an election results in no licensing. The most southern Delaware county, Sussex County, did exercise this option once prior to prohibition and was "dry" accordingly. There has been no attempt since then to exercise these options through either of the possible methods, so sale of all alcoholic beverages is legal throughout the state at the present time.

DISTRICT OF COLUMBIA — There are no local option provisions in the laws of the District of Columbia.

FLORIDA — Only intoxicating liquors are subject to the local option laws of Florida. Such liquors are defined as "all liquor and beverages, whether spirituous, vinous or malt, containing more than 3.2 per cent of alcohol by weight." Consequently, 3.2 per cent beer is legal in all counties in Florida. An election may not be held more often than once every two years in the same county. Seven of Florida's sixty-seven counties, with a population of 131,570 or 1.3 per cent of the total state population of 10,183,000 are dry for beverages containing more than 3.2 per cent of alcohol by weight.

GEORGIA — The local option laws of Georgia do not apply to beer. As to retail beer licenses, however, the governing bodies of counties and municipalities have broad power to decide whether or not they shall be issued within their jurisdictions. There is no appeal from their decisions. Under this provision, four of Georgia's 159 counties and an indeterminate and changing number of towns are dry for beer. The dry counties contain less than 0.5 per cent of the state population of 5,574,000.

HAWAII — There are no local option provisions in the Hawaii law. However, public hearings must be held by county liquor commissions on all applications for licenses. Protests against issuance may be filed by any registered voter of the precinct in which the licensed premises would be located or by any owner or leaseholder of real estate within five hundred feet of the premises. Their commission may grant or refuse the license in its discretion, except that if a majority of the registered voters of the precinct, or a majority of the eligible land owners or leaseholders shall have duly filed protests, the license shall not be issued. There are no areas dry for beer at present.

IDAHO — There are no local option provisions in the Idaho law. All alcohol liquors containing more than 4 per cent of alcohol by weight come under the provisions of the "Liquor Control Act" and are sold only through state stores or by special distributors designated by the State Liquor Dispensary.

ILLINOIS — The act legalizing the sale of beer on April 7, 1933, provided for local option elections in a city, village, incorporated town, township (outside the corporate limits of any municipality in such township) or road district; and in the city of Chicago in a precinct or a group of precincts. Elections may be called by petition and are held at the time of the regular elections every two years (odd-numbered years). After an election a subsequent election in the same political subdivision may not be held for 47 months. The total population living in dry areas is 736,000 or 6.42 per cent of the state's population of 11,462,000.

INDIANA — There are no local option provisions in the Indiana law. Indiana has 99 counties and a 5,490,000 state population.

IOWA — There are no local option provisions for sale of malt beverages in the Iowa law. Therefore sale of beer is legal throughout the state, which has 99 counties with a 2,913,000 population.

KANSAS — After repeal of constitutional Prohibition at the polls on November 2, 1948, Kansas enacted a local option law in 1949 on beverages above 3.2 per cent of alcohol but continued statewide sale of 3.2 beer. The law permits package sale only of beverages above 3.2 per cent in areas which voted to repeal on

November 2, 1948 or in subsequent local option elections. The question is submitted on petition at the general city election, not oftener than once in any four years in the same city. At the end of 1981 there were 299 cities in Kansas dry for the sale of beverages over 3.2 per cent comprising a population of 175,311 or .074 per cent of the state population of 2,383,000.

KENTUCKY — From 1933 until the enactment of a local option law in 1936, all 120 counties in Kentucky permitted the sale of alcohol beverages. Elections are called by petition and no election may be held in the same territory more often than once in three years, except where the territory previously voted wet.

The 1948 Legislature amended the local option law to provide that cities of 3,000 or more population located in dry counties could petition for elections and vote on their status independently of the county. Such an election is not subject to the three year limitation mentioned above.

At the end of 1981 eighty-four of Kentucky's 120 counties prohibited the sale of all alcohol beverages. Twenty-six counties are wet. In addition, there are ten counties dry except for a wet city in each. The total dry population of the state is 1,793,578 or 49 per cent of the total state population of 3,662,000.

LOUISIANA — Under the provisions of Act 17, passed in March 1939,

electors of a parish, ward or municipality could petition for an election to vote on one or both of two questions: (1) beverages containing not more than 6 per cent of alcohol by volume, and (2) beverages containing more than 6 per cent of alcohol by volume. As a matter of practice, both questions were consolidated and the vote taken on alcoholic beverages of more than ½ of 1 per cent alcohol.

The 1948 Legislature amended the local option law to provide that any ward, incorporated village or town could petition an election, while prohibiting elections on a parishwide basis, and eliminating beer containing not more than 3.2 per cent of alcohol by weight from the provisions of the local option act in future elections. Three propositions may be voted upon: (1) Sale of beverages containing more than 3.2 per cent of alcohol by weight and not more than six per cent by volume; (2) more than six per cent of alcohol by volume for consumption on the premises and (3) more than six per cent of alcohol by volume by package only.

The local option law was amended in 1974 authorizing the parishes of Beauregard, Washington, Rapides, Natchitoches, Red River, Grant, LaSalle, East Carroll, West Carroll, Bienville, Jackson and Winn prohibition of alcoholic beverages on a parish-wide basis, and the prohibition of beverages containing not more than three and two-tenths per cent alcohol by weight. Thus, in those parishes, four propositions may appear on the ballots: (1) Sale of beverages containing more than 3.2 per cent of alcohol by weight and not more than six per cent by volume; (2) sale of beverages containing more than one-half of one per cent alcohol by volume, but not more than three and two-tenths per cent alcohol by weight; (3) more than six per cent alcohol by volume for consumption on the premises; and (4) more than six per cent alcohol by volume by package only.

In all other parishes of the state, parishwide elections are prohibited. Elections must be called by wards and incorporated municipalities in each parish and only three propositions may be voted upon: (1) Sale of beverages containing more than 3.2 per cent of alcohol by weight and not more than six per cent by volume; (2) more than six per cent of alcohol by volume for consumption on the premises; and (3) more than six per cent of alcohol by volume by package only. No election may be held for the same subdivision oftener than once in every two years.

At the end of 1981, two parishes were dry. In addition, there were fifteen parishes which were partially dry. The total dry population was 151,400, or 3.5 per cent of the state population of 4,308,000.

MAINE — The Maine law provides for local option elections in every city, town, and plantation upon pe-

tition by voters. The eight questions which may be voted upon deal with: (1) State Liquor Stores and State Agency Stores; (2) sale of spirits and wine for consumption on the premises; (3) on premise sale of beer; (4) sale of beer in retail stores, and, four questions relative to Sunday sales and hours of sale.

There are 538 civil divisions (including cities, towns, plantations, and unorganized territories) in 16 counties in the State of Maine. The sale of beer is prohibited in 101 of these jurisdictions. These areas constitute a dry population of approximately 38,000 or 3.4 percent of the state population of 1,133,000.

MARYLAND — There are no local option provisions in the general law in Maryland except for the municipal corporation of Poolesville in Montgomery County. By virtue of local laws, however, the sale of beer and liquor is prohibited in parts of Baltimore, Frederick and Montgomery Counties. These totally dry areas have a population of 65,534 or 1.5 per cent of the state population of 4,263,000.

MASSACHUSETTS — Beer was legalized in Massachusetts on April 7, 1933. The Commonwealth of Massachusetts provides for the holding of local option elections by initiative petition of at least 10 percent of the total number of voters in the previous election within the city or town.

Of the 351 cities and towns in Massachusetts' 14 counties, a total of 20 communities are completely "dry." These areas constitute a total dry population of 104,086 or 1.8 percent of the total state population of 5,773,000.

MICHIGAN — Beer was legalized April 27, 1933 and provision was made for county local option on spirituous, vinous and malt beverages. Since that time several counties have held local option elections and have voted to retain legal manufacture and sale in every instance, leaving the sale of malt beverages legal in all of Michigan's eighty-three counties.

MINNESOTA — Minnesota has 87 counties. The county option law for intoxicating liquor and strong beer was repealed at the 1965 legislative session. Therefore, liquor and beer sales are legal in all counties.

Municipalities issue licenses for the sale of intoxicating liquor, including strong beer, within their corporate limits. County Boards are now also authorized to issue "on-sale" liquor licenses within the unorganized or unincorporated area of the county to restaurants, as defined in Sec. 340.07 Subd. 14 (seating not less than 100 guests at one time), with the approval of the Liquor Control Commissioner.

The Minnesota law contains no local option provisions applicable to malt beverages containing 3.2 per cent or less of alcohol by weight but broad licensing powers are vested in the governing authorities of counties, cities and towns. These occasionally have refused issuance of licenses for 3.2 per cent beer and have regulated closing hours and Sunday closing.

Municipalities having a population of more than 10,000 persons, may by municipal election, establish a municipal liquor store for both "on sale" or "off sale" or both.

Any statutory city or any home rule charter city of the fourth class may by election restrict the issuance of liquor licenses.

MISSISSIPPI — Beer was legalized in all eighty-two counties on February 26, 1934, when the Beer and Light Wines Law became effective. The original local option law permitted a county to prohibit the sale of beer and wine but did not provide for a subsequent election to relegalize these beverages. This situation was remedied by the 1942 Legislature and elections may now be called in wet or dry counties. In addition, a law passed in March, 1950 now permits cities having a population of not less than 2,500 to vote separately on local option. Prior thereto, elections were on a county-wide basis only. The same city or county may not vote local option more than once in five years. Of the eighty-two counties in Mississippi at the end of 1981, twenty-nine are totally dry and fourteen counties are partially wet for beer. The total population in the twenty-nine dry counties is 470,832 or 18.6 percent of the state population of 2,531,000.

MISSOURI — Local option does not apply to "malt beverages" having an alcohol content of not more than 5 per cent by weight, leaving the sale of beer legal throughout the state. On-premise sale of beverages containing more than 5 per cent alcohol by weight (liquor by drink) limited to municipalities of over 500 population by local option and to qualified "restaurants" and "resorts." The state of Missouri has 114 counties with a population of 4,941,000.

MONTANA — Beer was legalized in Montana on April 7, 1933. Manufacture and sale are allowed up to an alcoholic content of 7 per cent by weight. Ale, porter and stout containing more than 7 per cent of alcohol by weight are sold by state stores. "Montana Liquor Beverage Code" provides for local option elections upon petition, on a county-wide basis, which applies to beer up to 7 percent. There are, however, no areas in Montana "dry" for beer.

NEBRASKA — The sale of beer is not subject to local option, but if a "Remonstrance Petition" objecting to the issuance of beer licenses, signed by 51 per cent of the votes cast in the last preceding state general election, is filed with local governing authorities of a town or municipality, no retail licenses may be issued for the ensuing license year. A total of 86 dry towns have a combined population of 12,821 or .008 per cent of the state population of 1,577,000. One small town is dry because of provisions in its land grant that no licenses for the sale of alcohol beverages shall ever be issued. One county (McPherson) is dry.

NEVADA — There are no local option provisions in the Nevada law.

NEW HAMPSHIRE — Beer was legalized in New Hampshire on May 2, 1933. The law provides that at each town biennial election, held in even-numbered years, the municipality shall vote for or against state stores, and for or against the sale of beverages (meaning beer, ale and wine) containing not more than 6 per cent of alcohol by volume. Of the 240 towns and cities in New Hampshire, 6 are dry for malt beverages. Their population is 1,027 or .11 per cent of the total state population of 936,000. A 1965 law provided for local option by petition rather than at each biennial election.

NEW JERSEY — Local option elections may be the subject of a petition for beer and wine or for all alcohol beverages; and also to determine Sunday sales. These elections can be held at the time of the general election if 15% of the voters are petitioned. No further election may be held on the same question in the same municipality prior to the general election in the fifth year thereafter. In addition, local authorities have extremely broad powers even to the extent of deciding not to issue licenses. New Jersey, therefore, has some areas dry due to the failure of local authorities to issue licenses, or from prohibiting referenda.

As a result of one or another of these actions, 41 of 567 municipalities in N.J. are dry. The total population living in dry areas is 165,165, or 2.2 per cent of the state population of 7,404,000.

NEW MEXICO — The local option law of the state provides that an election may be petitioned for in any county, or in any city having a population of more than 5,000. The vote is for or against all alcohol beverages. After an election, a subsequent election may not be held in the same county or city for two years. Roosevelt County is dry, with the exception of the city of Portales, as is Curry County with the exception of the city of Clovis. These dry areas include a population of 16,580 or 1.2 per cent of the total state population of 1,328,000.

NEW YORK — The sale of beer became legal on April 7, 1933. Local option is provided for by towns or cities by petition. The election may be on either one of two groups of questions, one having to do with all alcohol beverages and the other pertaining to alcohol beverages other than malt beverages. The election is held at the time of the general election.

After an election on either of the two groups, a subsequent election on the same group may not be held in the same city or town before the third subsequent general election. There are twenty-one towns in New York totally dry for beer. Their population aggregates 36,653 or 0.2 per cent of the state population of 17,602,000.

NORTH CAROLINA — Until 1947 North Carolina had no provision for local option on beeer. However, two counties, nine towns and one township had been made dry by legislative pact. The legislature of 1947 amended the law to provide that local option elections for wine and/or beer may be petitioned by counties. A subsequent county election in the same county may be held within three years. Under 1957 legislation municipalities of over 1,000 population in dry counties may petition elections for (1) on- and off-premise sales; (2) off-premise sales only; or (3) on-premise sales by Grade A hotels and restaurants only and off-premise sales by other licensees. This provision was extended in 1963 to permit elections in resort communities which have a "seasonal" population of over 1,000 (i.e. for a period of six weeks in the year).

In 1971 through legislative enactment and a recodification of North Carolina GS 18 to GS 18A resulted in amending the 1957 legislation to permit municipalities of 500 population or more in Dry counties the right to petition for a vote.

Again in 1981, through legislative enactment, Chapter 18a was rewritten to become 18b. The provisions for local option were left virtually intact. Only minor changes in percentages of signatures required on a petition to call for an election were changed. New provisions were enacted to allow counties, municipalities and townships which operate ABC systems and have voted on the sale of mixed beverages to have legal sales of beer without elections.

At the close of 1981, there were 8 dry counties, 37 counties with legal sales in one or more towns and 55 counties with legal sales countywide. The legal sale of beer is now available to approximately 86% of the State's 5,953,000 citizens.

NORTH DAKOTA — There are no local option provisions in the North Dakota law.

OHIO — Sale of beer containing not more than 3.2 per cent of alcohol by weight may be voted upon at a general election on a previously filed petition containing signatures equal to 35 per cent of the total votes cast for governor and may be called in a municipality or a residential district consisting of two or more contiguous precincts in the residential portion of a municipality, or in a township or a part of a township outside of a municipality. No elec-

tions on 3.2 beer may be held in the same areas more often than once in each three years. An election on the sale of malt beverages containing more than 3.2 per cent and less than 7 per cent of alcohol by weight may be held in a municipality consisting of two or more contiguous election precincts, or in a township on petition of 35 per cent of the total number of votes cast in such area for governor. A special election is required, but may be held at the time of the general election.

OKLAHOMA — There are no local option provisions in the Oklahoma law. The sale of beer containing not more than 3.2 per cent by weight is allowed as a non-intoxicating beverage. Constitutional prohibition on alcohol beverages having an alcoholic content of more than 3.2 per cent by weight was repealed on April 7, 1959. The sale of beer is legal throughout the state.

OREGON — Malt beverages containing not more than 8 per cent alcohol by weight and wine containing not more than 21 per cent by volume may be sold in retail outlets. All other alcohol beverages may be sold only through state stores.

Local option elections may be petitioned in city or county areas and are held only on the regular November biennial election days. Three questions may be voted upon: (1) prohibition of the sale of all alcohol beverages; (2) prohibition of the sale of alcohol beverages containing over 14 per cent of alcohol by weight; (3) prohibition of alcohol beverages containing over 4 per cent of alcohol by weight.

Only one municipality with a population of 5,237 or 0.2 per cent of the total state population is dry.

PENNSYLVANIA — An election to mits, for beer; allow or prohibit the issuance of retail malt beverage licenses for on and off premise sale of beer may be held in any municipality or township on the date of the primary election preceding a municipal election, but not oftener in the same municipality than once in four years. Though Pennsylvania has state stores and a great bulk of spirituous liquors is sold through them, elections may be held also on the issuance of hotel, restaurant and club licenses for the on-premise sale of liquor. Such licenses include the right to sell malt beverages on- or off-premise, except in clubs where only on-premise sales are allowed.

There are 690 municipalities in Pennsylvania in which on-off premise retail sales of both beer and liquor and off-premise sale of beer and liquor are prohibited. Their combined population is 1,279,923 or 10.8 per cent of the state population of 11,871,000. There are a number of other municipalities where the issuance of retail beer licenses is prohibited, except 21 municipalities are dry for distributor and importing distributor licenses, but where liquor licenses are issued

to hotels, restaurants and clubs. Beer may be sold for on- and off-premise consumption by these licensees. Manufacturers, distributors, and importing distributors may sell and deliver malt beverages to the public, for off-premise consumption, in quantities of not less than a case, anywhere in the state, including the communities mentioned above where issuance of retail licenses is prohibited.

RHODE ISLAND — Under local option provisions, elections may be petitioned in any city or town. Elections are held on the day for the election of town or city officers. The vote may be on two questions: (1) the issuance of licenses of the sale of alcohol beverages; (2) the issuance of Retailers Class C licenses, authorizing the sale of alcohol beverages for consumption on the premises. An affirmative vote on the first question and a negative vote on the second would still permit retail licenses authorizing the sale of beer both on and off premises. Only one town in Rhode Island has voted dry. Barrington, with a population of 16,174 or 1.7 per cent of the state population of 953,000.

SOUTH CAROLINA — There are no local option provisions in the South Carolina law.

SOUTH DAKOTA — Beverages higher than 3.2 per cent alcohol content are subject to local option provisions. Any municipality upon petition may vote on whether these beverages shall be sold in that municipality by licensed on-premise dealers or by the municipality itself, or the sale prohibited.

TENNESSEE — Local option provisions do not apply to beer containing not more than five percent (5%) alcohol by weight. However, county courts and city councils frequently, with or without an advisory election, invoke a state law allowing local discretion in prohibiting the issuance of beer permits within up to 2,000 feet of a school, church or other place of public gathering. In addition, counties may also prohibit the sale of beer within 300 feet of a one or two family dwelling located in unzoned, unincorporated areas. Further, incorporated municipalities have been granted authority to set zones for the sale of beer or to prohibit its sale entirely.

At the end of 1981, 41 of Tennessee's 95 counties were totally wet (county and all cities), 47 counties were wet with one or more dry cities, 6 counties wet (unincorporated areas) with all cities dry, and 1 county was dry with its only city wet. Of the state's 333 cities, 228 are wet and 105 dry. Of the state's total population of 4,612,000, more than 98% of the population live in wet areas.

TEXAS — Local option elections may be called by petition on any county, justice's precinct, or incorporated town or city to vote on one of the issues permissible in the area at the time. The issue submitted may be "for" or "against" the legal sale of one of the following: (a) beer (4 per cent or less alcohol by weight); (b) same as (a) but for off-premise consumption only; (c) beer and wine (not over 14 per cent alcohol by volume); (d) same as (c) but for off-premise consumption only; (e) all alcohol beverages; (f) all alcohol beverages for off-premise consumption only; or (g) mixed beverages (on-premise only).

No subsequent election may be held oftener than once a year on the same issue in the same political subdivision.

As of January, 1982, of 254 counties in Texas, 74 were totally dry. In addition, 180 counties are partially wet. The total population living in dry areas as of January, 1982 was approximately 1,820,595 — about 12.3% of the population of 14,766,-000.

UTAH — There are no local option provisions in the Utah law. However, cities and towns and the counties outside of the corporate limits of any city or town have the power to "license, tax, regulate or prohibit the sale of light beer at retail." Light beer contains not more than 3.2 per cent of alcohol by weight. The town of Blanding in San Juan county and the rural areas of Cache County are dry for beer. The population involved is 25,000 or 1.6 per cent of the total state population of 1,518,000.

VERMONT — The law provides that each town in the state may vote annually if a petition of not less than 5 per cent of the registered voters of the town has been filed. The relevant matters are: (1) the sale of malt beverages containing over 1 per cent and not over 6 per cent of alcohol by volume; and (2) the sale of spirituous liquors. Of 246 towns and villages in the state four are dry for beer. As of the end of 1981, the total dry population was 2,711 or .5 per cent of the state population of 516,000. Reporting of local voting will be to the Secretary of State's office not to Liquor Control Board.

VIRGINIA — The Virginia Code has been amended to delete provisions for local elections on the sale of beer above 3.2% of alcohol by weight. Henceforth, licensees in a particular locality may sell beer under appropriate beer licenses issued by the Virginia Alcohol Beverage Control Commission, even though the voters may have voted previously against such sales. Therefore, at the end of 1981, virtually all of Virginia's 5,430,300 citizens were able to purchase beer within their locality.

WASHINGTON — An election as to whether the sale of liquor will be permitted, or the more limited question of whether restaurants will be permitted to serve liquor, may be petitioned for in any "local option unit." Such a unit consists of an unincorporated city, a town, or all of a county that is not included in an incorporated city or town. The petition must have a number of signatures equal at least to 30% of the number voting in the last general election. The decision in such an election is by majority vote.

There are no dry areas for beer in the state.

WEST VIRGINIA — There are no local option provisions affecting nonintoxicating beer containing not more than 4.2 per cent of alcohol by weight or 6% by volume. Upon petition of 25 per cent of the qualified electors, county or municipal elections may be held to determine whether such liquor (alcohol beverages over 4.2 per cent by weight or 6% by volume) shall be sold in the locality. If a county or municipality votes dry then the West Virginia Liquor Control Commission closes all state stores and agencies therein. Sale of liquor on premises is illegal in West Virginia. There are no dry areas for beer in the state.

WISCONSIN — The local option law differentiates between fermented malt beverages containing 5 per cent or less of alcohol by weight, and "intoxicating liquors" including malt beverages with an alcoholic content greater than 5 per cent. Under the malt beverage law local option elections may be petitioned for in any city, village or town. Such election must be held at the spring elections. The result is effective for two years and thereafter until changed in another election.

Two questions are voted upon: (1) the issuance of Class B licenses for the retail sale of beer for consumption on or off the premises; (2) the issuance of Class A licenses authorizing the sale of retail beer by the package. In addition, the various municipalities have a right to refuse to issue licenses.

There are 72 counties in Wisconsin. In 35 of these counties there are municipalities where the sale of malt beverages is prohibited. In 81 towns and villages sale is prohibited by referendum and in 35 towns by acts of local officials. These dry areas include a total population of 85,158 or 1.8 per cent of the state population of 4,742,000.

WYOMING — There are no local option provisions in the Wyoming law. Population is 492,800.

STATE AGENCIES IN CHARGE OF ADMINISTERING ALCOHOL BEVERAGE LAWS

ALABAMA
Alabama Alcoholic Beverage Control Board, Montgomery

ALASKA
Alcoholic Beverage Control Board, Anchorage

ARIZONA
Department of Liquor Licenses and Control, Phoenix

ARKANSAS
Department of Finance Administration, Alcoholic Beverage Control Division, Little Rock

CALIFORNIA
Department of Alcoholic Beverage Control, Sacramento

COLORADO
Department of Revenue-Liquor Enforcement Division, Denver

CONNECTICUT
Department of Liquor Control, Hartford

DELAWARE
Delaware Alcoholic Beverage Control Commission, Wilmington

DISTRICT OF COLUMBIA
Alcoholic Beverage Control Board, Washington, D.C.

FLORIDA
Department of Business Regulation, Division of Alcoholic Beverages & Tobacco, Tallahassee

GEORGIA
Alcohol & Tobacco Tax Unit, Department of Revenue, Atlanta

HAWAII
Liquor Commission for each county, Honolulu, Hawaii, Kauai, and Maui

IDAHO
Commissioner Law Enforcement, Boise (beer laws)
State Liquor Dispensary, Boise (liquor laws)

ILLINOIS
Illinois Liquor Control Commission, Springfield

INDIANA
Indiana Alcoholic Beverage Commission, Indianapolis

IOWA
Iowa Beer and Liquor Control Department, Ankeny

KANSAS
Alcoholic Beverage Control, Department of Revenue, Topeka

KENTUCKY
Department of Alcoholic Beverage Control (Malt Beverage Unit, and Distilled Spirits Unit), Frankfort

LOUISIANA
Commissioner of Alcoholic Beverage Control, Department of Public Safety, Baton Rouge

MAINE
State Liquor Commission, Bureau of Alcoholic Beverages, Augusta

MARYLAND
Comptroller of the Treasury, Alcohol & Tobacco Tax Division, Annapolis

MASSACHUSETTS
Alcoholic Beverages Control Commission, Boston

MICHIGAN
Michigan Liquor Control Commission, Lansing

MINNESOTA
Department of Public Safety, Minnesota Liquor Control Commissioner, St. Paul
(Licensing and regulation of sales 3.2 beer are under jurisdiction of local governing bodies)

MISSISSIPPI
Chairman of the State Tax Commission, Jackson

MISSOURI
Department of Public Safety, Supervisor of Liquor Control, Jefferson City

MONTANA
Department of Revenue, Liquor Division, Helena

NEBRASKA
Nebraska Liquor Control Commission, Lincoln

NEVADA
Nevada Department of Taxation, Carson City

NEW HAMPSHIRE
State Liquor Commission, Concord

NEW JERSEY
Division of Alcoholic Beverage Control, Newark

NEW MEXICO
Department of Alcoholic Beverage Control, Santa Fe

NEW YORK
New York State Liquor Authority, New York City

NORTH CAROLINA
Board of Alcoholic Control, Raleigh

NORTH DAKOTA
Alcohol Beverage Control Div., State Treasurer, Bismarck (Wholesale licenses; taxes)
Attorney General, Bismarck (Retail Licenses; enforcement)

OHIO
Department of Liquor Control, Columbus

OKLAHOMA
Oklahoma Tax Commission, Oklahoma City (3.2 beer)
Alcoholic Beverage Control Board, Oklahoma City (over 3.2 beer)

OREGON
Oregon Liquor Control Commission, Portland

PENNSYLVANIA
Pennsylvania Liquor Control Board, Harrisburg

RHODE ISLAND
Liquor Control Administration, Department of Business Regulation, Providence

SOUTH CAROLINA
South Carolina Alcoholic Beverage Control Commission, Columbia

SOUTH DAKOTA
Department of Revenue, Pierre

TENNESSEE
Department of Revenue, Nashville

TEXAS
Texas Alcoholic Beverage Commission, Austin

UTAH
Utah Liquor Control Commission, Salt Lake City

VERMONT
Department of Liquor Control, Montpelier

VIRGINIA
Alcoholic Beverage Control Commission, Richmond

WASHINGTON
Washington State Liquor Control Board, Olympia

WEST VIRGINIA
Alcohol Beverage Control Commissioner, Non-Intoxicating Beer Commissioner, Charleston (3.2 beer)

WISCONSIN
Department of Revenue, Excise Tax Bureau, Madison

WYOMING
Wyoming Liquor Commission, Cheyenne

THE FEDERAL GOVERNMENT

Federal distilled spirits taxes were first collected by the U.S. government in 1791. Almost two hundred years later, the federal government derives a large amount of revenue from beverage alcohol. It is therefore not surprising that the Department of Treasury plays a key role in the regulation of the beverage alcohol industry as well as the collection of alcohol-related revenues. The end of Prohibition marked enactment of the Federal Alcohol Administration Act of 1935 and placed responsibility for federal alcohol regulatory procedures within the Bureau of Alcohol, Tobacco and Firearms, a national bureau of the Department of the Treasury. The Bureau is large, employing almost four thousand persons. The Bureau's budget is approximately $1.5 million annually.

Although the states have primary responsibility for controlling and suppressing illicit beverage alcohol production within their jurisdictions, interstate "moonshining" activities are investigated by the Bureau. For example, in 1979 the Bureau seized forty illegal distilleries containing 25,507 gallons of mash and 1,803 gallons of non-tax paying distilled spirits. As a consequence of these activities the Bureau remanded forty criminal cases for prosecution.

Inspectors employed by the Bureau carry out routine surveillance at breweries, distilleries and wineries to ensure prompt collection of federal alcohol excise taxes. Four to five thousand on-site inspections are conducted annually. The Bureau has authority to issue original alcohol permits as well as amend and terminate permits for the production and distribution of beverage alcohol. Each year the Bureau audits tax returns for over one thousand distilleries, breweries and wineries.

During recent years the Bureau has undertaken a number of activities that may have significant impact upon regulatory practices and policies for the beverage alcohol industry. For example, the Bureau, in concert with the Food and Drug Administration, is exploring regulatory standards for ingredient labeling of beverage alcohol. Public awareness campaigns highlighting certain health hazards such as the fetal alcohol syndrome have been developed and promulgated by the Bureau. The Bureau is playing an active role in facilitating the development of viticultural areas. This procedure will specifically designate grape growing regions on the basis of geo-

Table 12. PER CAPITA CONSUMPTION OF TOTAL
ALCOHOL AND OF SPIRITS, BEER AND WINE IN 1981[a]

Total Alcohol (litres)		Spirits (pure alcohol) (litres)	
1. Luxembourg	16.8	1. Luxembourg	5.8
2. France	14.9	2. East Germany	4.7
3. Portugal	13.0	3. Hungary	4.6
4. Spain	12.7	4. Poland	4.3
5. East Germany	12.7	5. Czechoslovakia	3.6
6. West Germany	12.5	6. U.S.S.R.	3.3
7. Hungary	12.3	7. Canada	3.3
8. Switzerland	11.1	8. United States	3.2
9. Czechoslovakia	10.8	9. Spain	3.0
10. Austria	10.8	10. West Germany	2.9
11. Italy	10.4	11. Finland	2.8
12. Belgium	10.3	12. Netherlands	2.5
13. Australia	10.2	13. Sweden	2.5
14. Denmark	9.9	14. Iceland	2.2
15. New Zealand	8.9	15. Belgium	2.1
16. Canada	8.8	16. Switzerland	2.1
17. Netherlands	8.8	17. Yugoslavia	2.0
18. United States	8.6	18. France	2.0
19. Romania	7.7	19. Romania	2.0
20. Bulgaria	7.7	20. Bulgaria	2.0
21. Yugoslavia	7.6	21. New Zealand	2.0
22. United Kingdom	7.3	22. Italy	1.9
23. Ireland	7.2	23. Japan	1.9
24. Poland	6.6	24. Ireland	1.9
25. Finland	6.5	25. United Kingdom	1.7
26. U.S.S.R.	6.2	26. Norway	1.6
27. Japan	5.9	27. Denmark	1.6
28. Sweden	5.4	28. Austria	1.5
29. Norway	4.2	29. Australia	1.1
30. Iceland	3.7	30. Portugal	0.9
31. Mexico	2.7	31. Israel	0.9
32. Israel	2.1	32. Mexico	0.7
Average	8.9		2.5
Median	8.8		2.1

[a] Figures for East Germany, Yugoslavia, Romania, Bulgaria, U.S.S.R., Iceland, Mexico and Israel, from Dutch Distillers Association. Others as noted in Part I.
[b] Includes cider.
[c] Includes sake.

Beer (litres)		Wine (litres)	
1. West Germany	147.0	1. France	90.0
2. Czechoslovakia	140.1	2. Portugal	88.0
3. East Germany	137.5	3. Italy	74.0
4. Australia	134.1	4. Spain	60.0
5. Denmark	131.1	5. Switzerland[b]	53.2
6. Belgium	124.2	6. Luxembourg	40.2
7. Luxembourg	123.0	7. Austria	35.1
8. New Zealand	117.7	8. Hungary	34.0
9. Ireland	116.4	9. Romania	28.9
10. United Kingdom	111.5	10. Yugoslavia	28.2
11. Austria	104.8	11. West Germany	24.6
12. United States	93.0	12. Bulgaria	22.0
13. Netherlands	89.5	13. Belgium	21.0
14. Hungary	89.0	14. Australia	18.3
15. Canada	86.5	15. Denmark	16.1
16. Switzerland	71.0	16. Czechoslovakia	16.0
17. Bulgaria	60.9	17. Japan[c]	14.6
18. Finland	57.2	18. U.S.S.R.	14.5
19. Spain	56.0	19. New Zealand	14.3
20. Sweden	46.1	20. Netherlands	13.0
21. Norway	45.0	21. United Kingdom[b]	12.2
22. Romania	45.0	22. Sweden	9.7
23. Yugoslavia	44.2	23. East Germany	9.5
24. France	44.0	24. Canada	9.4
25. Japan	40.0	25. Finland	8.8
26. Mexico	40.0	26. United States	8.2
27. Portugal	37.0	27. Poland	7.5
28. Poland	28.6	28. Iceland	6.3
29. U.S.S.R.	23.5	29. Ireland	5.0
30. Italy	17.9	30. Israel	4.5
31. Iceland	14.0	31. Norway	4.2
32. Israel	14.0	32. Mexico	0.3
	75.9		24.7
	66.0		15.3

Courtesy of the Brewers Association of Canada

graphic and climatic factors. America may soon have officially designated vineyards whose names will be restricted on labels and in advertising a wine's "appellation" of origin.

The Bureau cooperates with state agencies designated to regulate beverage alcohol production and sales. Federal regulators have relied upon state authorities for maintaining industry compliance to regulatory standards, consistent with the public mandate for local control following repeal of prohibition.

Beverage Alcohol Consumption throughout the World:
A Comparative Analysis

Table 12 shows per capita consumption of total alcohol, and of spirits, beer, and wine throughout the world during 1981. The United States ranked eighteenth in total alcohol consumption, eighth in consumption of spirits, twelfth in consumption of beer, and twenty-sixth in consumption of wine.

Changes in patterns in per capita consumption in terms of absolute alcohol between 1950 and 1981 are shown in Table 13. Canada showed a 100 percent increase and the United States had a 55 percent increase. The only major countries to show a decrease were Portugal and France.

Table 13. CHANGE IN PER CAPITA CONSUMPTION IN TERMS OF ABSO-
LUTE ALCOHOL BETWEEN 1950 AND 1981

Country	*1950* litres	*1981* litres	Increase
Netherlands	2.07	8.75	323%
West Germany	3.29	12.52	281
Finland	2.23	6.50	191
Denmark	3.78	9.92	162
Hungary	4.91	12.34	151
Austria	4.79	10.75	124
Poland	3.11	6.63	113
Canada	4.41	8.80	100
Czechoslovakia	5.48	10.84	98
Ireland	3.69	7.18	95
New Zealand	4.57 (1953)	8.86	94
Luxembourg	8.71	16.77	93
Norway	2.17	4.16	92
Belgium	5.40	10.31	91
United Kingdom	3.94	7.31	86
Japan	3.60 (1962)	5.94	65
United States	5.52	8.55	55
Australia	6.62	10.19	54
Sweden	3.94	5.44	38
Switzerland	8.77(est)	11.11	27
Spain	10.60 (1962)	12.72	20
Italy	9.49	10.42	10
Portugal	13.21 (1962)	13.03	− 1
France	18.73	14.94	−20

Average increase (excluding Japan, New Zealand, Portugal and Spain)   108%

Courtesy of the Brewers Association of Canada

PART II

Alcohol: Biology and Behavior

CHAPTER 10

The Odyssey of the Alcohol Molecule: To the Brain and Beyond

In an earlier and more literary era, the brain was sometimes likened to a loom upon which the shuttle of experience, sensation and perception wove a rich tapestry of being. This poetic analogy of brain function permits a graphic image of alcohol's elusive actions. Perhaps intoxication occurs as alcohol permeates the tapestry, alters the shapes and nuances of the pattern for a while, and then evaporates, leaving the pattern of the tapestry as it was before. Contemporary models of brain function usually are based on computers. Computers not only mimic but transcend certain capacities of the human brain, and the analogies are compelling. Yet the most sophisticated computer models tell us little more about the neural processes involved in intoxication than the tapestry analogy.

Current ignorance about the neurobehavioral effects of alcohol is balanced by extensive information about its molecular structure and the rules that govern its absorption, metabolism and excretion. These physiological processes have been documented in detail and the way in which each may enhance or attenuate intoxication is known. As we follow the alcohol molecule on its path to the brain, we will see how intoxication is modulated by these processes and their complex interactions. We will learn the biological basis for the hydrophilia (affinity for water) of the alcohol molecule and how that contributes to its intoxicating properties, only to be reminded again how little is known about the behavioral phenomena of alcohol-philia (affinity for alcohol) or alcohol aversion.

THE ALCOHOL MOLECULE

The alcohol molecule is comprised of three basic elements: carbon, hydrogen and oxygen. The way in which these elements combine to form alcohol is shown below. The alcohol molecule contains two carbon atoms, six hydrogen atoms and one oxygen atom. Five of the six hydrogen atoms are attached to the two carbon atoms, which occupy a central position in this configuration. The sixth hydrogen atom is attached to the single oxygen atom, and these in turn are bound to an atom of carbon. The combination of oxygen and hydrogen is called a hydroxyl group. The etymologist will immediately recognize that the root of hydroxyl is the Latin root for water. Hydroxyl groups have a particular affinity for water. Because alcohol molecules contain a hydroxyl group (which is also the basic portion of the water molecule), there is a strong attractive force between alcohol and water. Alcohol and water are good mixers in every solution, including the blood. It is this aspect of the molecular structure of alcohol that makes it so easily soluble in water and determines its progress through the body. Alcohol goes wherever water goes and it is rapidly distributed throughout the fluid compartments of the body.

The progress of alcohol through the human body is a fascinating story. Once alcohol enters the body, it easily penetrates the brain and other body tissues. It is also readily metabolized to the simple substances of carbon dioxide and water and eliminated. Some of the unpredictability of the nature and quality of intoxication can be attributed to these interacting physiological processes.

Figure 1. ALCOHOL MOLECULE

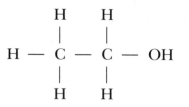

C—CARBON ATOM
H—HYDROGEN ATOM
OH—HYDROXYL GROUP

Alcohol enters the blood primarily from the small intestine, rather than the stomach. Once alcohol is absorbed into the bloodstream, it is rapidly circulated throughout the body. It is in the bloodstream that the hydroxyl characteristic of the alcohol molecule is first important, because water is the major constituent of blood. Water in the bloodstream exerts a powerful attractive force on the substances dissolved in it, in much the same way that gravity attracts masses to the earth. The strength of water's attractive forces depends upon the way in which molecules separate or dissociate when dissolved in water. Alcohol is bound to water through a process called hydrogen binding.* The attraction between water and the hydroxyl group of the alcohol molecules is strong enough to solubilize alcohol, but not so strong that alcohol is irrevocably bound to water. In order to enter brain cells and cells of other body tissues, alcohol must be able to "escape" from water in order to enter other substances that form the lining of the blood vessels and the walls of nerve cells.

Lipids are one of the major constituents of all cell walls or cell membranes. To travel from the bloodstream into the brain, and into the nerve cell, alcohol has to pass through cell walls and membranes composed of lipids. Once inside the nerve cell, the alcohol molecule must pass through other cell barriers and out again to return to the bloodstream.

The alcohol molecule is unique in the ease with which it can enter nerve cells and return to the blood. The passage of most other substances into the brain is prevented by barriers formed by the cells that line the interior of the blood vessel, and those that comprise the walls of the nerve cell. This physiological system, usually referred to as the blood-brain barrier, actively excludes the entry of certain substances, particularly large molecules, into the brain. The blood-brain barrier is adaptive, since it insures that potentially harmful compounds do not reach the cells of the brain.

For many substances, passage into the lipid compartment of the cell wall is the equivalent of entering a room with one door that locks after entry. In contrast, the lipid solubility of alcohol permits alcohol to enter, but the forces that attract alcohol to lipid molecules

*Not all substances dissolve in water through a process of hydrogen binding as does alcohol. Many substances are attracted to water through a process called ionic binding. Ionic binding involves the process of ionization in which the substance separates into negatively and positively charged atoms and molecules.

are not powerful enough to permanently entrap the alcohol molecule in the lipid phase. This is extremely important, for if alcohol were held or trapped in the lipids that constitute cell walls, it could not diffuse into water on its way to and from the nerve cell. Many substances are trapped in lipids for hours, days or even permanently. When the substances are drugs, their effects may be adverse because they persist for such a long period of time.

Alcohol is a molecue uniquely suited to the physiological terrain through which it must pass to reach the brain, produce intoxication, and leave again as easily as it came. The evanescent quality of intoxication, its ever-changing, often contradictory moods, seems a fitting parallel to the biological odyssey of the alcohol molecule.

THE PATHWAY TO THE BRAIN

The rate at which the alcohol molecule reaches the brain is not solely determined by its rapid interchangeability with water and lipids. A number of factors in the body and in the nature of the alcohol beverage itself may retard or accelerate the transport of alcohol into the bloodstream. These entry control factors each contribute to the relative unpredictability of alcohol effects.

The presence or absence of food in the stomach is one critical determinant of the rate of alcohol absorption. Before alcohol can be absorbed into the bloodstream it must enter the small intestine from the stomach. Passage from the stomach is controlled by a muscular valve called the pyloris. The pyloric valve remains closed until food in the stomach has been acted upon by acids and enzymes secreted by the stomach wall. If alcohol is ingested with food, it will take longer for the alcohol to reach the small intestine because the pyloric valve will remain closed. This is the basis of the well-known fact that eating while drinking slows down the rate of absorption of alcohol.

Alternatively, if alcohol is consumed when the stomach is empty, it may be emptied into the small intestine very rapidly. The optimal alcohol concentration to facilitate rapid stomach emptying corresponds to that of distilled spirits, 86 proof or 43 percent alcohol. This concentration of alcohol passes most readily from the stomach into the small intestine, where it is absorbed into the bloodstream. However, if too much alcohol is consumed too rapidly, resulting in a high concentration of alcohol in an empty stomach, the pyloric valve may go into spasm. Under these conditions, vomiting usually occurs and alcohol does not enter the intestinal tract. This is one of the body's many protective devices to prevent toxic substances from leaving the stomach and entering the intestinal tract and the bloodstream.

There is a pervasive misunderstanding about the intoxicating qualities of different types of beverage alcohol. Many believe that beer, for example, is less intoxicating than bourbon. In fact, this is dependent upon the amount of either beverage consumed. It has recently been shown that consumption of an equal amount of *alcohol* in the form of wine, beer, or whiskey will produce equivalent behavioral effects and equivalent levels of alcohol in the blood. A twelve-ounce glass of beer, a six-ounce glass of wine and one mixed drink containing an ounce and a half of 86 proof distilled spirits all contain the same amount of pure alcohol.

Another factor that facilitates the absorption of alcohol from the small intestine is whether the beverage mixer consumed with the alcohol, or the alcohol itself, is carbonated or not. Carbonated beverages are absorbed more readily than noncarbonated beverages. The rapid intoxication produced by champagne, in contrast to a still wine with an equivalent alcohol concentration, is due to the fact that champagne is carbonated.

The rate of alcohol absorption from the small intestine is also influenced by the congener content of the beverage. Congeners are substances that may be added to alcohol or evolve through the process of fermentation and aging. Low congener beverages, such as vodka, are absorbed more rapidly than high congener beverages, such as brandy. Distilled spirits, which have a high congener content, will be absorbed more slowly than either wine or beer. The congener content of distilled spirits has been shown to increase as the beverage is aged. For the most part, congeners are complex organic molecules, and the exact composition of many of these substances has not been precisely determined. These congeners contribute to the unique color and flavor of different brands of bourbon, whiskey and gin. Beverages of the same proof can vary in congener content. A "light" scotch contains fewer congeners, but the alcohol content is usually unchanged.

The possible contribution of congeners to alcoholism and alcohol-related medical problems has been an issue of recurrent speculation. Some people believe that drinking two or more types of alcohol (for example, scotch and gin) is more likely to produce drunkenness, sickness and hangover than the use of a single beverage alone. Since the fundamental ingredient of all alcoholic beverages is alcohol, does mixing of congeners account for this alleged phenomenon? It is unlikely that the congeners have very much to do with the consequences of social drinking. It is probably a myth that mixing one's liquor produces bad effects, but belief in a myth can determine the outcome. In fact, expectancy about how alcohol affects feeling states and behavior is one important determinant of alcohol's effects. However, it is important to recognize that it is alcohol and not the color, taste or smell imparted by congeners that affects brain function and ultimately behavior.

A number of techniques have been devised to measure the concentration of alcohol in the body. The resulting measurement is usually referred to as the *blood alcohol level*. Blood alcohol levels are usually expressed as milligrams of alcohol per hundred milliliters of blood (mg/100 ml). Many states define the legal limit of intoxication in terms of the concentration of alcohol in the blood. Levels of 100 mg/100 ml and above are often the basis for determining intoxication.

Blood alcohol levels may be assessed by taking a blood sample and measuring the alcohol concentration. A second and more common method is to measure the concentration of alcohol in expired air, using a breathalyzer device. Approximately 2 percent of alcohol ingested is excreted directly in the breath. Because this ratio is so constant, the amount of alcohol in the bloodstream can be inferred by measuring the amount of alcohol present in a sample of expired air from the lungs. This technique has been employed widely by police departments to measure alcohol concentration in blood of individuals who may be charged with drunken driving or other offenses.

The breath alcohol measuring technique is simple to use. The individual who takes the test inhales and then exhales as forcibly as possible into a tube, which is connected to a cylinder that holds a precise volume of the air. The cylinder is kept at body temperature (by a heating coil), and the warm air is subsequently forced into a chemical solution. The alcohol in the air is dissolved in the solution and changes the color of the solution. The amount of color change is proportional to the amount of alcohol in the breath. After the solution changes color, the amount of color change is measured by a device that is sensitive to light. This instrument is similar to the photocell, which is commonly used in exposure meters for cameras.

The operator can easily read the degree of color change and calculate the concentration of alcohol in the breath and blood. The newer instruments for measuring alcohol in the breath may utilize gas chromatographic or infrared devices and they also have electrical circuits that not only carry out the computation, but also print out the result in terms of milligrams of alcohol per milliliter of blood, as well as the date and time of day. Such reports are gaining more and more credence as admissible evidence in court procedures and litigation.

Can the blood alcohol level be predicted from the amount of alcohol consumed? Only approximately, because food in the stomach, carbonation of the alcohol vehicle, the congener content and the concentration of the beverage alcohol all combine to affect the rate at which alcohol is absorbed from the small intestine and consequently the blood alcohol level at any point in time.

Still another factor that contributes to the unpredictability of blood alcohol levels is the difference in the concentration of protein, lipids (fats), and water in the human body. Although alcohol is readily soluble in fat, it is more soluble in water. Persons who have relatively large amounts of body fat will have higher blood alcohol levels than lean individuals after drinking the same amount of alcohol. For example, two men who have the same body weight may have very different blood alcohol levels after drinking exactly the same amount of alcohol. A lean, muscular man is likely to have a lower blood alcohol level than an obese man. A female weighing exactly the same as a male will invariably have a higher blood alcohol level even though she may consume the identical amount of alcohol as the male. This is because females usually have a higher proportion of fat in their bodies than males. Thus the popular notion that females tend to become more intoxicated than males after drinking approximately the same amounts of alcohol is often true.

Despite all of these unpredictable differences, it is estimated that the usual legal limit of intoxication, 100 mg/100 ml, often occurs after consumption of about four to five ounces of distilled spirits (43 percent, or 86 proof) on an empty stomach over a period of one to two hours. This estimate is only approximate.

Does the blood alcohol level correlate with the behavioral changes associated with intoxication? It is reasonable to assume that the level of intoxication is a direct function of the concentration of alcohol in the blood. Alcohol remains in the body for a finite period. The ab-

Figure 2. TIME COURSE OF ALCOHOL DECAY

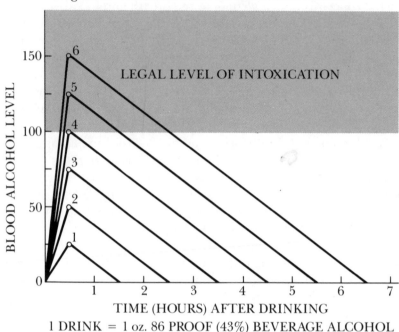

sorption, peak effect, and decay of alcohol concentrations in the blood form a curve through time as is shown below. The rate of alcohol disappearance from the blood is estimated to be about one ounce an hour.

Alcohol continues to affect the brain as long as it remains in the bloodstream. Once drinking stops, the effective concentration of alcohol in the blood is decreased slightly each time the blood passes through the liver, where alcohol is metabolized. The blood completes one cycle of circulation through the body about every three minutes.

It was once believed that alcohol was metabolized by the brain as well as by the liver. The most recent evidence suggests that if any breakdown of alcohol does occur in brain tissue, it accounts for an infinitesimally tiny change in effective alcohol concentration in the blood.

Almost nothing is known about the way alcohol alters the function of brain cells to produce intoxication. Hypotheses range from interference with crucial chemical substances produced in the brain to disruption of nerve cell walls or membranes. Since progressive intoxication may involve loss of cognitive faculties, then motor control, and eventually interfere with respiration, it was once believed that the sites of action of alcohol in the brain paralleled the observable functional disruptions. Perhaps alcohol first affects the cortical centers believed involved in cognitive function, then progresses to lower brain centers involved in motor control, then respiration. Despite the attractive logic of this hypothesis, evidence from electrophysiological studies in animals does not appear to support it. Alcohol intoxication in man has been shown to produce discernible changes in brain activity as measured by electroencephalography (EEG). However, the EEG patterns commonly reported are not uniquely associated with any particular behavior change.

How alcohol and other drugs affect the brain to produce intoxication is an enigma now far from resolution. Human intelligence has achieved near-miraculous advances in medicine and science. Yet it may be that exploration of the universe will prove less of a challenge than the analysis of the function of the human brain.

ALCOHOL METABOLISM

The next major step in the odyssey of the alcohol molecule is the process by which it is broken down or metabolized, prior to its excretion from the body. About 92 to 94 percent of alcohol ingested is metabolized and excreted; the remainder is excreted directly in breath and urine. We have followed the molecule in its passage through the pyloris, and its absorption into the bloodstream. The molecule has passed through the lipid constituents of the cell wall of the brain into the cell body and out again to continue on its course. Imagine the alcohol molecule as a small boat traveling along the bloodstream, borne inexorably toward a point of certain dissolution and destruction. Alcohol is broken down rapidly, and the rate of metabolism is influenced by the same factors that initially determined its rapid absorption and distribution through the body, i.e., its water and lipid solubility and rapid interchangeability. However, the biological processes involved in alcohol's entry into the body are relatively simple in comparison to the very complicated processes involved in its inactivation and excretion.

The bloodstream carries the alcohol molecule from the small intestine to the liver and then on to the rest of the body tissues including the central nervous system and the brain. In the first pass through the liver, noxious products that may have been absorbed from the small intestine into the bloodstream are detoxified. A significant amount of alcohol is removed from the bloodstream on the first passage of the alcohol molecule through the liver detoxification process. The liver removes toxic agents through the actions of a series of enzymes. Enzymes are necessary for certain chemical reactions to occur but are not broken down in the process, i.e., enzymes are catalysts. The enzyme responsible for the breakdown of alcohol is alcohol dehydrogenase.

Why does alcohol dehydrogenase exist in the human liver? Was the cocktail-hour ritual anticipated in the earliest stages of evolutionary development? In fact, a small amount of alcohol is generated in our gastrointestinal tract by fermentation processes of normal bacterial flora. Using very sensitive measurement techniques, it is possible to demonstrate that everyone has some free alcohol circulating in his or her bloodstream at all times, whether he or she drinks or not. If there was no system for the removal of this internally generated or endogenous alcohol, conceivably it could accumulate and produce many adverse effects.

Comparative biologists tell us that alcohol dehydrogenase is ubiquitous in nature. In mammals, the very highest concentration of alcohol dehydrogenase is found in the liver of the horse. This situation is dramatized in the old cliché, "If God did not intend the horse to drink, why does this animal have such a high level of alcohol dehydrogenase?" The answer lies in the fact that most mammals, including man, produce some alcohol in their own gastrointestinal tract as a function of the fermentation of microorganisms which are ordinarily present in the gut. A ruminant, such as the horse, produces a great deal of alcohol by fermentation in the gastrointestinal tract, and hence the evolution of an efficient mechanism for removing alcohol from the body.

Alcohol dehydrogenase is the most efficient enzyme for alcohol degradation and it is responsible for the first step in the breakdown of alcohol. The level of alcohol dehydrogenase activity sets the pace for all subsequent steps in the elimination of alcohol.

Alcohol is converted to a substance that can easily enter the same chemical pathways that derive energy from sugar. Alcohol, like sugar, contributes energy to the body. In fact, alcohol yields the same number of calories (7 per gram) as sugar. But even though alcohol breakdown in the body yields energy, it has often been described as "an empty calorie." The reason is that alcohol, unlike ordinary foods, does not contain important vitamins and minerals, which are necessary for the body to function properly. A diet that consists mainly of alcohol can cause severe nutritional and metabolic deficiencies.

When alcohol is metabolized in the liver, both enzymes and substances called co-enzymes are required. Enzymes like alcohol dehydrogenase regulate the rate or speed at which metabolism occurs, and therefore are called catalysts of chemical reactions. Co-enzymes are involved in the transport of atoms from the alcohol molecule and make possible its conversion to a substance that can be subsequently broken down in a manner similar to sugar. Co-enzyme transport mechanisms operate rather like freight cars, which transport products from one site to another. During alcohol breakdown in the liver, these transport mechanisms are diverted from jobs they ordinarily perform. The consequences of this metabolic diversion are analogous to what might occur if all the freight cars in a given region were diverted from carrying essential products to less essential items. One could envision a breakdown in the normal economy of a city or state if this were to occur. Similarly, diversion of co-enzymes for alcohol metabolism results in a disruption of the physiological economy of the liver when large amounts of alcohol are consumed. One consequence is the impairment of other activities carried out by the liver, such as the detoxification of noxious metabolic by-products and drugs. It is for this reason that ingestion of large amounts of alcohol and other drugs (e.g., sleeping pills) is more dangerous than taking either alone. Polydrug abusers who also abuse alcohol are at a higher risk for adverse consequences of intoxication.

Although alcohol dehydrogenase acts as the master control for the rate at which alcohol is eliminated, several other factors also influence the rate of degradation. One factor is the amount of stored sugar in the liver. If the drinker has been fasting, or has eaten poorly, the stores of sugar in the liver may be depleted, with a consequent slowing of the rate of alcohol metabolism. Fasting for twelve to twenty-four hours can significantly slow down the rate of alcohol metabolism in man. One practical consequence of this physiological fact is that many problem drinkers learn that they can prolong the duration of alcohol intoxication by not eating.

Since drinking alcohol on an empty stomach also facilitates rapid absorption, not eating produces a higher level of intoxication, which lasts longer. In the jargon of the street, not eating when drinking "gives a bigger bang for the buck." It is this factor, rather than the calories in alcohol, which probably contributes to the marginal food intake of many chronic alcoholic individuals. Chronic alcoholism in combination with poor nutrition can result in a number of very serious medical consequences, which will be discussed later.

It is somewhat ironic that although it is easy to slow down the breakdown of alcohol, there is no effective way to speed up alcohol metabolism once intoxication has occurred. The search for an effective sobering agent continues, but most familiar techniques are of little value. The time-honored notion that strong black coffee and a cold shower can speed sobriety is erroneous, since these do not alter the rate of alcohol metabolism. Although such maneuvers may stimulate the drinker, the amount of alcohol circulating in the body remains essentially unchanged and the drinker is just as intoxicated. Sobering agents are discussed more fully in Chapter 17, "Health Consequences of Alcohol and Drug Combinations."

The experience of many social drinkers suggests an exception to the general rule that the rate of alcohol metabolism is unalterable. Many have noticed that the more they drink, the more they are able to drink without incapacitating intoxication. This observation is valid and reflects a general pharmacological phenomenon known as "tolerance." Alcoholics may develop enormous tolerance for alcohol.

As an individual becomes tolerant of alcohol, more and more alcohol is required to produce the same subjective and behavioral effects. The development of alcohol tolerance as a function of drinking experience through time does have a metabolic basis. A number of studies have shown that the rate of alcohol metabolism can be accelerated in man as a function of exposure to alcohol. Figure 3 shows a graph of the concentration of alcohol in the blood of a heavy drinker before and after a period of high alcohol intake. Measurements of blood alcohol were made after equivalent doses of alcohol were administered. Notice that there is a more rapid decrease in blood alcohol levels after seven days of heavy drinking. In fact, the rate of

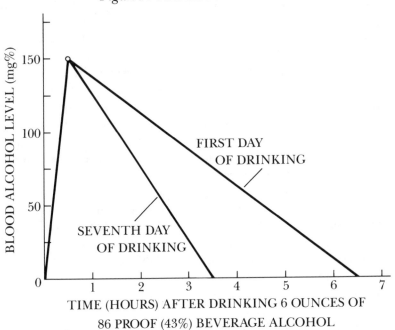

Figure 3. ALCOHOL TOLERANCE

FIRST DAY
OF DRINKING

SEVENTH DAY
OF DRINKING

BLOOD ALCOHOL LEVEL (mg%)

TIME (HOURS) AFTER DRINKING 6 OUNCES OF
86 PROOF (43%) BEVERAGE ALCOHOL

decrease is almost 50 percent greater. Although the mechanisms accounting for metabolic tolerance are unknown, alcohol-induced enhancement of enzyme activities in the liver, which are involved in the breakdown of alcohol, appear to be the major contributors. One consequence of metabolic tolerance for alcohol is that heavy drinkers drink progressively more and more in order to achieve the blood alcohol levels they formerly were able to obtain after drinking lower doses of alcohol.

Because major changes may occur in liver enzyme activities which affect the rate of alcohol metabolism, some scientists have wondered if genetic differences in enzyme activities in the liver could account for the different degrees of tolerance that people have for alcohol. For example, do individuals with a low tolerance for alcohol have an enzyme deficit or a low enzyme activity? Conversely, do people with a high alcohol tolerance have high levels of liver enzymes or especially efficient enzyme systems?

These kinds of questions prompted studies of the liver enzyme activities of certain genetically defined populations who appeared to be at unusually high risk for the development of alcohol problems. Several investigators examined American Indians with documented histories of severe drinking. It was found that liver enzyme activity in Indians did not differ from Asian and Caucasian populations, and metabolic factors could not account for differences in alcohol tolerance. At present, there is no evidence that any genetic process involving regulation of the way alcohol is metabolized critically affects the development of alcohol problems or alcoholism. Although recent studies suggest that genetic factors may increase the risk of developing alcoholism, these genetic factors do not appear to be related to the capacity to metabolize or break down alcohol.

It has long been known that the rate of alcohol metabolism in *abstinent* alcoholics and normal drinkers is not significantly different. Although sustained exposure to alcohol can induce a transient increase in alcohol metabolism, this alcohol-induced acceleration does not persist indefinitely after cessation of drinking.

CONCLUSIONS

This concludes the biological odyssey of alcohol to and from the brain. We have followed the journey of the alcohol molecule from its absorption into the bloodstream through the blood-brain barrier into the lipid compartments of cell membranes and finally to its metabolic breakdown in the liver and eventual excretion.

The unique hydrophilic qualities of alcohol make it a perfect molecule for rapid entry into and exit from the brain. It is ubiquitous in nature and is universally produced by fermentation, even in such improbable sites as human and other mammalian gastrointestinal tracts. As a function of its ubiquity, a superbly efficient biochemical system has evolved to detoxify and rapidly eliminate alcohol from the body. This system has the flexibility and adaptability to increase its efficiency as a function of exposure to alcohol through time.

Although alcohol can gain rapid access to the brain and alter the quality of our experience and fantasy, this change cannot be sustained. The evanescent effects of psychoactive agents, including alcohol, have been succinctly described by Dr. Daniel X. Freedman, Professor of Psychiatry at the University of California School of Medicine:

In enhancing our purposes with chemicals, we seem nevertheless, to be intrinsically limited in the extent to which we can endure chemical pleasures. I have often wondered whether the fact that drugs rarely sustain their initial allure and promise was a mischievous and malevolent trick of nature. Whatever the problems of design, there do appear to be intrinsic biological and biosocial limits upon the extent to which the chemically-induced moment of pleasure can be sustained and woven into a fabric of viable social life. We can readily give up our lease on reality, but apparently it is difficult to gain ownership of paradise!

CHAPTER 11

Alcohol, Sex and Aggression

Science has yet to discover the essential biological mechanisms that regulate the most commonplace human drives and behaviors. For example, the intricate processes that control hunger and satiation remain an enigma, despite knowledge about the requirements for good nutrition. Sleep, once thought to be a necessary restorative for weariness and fatigue, is now believed also to serve a reprogramming and information storage function. The scientific study of sexual behavior is a very recent phenomenon and systematic investigation of aggressive behavior is just beginning. Thus, it is not surprising that the relationships between alcohol use and sex and aggression are shrouded in folklore and ignorance.

Alcohol does affect both sex and aggression. There is increasing evidence that some facets of these behaviors are regulated by the same brain hormone systems. It is now possible to measure hormone levels and trace the pattern of episodic fluctuations. Many hormones are secreted in irregular pulses or surges. The frequency and amplitude of these hormonal pulses may change as a function of some external stimulus such as alcohol, an erotic film or an aggressive provocation. Changes in the secretory activity of one hormone are controlled by changes in other hormones, which in turn influence the activity of still other hormones. It is difficult to conceive of a more perfectly orchestrated, intricately balanced system than this symphony of neuroendocrine hormones. Examination of the covariation between neuroendocrine hormones, sex and aggression, and how these patterns are modulated by alcohol may someday clarify these fundamental biological processes.

ALCOHOL AND THE BIOLOGY OF SEX

Sophistication in neuroendocrinology is not one of the virtues commonly attributed to Shakespeare. Yet, less than a decade ago, *Macbeth* was one of two citations on alcohol effects on sexual function in a leading text on pharmacology. There has been a rapid expansion of knowledge in this area in the last few years, but Shakespeare's summary remains one of the most concise and accurate [Macbeth, Act II, Scene 3]:

> MACDUFF: *What three things does drink especially provoke?*
> PORTER: *Marry, sir, nose painting, sleep and urine. Lechery, sir, it provokes and unprovokes. It provokes the desire, but takes away the performance: therefore much drink may be said to be an equivocator with lechery; it makes him and it mars him; it sets him on and it takes him off; it persuades him and disheartens him; makes him stand to and not stand to; in conclusion equivocates him in a sleep, and giving him the lie, leaves him.*

This passage describes several commonly experienced effects of alcohol intoxication which have a clear and predictable physiological basis. "Nose painting" refers to the enlargement and reddening of the nose that occurs in some alcohol abusers, and which cartoonists often use to depict the alcoholic. This reddening and enlargement is due to increased blood flow to the nose, as well as to the skin elsewhere on the body. The sensation of warmth after alcohol consumption reflects the increased blood flow to the skin. Paradoxically, this sensation of increased warmth is actually associated with a *loss* of body heat. As peripheral blood vessels dilate, the expanded surface area releases more heat, just as large windows lose more heat than small windows. Thus, although one may feel warmer after drinking, the actual heat loss may lower body temperature. This seldom-appreciated fact can be added to the list of perils of drinking alcohol to combat an Alaskan winter's night.

The porter's observation that alcohol provokes urine is familiar even to the occasional imbiber. Alcohol does not produce more urine as the result of increased fluid volume in the body; rather, alcohol suppresses the effects of a hormone that controls urine output from the kidneys. This hormone (vasopressin) is secreted by the pituitary gland in the brain. Under ordinary conditions, vasopressin inhibits excessive urine output (diuresis). However, when alcohol suppresses the urine inhibitor vasopressin, this leads to a concomitant increase in urine output. Curiously, alcohol inhibits vasopression only when blood alcohol levels are rising and not when blood alcohol levels are stable or falling.

Just as the ways in which alcohol affects peripheral blood flow and urination are complex upon close examination, so too is the effect of alcohol on sex. The porter's comment succinctly summarizes alcohol's almost contradictory effects on sexual behavior. How does alcohol act to increase sexual desire but diminish sexual performance in males? The same brain structures that regulate peripheral blood flow and control urine volume output also control the secretory patterns of hormones that influence sexual function. It has recently been discovered that alcohol directly affects these hormones and their regulatory interactions.

One important sex hormone affected by alcohol is the male hormone, testosterone. Testosterone is produced by the testes and plays an important role in human sexual development and sexual function. One of its most important functions is to permit normal development of male secondary sexual characteristics during puberty. Without testosterone, normal male hair growth and development, deepening of the voice, as well as the development of the external genitalia, will not occur. But even after puberty, testosterone remains important for male sexual function. Sufficient levels of testosterone are necessary for both sexual performance (penile erection) and for fertility.

Alcohol and Fertility

Although penile erection is necessary for ejaculation, normal fertility is dependent on the "accessory" male sexual glands. The "accessory male" sexual glands produce fluids that are necessary for the sperm cell to survive. These fluids also determine sperm motility, which affects whether or not the ejaculate reaches and fertilizes the ovum.

The fluid in which the sperm cells are discharged, the seminal fluid, contains large amounts of sugar called fructose. Fructose is the sugar ordinarily found in many fresh fruits, but in the human body this sugar is produced only by the seminal vesicles of the male. Fructose secreted into the seminal fluid provides a very potent source of energy, which is necessary for sperm cell mobility. The sperm cell uses fructose as a rapidly available energy source for movement of its flagella (the tail-like appendage of the sperm cell). When testosterone levels are low, fructose production in the seminal vesicles is decreased and sperm motility is diminished. Thus, heavy alcohol consumption may not only reduce sexual potency in males, but it may also act to decrease fertility.

Alcohol and Sexual Performance

The porter in *Macbeth* comments that too much drink "makes him stand to and not stand to." The porter may have been referring to the fact that although men can achieve an erection when heavily intoxicated, they often fail to maintain the erection. Failure to maintain penile erection can lead to severely compromised sexual function. Kinsey observed, in his early reports on male sexuality, that the most frequent cause of male impotence was alcohol abuse or a recent history of excessive alcohol intake. Although male impotence is multiply determined, recovery from alcohol abuse problems is often accompanied by improved sexual potency.

Does alcohol affect testosterone levels in a way that adversely affects maintenance of male penile erection? Recent studies of male alcoholics have shown that alcohol directly suppresses testosterone levels. Moreover, the magnitude of fall in testosterone levels is related to the amount of alcohol consumed. The more alcohol ingested and the higher the blood alcohol levels, the lower the testosterone levels. Often abrupt cessation of drinking is followed by a rebound increase in testosterone levels. The alcoholic men studied did not have evidence of liver disease, a condition which might otherwise account for changes in testosterone levels.

Alcohol suppresses testosterone production in all men, not just in alcoholics. When healthy young adult men drank moderate amounts of alcohol in an ordinary cocktail-party setting, plasma testosterone levels fell as blood alcohol levels rose. When blood alcohol levels were highest, testosterone levels were lowest. Thus, even moderate amounts of alcohol can suppress testosterone levels, regardless of past drinking history of the individual. These findings are consistent with observations that both alcoholic men and occasional social drinkers may suffer from impotence if they drink enough alcohol before attempting sexual intercourse.

Alcohol, Sexual Desire and Hormones

But if alcohol suppresses male testosterone levels, and thereby compromises penile erection, how can this be reconciled with Shakespeare's observation recounted by many others, that men may experience increased sexual desire during heavy intoxication? To understand this paradox, we need to consider the nature of the interaction between the brain and the testes. The brain regulates the pituitary gland, and the pituitary in turn regulates the testes. A diagram depicting the interrelationships between brain, pituitary, and testes is shown in Figure 4. Testosterone production is preceded by a sequence of changes in hormonal levels, and these hormones also may have some behavioral effects. The brain controls the production of a hormone (LHRH) that stimulates the pituitary gland to produce a second hormone (luteinizing hormone, or LH), which travels in the bloodstream to stimulate production of yet a third hormone, testosterone.

Alcohol lowers testosterone levels by a curious and complex mechanism. One component of this mechanism involves the neural regulation of testosterone production. Testosterone exerts a "feedback" effect on the brain and the pituitary, which is analogous to the way in which a governor regulates the speed of an engine. When testosterone levels are high, the brain and the pituitary are signaled to reduce production of the hormones that stimulate testosterone production. When testosterone levels are low, the brain and pituitary are signaled to produce more of the hormone that stimulates testosterone production. Consequently, when testosterone levels decrease following alcohol intake, the brain and pituitary are instructed to produce more hormones that stimulate increased production of testosterone by the testes. Circulating levels of LH, the pituitary hormone that regulates testosterone production, are significantly increased when men are most intoxicated and when their testosterone levels are lowest.

The high circulating level of this pituitary hormone, LH, has a direct effect on behavior, in addition to stimulating the testes to produce testosterone. Luteinizing hormone also stimulates brain cells

Figure 4.

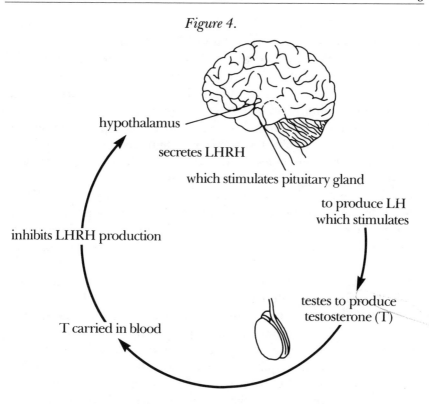

hypothalamus

secretes LHRH

which stimulates pituitary gland

to produce LH
which stimulates

inhibits LHRH production

testes to produce
testosterone (T)

T carried in blood

that have a crucial role in the regulation of both sexual and aggressive behavior. When a surge of luteinizing hormone occurs after alcohol-induced suppression of testosterone, this LH surge affects both
the testes and the brain. Some scientists speculate that increased
sexual desire following alcohol intake by males is related to the increase in luteinizing hormone secretion. Research shows episodic
surges of LH occur at the same time that young men reported feeling sexually aroused when they were shown erotic stimuli.

It now appears that when alcohol reduces testosterone levels, this
signals the brain and the pituitary to produce more luteinizing hormone. The subsequent surge in luteinizing hormone levels stimulates more testosterone production and also stimulates increased
sexual desire through its direct action on the brain.

The mechanism by which alcohol reduces testosterone has only
recently been clarified. The brain and the pituitary are not the primary sites of alcohol's effects on male sex hormones; rather, alcohol

suppresses testosterone production in the testes. Recent studies have shown that the testes metabolize and break down alcohol in a manner similar to that which occurs in the liver. Although the liver is the prime organ for metabolizing alcohol, the testes also possess the necessary enzymes to oxidize alcohol. Thus, when men drink alcohol, some of the alcohol is broken down by the testes.

The enzymes used to metabolize alcohol in the testes are also crucial for the production of testosterone. When alcohol is metabolized in the testes, these enzymes are diverted from their normal role and testosterone production is decreased. Researchers have recently identified the specific enzyme co-factor (a substance called NAD) that is depleted in the testes during alcohol oxidation. At present, there is no known way to administer this substance to humans to overcome the suppressive effects of alcohol on testosterone production.

Other drugs also affect male sexual behavior and testosterone levels, but through different biological mechanisms. For example, opium derivatives, such as morphine and codeine and heroin, also suppress male testosterone levels and also impair sexual function. But unlike alcohol, opiates depress rather than increase sexual desire. One reason that opiates and alcohol affect sexual desire differently is because these drugs have opposite effects on the pituitary and gonadal hormones. Unlike alcohol, opiates inhibit testosterone production by first inhibiting LH, the pituitary hormone that stimulates testosterone production. This decrease in luteinizing hormone is consistent with reports by opiate addicts that opiate drugs decrease their desire for sexual activity. Over time, opiates also decrease an addict's sexual performance or potency.

Alcohol and Male Feminization

In addition to decreasing testosterone and potency and increasing sexual desire, alcohol also has other effects on male sexuality. In 1926, an astute clinician noted that male alcoholics with liver disease developed enlargement of the breasts (gynecomastia) and shrinkage of the testes (testicular atrophy). These observations were subsequently confirmed in many clinical studies. For years, it was believed that alcohol-induced liver disease was the cause of breast enlargement and testicular atrophy. Only within the past decade has it been shown that male alcoholics whose liver function is normal can also develop testicular atrophy and gynecomastia. Thus, liver disease associated with alcoholism is not necessary for feminization to occur in male alcoholics. Precise details of how alcohol produces severe and irreversible feminization in some male alcoholics are unknown, but prolonged suppression of testosterone levels is one important contributing factor.

ALCOHOL AND THE BIOLOGY OF AGGRESSION

It has been long known that many components of sexual and aggressive behavior are controlled by the *same* hormonal and central nervous system processes. Although alcohol intoxication is not invariably associated with aggressive behavior, brawls in a barroom differ quantitatively and qualitatively from disputes in a tearoom. Since alcohol can affect sexual behavior and change hormone levels that are important in the regulation of male sexual function, this may have important implications for understanding how alcohol affects aggression.

Scientists have identified minute regions of the brain that contain nerve cells that regulate both aggressive and sexual behavior in animals. These brain cell aggregates are quite remarkable because they can "sense" sex hormones circulating in the blood. These cells are called steroid receptors, because they can bind with the sex steroid hormones, such as testosterone. This binding affects activity of the steroid receptor cell itself and the activity of other nerve cells that these cells regulate. Certain steroid receptor cells have the remarkable capacity to convert one sex steroid hormone to a dramatically different hormone. For example, testosterone can be converted to estradiol. Testosterone is the male sex hormone, whereas estradiol is the most prominent female hormone of the estrogen family. It has been discovered that sex steroid action on the brain can affect very complex sexual behaviors such as courting, grooming and coitus.

The brain centers that regulate sexual activity also are important in the regulation of aggressive behaviors. The range of aggressive behaviors in animals that are affected by sex steroid action on the brain includes predatory functions necessary for survival, as well as spontaneous killing behavior which has no known survival role.

There has been a concerted effort to discover if aggression is associated with any particular hormone. Because the regulation of any single hormone is so complexly related to many other hormones, no simple answers have emerged. Research in primates has shown that dominance hierarchies often parallel testosterone levels, i.e., the alpha male who leads the troop has higher testosterone levels than subordinates. Measurements of testosterone in primates, before and after a fight, have shown that testosterone falls after a defeat and increases after a victory. Fascinating as these observations are, there are also many exceptions which prevent any definite conclusions about the role of a hormone in aggression.

Consider a familiar example. Male animals are castrated to decrease sexual behavior and there is usually also a striking decrease in aggressive behavior. Castration, as well as alcohol, lowers testosterone levels. Most pet owners know that castration or spaying of a male dog or cat produces these changes in general behavior. Animals that were aggressive or predatory tend to become gentle and docile. In man, castration was once used to produce eunuchs valued as asexual guardians of the harem. Their relative docility or aggressivity is not a matter of scientific record.

The astute reader will notice an apparent contradiction. If alcohol intoxication is often associated with increased aggression and lower testosterone, but decreased testosterone in turn seems related to docility, how can these be reconciled? This is one of the many unresolved questions about the interactions between alcohol, behavior and neuroendocrine hormones.

Whatever the contributing biological factors, the relationship of alcohol and aggression in humans is amply documented. It is commonly believed that drinking enhances courage and alleviates fear and apprehension in the face of danger. Although alcohol may have a salutory effect in promoting courage and aggression in certain life-threatening situations, in most instances, alcohol-related aggression is destructive to the aggressor as well as the victims.

Alcohol-related aggression in this society exacts an enormous toll. Over 50 percent of highway fatalities are associated with alcohol abuse or alcoholism. Studies of alcohol effects on driving behavior have consistently demonstrated that alcohol impairs skill and judgment. However, these effects on performance may not be the most crucial consequences of alcohol intoxication on driving behavior. There is increasing evidence that alcohol intoxication is associated with a great increase in risk-taking and overt aggression while driving. The ordinarily courteous and conservative driver may become quarrelsome and drive aggressively after drinking. Recent police statistics indicate that intoxicated individuals, involved in minor accidents, frequently leave their vehicles and verbally or physically attack the other driver.

Another example of alcohol-induced aggression is the high incidence of homicide associated with alcohol abuse. It is estimated that a large proportion of perpetrators (and victims) of homicide were drinking heavily before commission of a crime. In societies where male drinking and macho behavior is condoned, the vast majority of all homicides occur in the context of heavy drinking. Less dramatic, but more frequent, is the common observation that intoxicated males in barroom settings frequently become quarrelsome and aggressive. Alcohol intoxication in the social drinker or sustained inebriation in an alcoholic is often accompanied by the enhanced probability of the emergence of aggression and belligerence.

Although the evidence is far from complete, the data suggest that alcohol may directly alter the biological mediators of sex and aggression through its actions on the brain, the pituitary and the gonads. But biology alone does not determine human behavior, and expectations about alcohol's effects can also influence sexuality and aggression.

ALCOHOL AND EXPECTANCY

The power of expectations about alcohol effects has long been appreciated by those bartenders who "water the drink" to increase their profit margin. The diluted alcohol may be just as satisfying to the customer as an unadulterated drink and far more satisfying for the balance sheets. Social scientists have also made a convincing case for the importance of expectations about alcohol on subsequent behaviors. One dramatic example is the finding that people who believed they were given alcohol acted more aggressively than people who believed they were given tonic (an alcohol placebo). Expectations changed behavior independently of whether the subjects had actually received alcohol or not.

In a model study of the influence of expectancy about alcohol's effects on aggression, forty-eight people were told they would receive alcohol and another forty-eight people were told they would receive tonic. In fact, half of the people in *each* group were given alcohol and the other half were not. The measure of "aggression" was the intensity and duration of electric shocks administered to a person in another room in the context of a "teaching exercise." People who expected alcohol administered longer and more intense shocks than people who expected to consume tonic, regardless of the actual alcohol content of their beverage. Those who expected tonic, and actually received alcohol, were indistinguishable from those who expected and also received tonic. The expect-tonic, receive-alcohol group were also less aggressive than those who expected alcohol and actually received tonic. No subject could taste the difference between vodka and the tonic placebo at a rate better than chance. Consequently expectancy rather than any drug discrimination factors appeared to influence the expression of aggression.

If the belief that alcohol has been consumed is just as effective as alcohol itself in increasing aggressive behavior in social drinkers, this strongly suggests that other aspects of intoxication are modulated by expectation. It is often argued that alcohol may provide an acceptable excuse for aggression and other deviant behaviors that are usually considered unacceptable. Alcohol intoxication is often "blamed" for a variety of types of asocial acting out, when in fact, it may merely set the stage.

Alcohol is commonly believed to increase sexual arousal, and this expectation may be as significant a factor in reports of sexual feelings as the hormonal accompaniments. This possibility was examined in a study similar to the study of alcohol, aggression and expectancy. Sexual stimuli were erotic films. Half of the subjects, all of whom were men, were told they would receive alcohol and half were told they would receive tonic; whereas, in fact, half of the alcohol-expectancy group received tonic and half of the tonic-expectancy group received alcohol. Physiological indices of sexual arousal involved a measure of penile erection, i.e., the tumescence of the penis. Although the dose of alcohol was too low to affect penile erection directly, subjects who believed they had been given alcohol showed significantly more sexual arousal than subjects who believed they had consumed tonic. Similar results were obtained when subjects listened to tape recordings of erotic narratives. Again, those who believed they had consumed alcohol became more sexually aroused than subjects who believed they had been given tonic.

When subjects were asked to rate the relative stimulating effects of erotic slides, those who believed they had consumed alcohol rated the slides as more sexually stimulating, whether they had actually consumed tonic or alcohol. Several studies have now confirmed the general finding that men who believed they had drunk alcohol became significantly more aroused by erotic stimuli than those who believed they had drunk tonic, regardless of the actual content of their drinks. Moreover, men's reports of sexual arousal and penile tumescence were usually positively correlated.

Sexual responsivity in women is affected differently by low doses of alcohol. In women, unlike men, very low doses of alcohol have been shown to *reduce* a physiological measure of arousal, vaginal pressure pulse scores. However, women's reports of sexual arousal increased with increasing alcohol dosage. In one study, women were given four different doses of alcohol in random order and were shown an erotic and a neutral film in each weekly session.

Women were asked to rate the extent to which films were sexually arousing and enjoyable, and were also monitored for their physiological responses through measurement of vaginal pulse pressure. Subjective reports of sexual arousal increased with increasing intoxication, whereas physiological measures of arousal declined. The lowest dose of alcohol (.05 gm/kg) was essentially a placebo, since it yielded no measureable blood alcohol level. Physiologically measured sexual stimulation increased during the erotic film but not during the control film when a nonalcohol placebo was given.

This apparent dissociation between sexual arousal and physiological responsivity is interesting and suggests that alcohol's alleged aphrodisiacal powers are complex indeed. To the extent that sexual gratification may be related to vaginal pulse pressure, traditional sophomoric advice to male seducers may not apply to women today. "Liquor" may not be "quicker" from the women's vantage point.

Alternatively, the apocryphal sophomores' approach to low-budget seduction may in fact be effective. The prototypical fraternity Don Juan reports, "We drank vodka with the orange juice, but gave the girls plain orange juice. Of course they thought they were getting vodka with orange juice and we were able to save money and get them drunk at the same time." Thus may chauvinism and penury combine to exploit the role of expectancy in the effects of alcohol intoxication — a strategy sanctioned to some extent, by the dispassionate data of social science. However, since biological studies clearly show that alcohol impairs male sexual performance, diluted alcohol plus expectancy may be the best formula after all. Empirical studies of seduction (success or failure) and sexual performance and gratification under alcohol and placebo control conditions are not yet available. Candy cannot be dismissed out of hand!

We are just beginning to appreciate the complex changes in behavior that a simple molecule such as alcohol can produce. Why humans seek these changes remains a fascinating question. It is likely that prehistoric man accidentally discovered that alcohol enhanced sexual desire and simultaneously enhanced aggression. Perhaps this chance occurrence in some way facilitated acquisition of a mate and subsequent procreation. Transmutation of such primative behavioral processes to contemporary behaviors associated with sex and aggression is subtle and extraordinarily complicated. Yet, if alcohol does produce changes in very basic biological drives, perhaps some individuals abuse alcohol to enhance such effects. The mood states associated with alcohol intoxication are even more ephemeral, and their possible relation to biological processes remains a mystery.

ALCOHOL AND FEMALE SEXUALITY

Except for the brief review of studies on alcohol and expectancy, the female half of the sexual equation has been blatantly ignored in the foregoing discussion. This omission reflects the information currently available rather than deliberate neglect. Sadly, research on alcoholism and alcohol effects has tended to ignore women until very recently. It has been generally assumed that conclusions drawn from studies of men apply equally to women. This may be true for some basic physiological functions such as gastrointestinal and liver function, but the applicability to behavioral dimensions is not known.

Questions about alcohol and female sexuality have been avoided with the assiduousness of a Victorian matron. But scientific chauvinism is not entirely at fault. Study of the effects of alcohol and other drugs on sexuality in men provoked congressional displeasure as recently as 1976. A Midwestern congressman, running for re-election, declared that studies of drug effects on sexual arousal were "offensive to the sensibilities of most Americans."* He persuaded his congressional colleagues to stop federal funding for a research project in his home district that addressed these issues. Special legislation was enacted as a rider to the House Appropriations Bill. Politics and prudery combined to publicize the congressman's carefully orchestrated pronouncements of outrage and he was reelected. So less is known than might have been about the effects of alcohol on male or female sexuality. Only a few intrepid investigators have pursued this line of research, while endeavoring to evade the unwelcome attention of their local golden fleecers.

Although evaluation of the effects of alcohol on pituitary and gonadal hormones is far less likely to excite prurient or political interest, this area also has been neglected until very recently. Preliminary data from studies in women and in female primates suggest that single episodes of intoxication do not suppress either luteinizing hormone or estradiol. Estradiol is the female gonadal hormone equivalent to testosterone, the male gonadal hormone, and the pathways for biosynthesis are very similar for both. Although the generality of these findings remains to be established, it now appears that acute alcohol intoxication has less severe hormonal consequences for women than for men. However, chronic alcohol abuse may have equally serious effects in men and women as inferred from the clinical literature. Alcoholism in women is associated with a number of disorders of reproductive function ranging from complete cessation of menses to structural damage to the ovaries. These disorders are discussed more fully in Chapter 13 on alcohol and health. The importance of learning about alcohol's effects on female procreative function and sexuality is obvious.

*Congressional Record, April 13, 1976, H 3299.

CHAPTER 12

Alcohol and Mood: The Illusion of Happiness

Anyone who has ever thought, "I really need a drink today," knows that alcohol works and that it doesn't work. The transient changes in feelings one has while drinking are seldom simply good or bad. A pleasant initial sensation of warmth, relaxation and conviviality may suddenly dissolve into sadness, guilt, anger and remorse. It is very difficult to predict the effects of alcohol from time to time or even from drink to drink. The third or fourth drink rarely has the same effects as the first drink. As intoxication increases, feelings of despondency and uneasiness may also increase. Although heavy drinking almost ensures the dissolution of any positive mood change, limited drinking does not guarantee that the initial glow will remain. As more sensitive measures of mood have been developed, it has been possible to trace fluctuations in feeling states over a few hours. These measures of mood in social drinkers have confirmed that even the pleasures of mild intoxication are evanescent indeed. After one or two drinks, the blood alcohol curve gradually rises over sixty to ninety minutes, and so do reports of good feelings. But once the blood alcohol curve reaches a peak and begins to decline, reports of dysphoria and anxiety increase and often exceed pre-drinking levels. A look at this process in the microcosm of a single drinking occasion invites a reexamination of some time-honored assumptions about alcohol and mood.

IMAGES AND ILLUSIONS

The belief that alcohol intoxication leads to feelings of well-being, contentment, happiness or even euphoria is a central theme in classical mythology. An idealized conception of the ancient bacchanalian revels has been conveyed through the centuries in paintings, sculpture and allegorical accounts. Bacchus, the giver of wine, is most often depicted as a jovial cherub or an engaging youth. The Bacchus-Dionysus of Roman and Greek legend was a healer, a comforter, a giver of gifts to mankind, and above all a celebrant in idyllic pastoral scenes. Exuberant orgies, lush grapes and wine in ornate cups are only a few of the images we associate with Bacchus. The vapid, dissolute Bacchus depicted by Michelangelo is a more sophisticated image, probably based on the sculptor's observations of the effects of chronic intoxication rather than on traditional fantasy.

In Western culture the association of wine with pleasure survived the disappearance of Bacchus as a diety. Many Dutch and Flemish painters of the seventeenth and eighteenth centuries showed drinking as a central part of family celebrations. Drunkenness at village dances and festivals was convincingly shown by painters such as Jan Steen, Hobbema, and Brueghel. These festive themes have found a modern expression in contemporary advertising. Celebrant groups and attractive young couples, faces suffused with determined gaiety or intense romantic communication, lift their glasses to the reader of almost every glossy magazine.

What we believe alcohol does today is essentially what has always been believed. Only the dominant images and the vocabulary have changed over the centuries. Today's mythmakers tell us that each glass of alcohol contains an implicit promise of happiness, relaxation and conviviality. The extent to which these promises are fulfilled depends on many factors in the drinking situation, the expectations and experience of the drinker and the amount of alcohol consumed. The promise is most likely to be fulfilled after one or two drinks, and the positive experience with alcohol of most normal drinkers conforms to the usual expectancy and validates the myth. It is less generally appreciated that the pleasures of alcohol are not invariant, and that chronic heavy drinking may result in increased tension, sadness and depression.

Until very recently, the kinds of effects alcohol exerted on mood were seldom questioned. Drinkers are consistently depicted as jovial and carefree, bubbling with good spirits. The forced jocularity and the strained "happy" expression of the chronic inebriate are sometimes mistaken as further evidence that alcohol brings unlimited cheer. But our society has tended to view unrestrained hedonism with mistrust and distaste. The language that evolved to describe alcoholism during the growth of the temperance movement indicts the alcoholic and prejudges the nature of the problem. Terms from a pre-Victorian era (e.g., libertine, debauchery, dissolute and intemperate) have as antonyms, sober, temperate, chaste and decent. Roget's entries in the section on moral practice are instructive on this point . . . his first compilation of the *Thesaurus* was published in 1852. Although language usage is often less elegant today, the disdain conveyed by common descriptions of alcoholism remains an integral part of our culture.

Contemporary society appears to disavow intemperance, to condone moderation and to be suspicious of those who use alcohol and other drugs to excess with the goal of achieving some variant of "happiness." Moreover, there has been little reason to question the conventional wisdom that happiness and drunkenness are often intertwined. Even Webster's, devoid of moral overtones and focusing on what "drunk" means, defines "inebriated" as implying such a state of intoxication that exhilaration or undue excitement results (inebriated revelers). Webster's further asserts that "drunk" is used in a figurative sense as implying excess of emotion (drunk with joy). Common sense argues that one doesn't drink something that is going to make one feel worse!! Or does it . . . ?

WHAT CLINICAL RESEARCH SHOWS

It was not until clinical researchers began to study alcoholic individuals during intoxication (instead of asking them about it later) that a more complex picture of the effects of alcohol on mood and feeling states began to emerge. The common belief that alcohol intoxication is the key to even a transient state of "happiness" has been challenged by clinical findings. Over the last decade, research has shown repeatedly that alcohol may produce profound dysphoria, depression, anxiety and tension *during* intoxication. Alcoholics and social drinkers alike have been shown to become despondent after heavy drinking.

Yet alcoholics usually report that alcohol makes them feel better, even wonderful, in a variety of ways. These statements were often made during sobriety, when the alcoholic described his or her recollection of a drinking experience. When scientists asked a group of alcoholic men to predict what a drinking experience would be like for them, they replied that they would become more adequate, more masculine, more socially adept, happier, and more relaxed and comfortable. Observations of these same men during chronic intoxication indicated that quite the opposite effects occurred. Most became profoundly depressed, anxious and belligerent.

When these same men were asked to recall the drinking experience, they remembered only their anticipated fantasy of intoxication. They did not recall the depression and anxiety that occurred during drinking. Whether this was a genuine failure of recall or a self-serving form of denial is unclear; but a discordance between the expectancy and the actual events during drinking has now been observed in several clinical studies. Whatever the explanation for this affect-amnesia, it is apparent that forgotten consequences are unlikely to influence future drinking behavior.

These findings point to two important factors that have long distorted information about alcoholism. First, recollections about intoxication are often inaccurate and bear little resemblance to actual events. Second, the effects of alcohol during intoxication may be more adverse than positive for the drinker. This new information greatly complicates an already complex problem. As long as the alcoholic was viewed as someone who enjoyed momentary gratifications with no regard for possible adverse consequences in the future, alcohol abuse seemed somewhat understandable. Problems could be dismissed, ignored, or delayed and the present was assumed to be a state of sublime inebriation. But if heavy drinking causes or increases depression and anxiety rather than relieving or reducing it, why do people continue to drink?

Some scientists speculated that perhaps a lacunar amnesia for events during intoxication was the culprit in perpetuating alcoholism. Perhaps, they argued, if problem drinkers realized exactly what effects alcohol had for them, they could learn to drink moderately or not at all. This reasoning led to some innovative studies that used videotapes of alcoholic patients during intoxication as an adjunct to therapy. The sober problem drinker was allowed to see himself as he was during alcohol intoxication, at his worst! This experience was then discussed and analyzed with the support and assistance of a therapist. Despite the self-evident rationale and refreshing directness of this approach, the results thus far have been little better than those achieved with conventional therapies.

Is it possible that alcoholics may drink, in part, to become unhappy, anxious and sick, rather than for the hedonistic reasons usually ascribed? This question will not be easily resolved but does call attention to some aspects of problem drinking that have been largely ignored. Some provocative and tantalizing hints come from both clinical and basic behavioral studies.

One curious fact is that the *first* drinking experience is often rather unpleasant in several ways. Nausea, vomiting, anxiety and depression are frequent consequences of heavy drinking for the alcohol-naïve individual. The alcoholic purportedly has a clearer recollection of his "first drink" than occasional social drinkers. Usually the first drinking experience is not one of euphoria, relaxation, and enhanced self-esteem but more often a "sick drunk" that is physically and emotionally disturbing. Such adverse consequences of drug intoxication in naïve users are not unique to alcohol. Opiate drugs invariably produce nausea, vomiting, anxiety and depression. Comparable first use effects have also been reported for a variety of other substances including tobacco (nicotine), barbiturates and even coffee.

If these initial alcohol and drug experiences are so seemingly unpleasant, why are they repeated? What contributes to the transition from initial exploratory drug use accompanied by adverse consequences to repeated and frequent use, probably also with adverse consequences? Does the drinker become tolerant to these effects of alcohol? Although tolerance may be one influential factor, tolerance to both the positive as well as to the negative effects of alcohol probably occurs. The chronic alcoholic can consume far higher doses of alcohol than the casual social drinker without appearing to be intoxicated or exhibiting gross motor discoordination. However, after enough alcohol is consumed, the negative somatic and emotional symptoms previously described invariably occur.

THE SEMANTICS OF HAPPINESS

The lay vocabulary relies on terms such as "reward" and "punishment" to denote the polar extremes of consequences for behavior.

These descriptive terms are generally understood and traditionally used to explain alcohol and drug effects. However, we have seen that the effects of alcohol (and of certain other drugs) are not exclusively or necessarily "rewarding" but may be somewhat "punishing" as well. To describe problem drinking as "reward seeking" or "punishment seeking" is merely to apply a label that does not help to analyze the complexities of the behavior involved.

Contemporary behavioral science uses a different language that analyzes behavioral events in terms that do not have any immediately evident or "surplus" meaning. The intention of this more precise vocabulary, derived from the pioneering conceptual and scientific advances of B. F. Skinner, is to permit the examination of events and the behavioral consequences without a priori judgment of their value or meaning. Of special relevance to this discussion is Skinner's concept of "reinforcement." A central premise of Skinner's approach to the experimental analysis of behavior is that behavior is maintained by its consequences. Any consequence that increases the probability of recurrence of a specific behavior is defined as a reinforcement. This concept is not limited by the confines of "reward" and "punishment" and does not require that positive or negative values be ascribed to events. Behavior and the attendant consequences can be examined in an effort to determine which consequences maintain a pattern of behavior under what conditions.

If, in fact, behavior is maintained by its consequences, and the consequences of heavy drinking are depression, anxiety, and often nausea and vomiting, perhaps it is not unreasonable to postulate that these may be part of the total reinforcing complex that maintains alcohol abuse. A similar argument may be advanced for certain other forms of drug abuse such as heroin addiction. Chronic heroin use, under research ward conditions, has also been shown to be associated with increased depression and anxiety rather than feelings of contentment and well-being or euphoria.

Are we to believe that the alcoholic drinks to become depressed and sick; that this is part of the reason for drinking? No answer is now available to confirm or refute that hypothesis, and even the most generously inclined skeptic may bridle at such manifest absurdity. However, there is accumulating evidence that these adverse effects of alcohol are frequent, and perhaps the most common consequences of heavy drinking. If, in fact, behavior is maintained by its consequences, then we fail to recognize and ignore this aspect of the consequences of alcohol intoxication at our peril.

Other even more dramatic examples of drug use lend some support to this notion. The cyclohexamine derivative, phencyclidine, is one case in point. Phencyclidine or PCP and related compounds, ketamine and sernalan, were developed as anesthetics for use in veterinary medicine because of their rapid onset and relatively brief duration of action. These compounds have proved highly effective for the immobilization of large animals for brief surgical procedures and other manipulations. Efforts to use phencyclidine as an anesthetic in human surgery proved disappointing because convulsions and psychological disturbances were frequently seen. Yet, within the last several years, phencyclidine has been increasingly abused by young people, even by grade-school children. The consequences of phencyclidine use appear to be aversive in the extreme. Stupor, psychotic reactions, hallucinations and severe anxiety are commonplace. At high doses, phencyclidine is life threatening and presents a serious problem in medical management. Yet, case reports indicate that some individuals inject phencyclidine to the point of loss of consciousness, recover, and reinject to achieve loss of consciousness again.

It is difficult to envision a drug that produces more aversive consequences. Phencyclidine would seem to be a drug that most people would try strenuously to avoid. Yet some aversive components of phencyclidine intoxication are shared with alcohol and some other abused drugs. In the vernacular, such drug use is rather like the proverbial "knocking one's head against the wall." Examples of repetitive self-destructive behavior are numerous and are not specific to alcohol and drug abuse. Masochism or comparable constructs, which imply that a psychodynamic or characterological trait is the culprit, offer a facile but probably untestable explanation for such behaviors. Since engaging in behaviors that have aversive consequences is not limited to humans, this tends to diminish the explanatory appeal of psychodynamic formulations.

In the early 1960s a chance observation led to a series of important experiments that showed that "aversive" stimuli may be reinforcing in subhuman primates as well as humans. It was found that a monkey trained to avoid a painful electric shock would later work to produce an electric shock under certain stimulus conditions. A series of elegant and elaborate experiments were designed to test the hypothesis that an aversive event, such as electric shock, could maintain response behavior leading to its administration. It is now well documented that under certain experimental conditions, monkeys will continue to self-administer a painful electric shock for months and even years. This finding has proved to be robust and repeatable. In some instances, the monkey will self-administer an electric shock of the same intensity that it previously learned to avoid. If the shock is disconnected, the monkey stops responding. As shock intensity is increased (over a range of 0–10 ma), the monkey's response rate on a shock presentation schedule also increases. In some studies, the presence of a green light signals the occasion for the monkey to avoid electric shock, whereas the presence of a red light signals the occasion for the monkey to self-administer a shock of the same intensity. The monkey learns to respond appropriately under both red and green stimulus conditions.

This phenomenon, sometimes called response-produced shock, indicates the importance of examining the actual consequences that maintain behavior rather than ascribing a priori values, positive or negative, to particular events. Electric shock is the favorite aversive consequence of generations of psychologists, and the capacity of this consequence to maintain avoidance behavior is familiar to every beginning student. One important implication of the response-produced shock studies is to demonstrate that an "aversive" event can also maintain response behavior in an animal model. This important basic research has shown that the same event may be a reinforcer (maintains behavior that leads to its repetition) or a punisher (suppresses behavior that leads to its repetition) depending upon the situation.

There are other examples of "aversive" control of behavior that attest to the generality of this phenomenon. Narcotic antagonists are drugs that induce opiate withdrawal signs in individuals and monkeys addicted to opiates. Everything we know about opiate withdrawal suggests that it is unpleasant. In opiate-dependent monkeys, administration of a narcotic antagonist is followed by vomiting, tremors, coughing, salivation and irritability. Yet opiate-dependent monkeys will work at an operant task to obtain injections of a narcotic antagonist that produces those very symptoms. "Meshuga monkeys"? or a disquieting parallel to alcoholic drinking? If the alcohol abuser hopes for joy and consolation in alcohol but instead finds dysphoria, anxiety, nausea and vomiting, to say nothing of disdain and abuse from others, and still continues to drink, something more subtle, intricate and curious is happening than "expectancy" explanations take into account.

Nausea, vomiting, depression and anxiety may be part of the constellation of events that maintain excessive drinking. Undoubtedly, these consequences are balanced by the occurrence of the expected positive consequences to some degree. However, the common assumption that alcohol is an unvarying euphoriant is clearly false. Considerably more systematic examination of the actual behavioral consequences of alcohol intoxication will be required before we can fully appreciate the entire spectrum of its behavioral effects.

AT LEAST IT'S DIFFERENT . . .

Undeniably alcohol does change mood and other subjective states, but the direction of change (up-down; better-worse) is less predictable, and this is also true for many other abused drugs. Polydrug use further illustrates the reinforcing properties of change, any change! Drugs with markedly different pharmacological properties often are used at the same time or in rapid succession. If an individual takes sedatives, then stimulants, then sedatives again, the objective would seem to be a rapid change in subjective state. The direction of the change (up-down) may be less important than the occurrence of change itself.

Albert Goldman's eloquent portrayal of polydrug use patterns in Lenny Bruce's world fits this descriptive framework:

> The night before, they wound up a very successful three-week run in Chicago at the Cloisters with a visit to the home of a certain hip show-biz druggist — a house so closely associated with drugs that show people call it the "shooting gallery." Terry smoked a couple of joints, dropped two blue tabs of mescaline, and skin-popped some Dilaudid; at the airport bar he also downed a pair of double Scotches. Lenny did his usual number: twelve 1/16th-grain Dilaudid pills counted out of a big brown bottle like saccharins, dissolved in a 1-cc ampule of Methedrine, heated in a blackened old spoon over a shoe-stuck lucifer and the resulting soup ingested from the leffel into a disposable needle and then whammed into the mainline until you feel like you're living inside an igloo.
>
> Lenny also was into mescaline that evening: not just Terry's two little old-maidish tabs, but a whole fistful, chewed up in his mouth like Gelu-

sils exposed with his tongue to the assembled company in the manner of a dyspeptic Seventh Avenue garment manufacturer, and then washed down with a chocolate Yoo-Hoo. [Goldman, 1971, p. 4]

An insatiable eclecticism appears to distinguish the professional polydrug user from the ordinary person whose pharmacological armamentarium may be limited to sleeping pills, diet pills, tranquilizers, alcohol, coffee and tobacco. Whatever the drug-use repertoire, the goal of contradictory multiple-substance use is unclear. If state change is the goal, then drug use and abuse may be just one special case of what, for lack of a better term, we refer to as "stimulus self-administration." The types of stimulus events that plausibly could fit into this rubric are endless. From child's play (to get dizzy by standing on one's head; holding one's breath; to get scared by playing scare games) to adult's play (sky-diving, hang-gliding, climbing hazardous cliffs) to socially sanctioned rituals (religious ecstasies; ceremonial self-deprecations) to socially condemned rituals, including drug and alcohol abuse.

Any concept that is so all-inclusive and generally applicable, necessarily lacks precision. But a concept such as "stimulus self-administration," used in this context, does connote that a change in mood and subjective state need not necessarily be "better" to be sought and preferred. "Worse" may suffice. "Different" is reinforcing.

And happiness, that long-cherished illusion of drinkers, prospective and past — should it give way to the prolix pronouncements of behavioral science? A sad prospect for the drinking ballads of the next generation. Or should Webster's be urged to update "drunk with joy" to include "drunk with sorrow," or, perhaps more economically, "drunk with dyspepsia"? Could some four thousand years of celebrants really be so wrong about alcohol and happiness? The answer is both yes and no, qualified by the amount of alcohol drunk, the drinking experience and expectancy of the drinker, the rate of drinking and the overall social context in which alcohol is consumed. A low dose of alcohol, resulting in blood alcohol levels of 30 to 50 mg/100 ml, often produces the anticipated pleasurable consequences. Some studies have shown that the effects of a low dose of alcohol given intravenously were difficult to distinguish from one marijuana joint. Subjects did not know in advance if they would be given saline (placebo alcohol) or real alcohol or marijuana or marijuana placebo that day. At low doses, both marijuana and alcohol appear to be mild intoxicants that facilitate feelings of conviviality and contentment. However, even at relatively low levels of intoxication, anxiety and irritability may increase as blood alcohol levels descend from peak values.

The definition of a high or low alcohol dose varies for social drinkers and alcohol abusers because of differences in behavioral tolerance for alcohol. Six to eight ounces of alcohol may lead to dysphoric mood changes in social drinkers, whereas in alcoholics, depression and anxiety are reported during chronic drinking when blood alcohol levels remain between 150 and 300 mg/100 ml. Thus, the point at which euphoria gives way to dysphoria cannot be related to a specific blood alcohol level. But the greater the relative degree of intoxication, the more likely that despondency and anxiety will occur.

The conclusion that the effects of alcohol on mood vary directly with the dose of alcohol was presaged by an archaic scheme advanced by an early temperance spokesman. In 1785, Dr. Benjamin Rush, respected Philadelphia physician (and signer of the Declaration of Independence), published his "Inquiry into the Effects of Ardent Spirits on the Human Body and Mind." To convey his argument graphically, he devised a moral and physical thermometer in which intemperate use of grog, rum, brandy and gin (on a scale from 0 to 70) led inexorably to a graded series of "vices, diseases and punishment" ranging from idleness, sickness and debt; to swindling, burglary and murder; madness and despair; to life imprisonment and gallows. In contrast, the temperate use of wine, cider and strong beer all led to "cheerfulness, strength and nourishment when taken only in small quantities, and at meals." Only milk, water and small beer insured "serenity of mind, reputation, long life and" — that much maligned term — "happiness."

Contemporary medical opinion on the implications of excess milk consumption for cholesterol may elicit as much rhetorical fervor as did ardent spirits in Benjamin Rush's day, whereas moderate alcohol use continues to receive social and medical approval. What Dr. Rush, as well as some contemporary authorities, failed to appreciate is that alcoholism can develop if enough beer, wine or cider are consumed. Beverages with low concentrations of alcohol do not necessarily protect against alcohol abuse. Even less commonly appreciated is that excessive consumption of water may also lead to intoxication and its attendant discomforts. As aphorisms go, *abjure excess* may still be valid for those who would pursue the illusion of happiness in alcohol.

ALCOHOL AND DEPRESSION

And what of those already sorrowful who turn to alcohol to try to cope with a disappointment, bereavement, or that special type of implacable sadness psychiatrists call depression? In these drinkers there may be no expectation of "happiness" through intoxication, only the hope that alcohol may briefly ameliorate an intrusive despondency. Objective studies of alcohol and depression are rare, but it appears that alcohol in high doses does not relieve depression. Psychiatric patients suffering from a depressive disorder reported feeling about the same after an intravenous alcohol infusion. The coincidence of suicide and alcohol intoxication also suggests that alcohol is not an effective antidepressant.

The depression-facilitating effects of alcohol are accentuated at high doses. Alcoholic men who were somewhat depressed during sobriety became tearful and despondent during sustained drinking. Dreams and reveries about disturbing emotional events, repetitive expression of self-deprecation and guilt occurred more often during severe intoxication. It is as if alcohol gives access to memories and fears that are less available during sobriety. Astute clinicians first noted this enhanced remembrance for painful experiences during intoxication while treating alcoholic patients with psychotherapy. Sobriety was followed by amnesia for the emotionally painful revelations.

Studies such as these are especially relevant to the recent finding that depression may develop as a *consequence* of alcohol problems. Resolution of the drinking problem often was accompanied by a significant alleviation of depression. Alcoholism has often been explained in terms of an antecedent depression and both conditions often co-exist. However, careful longitudinal studies have shown this sequence is not the only possible pattern. The patient with alcoholism or other drug abuse problems may be surprised to feel considerably less depressed during a drug-free period.

Much remains to be learned about the biology of depression and how it may relate to the effects of alcohol and other drugs on the brain. The analogy between dysphoria and anxiety associated with severe intoxication and the symptoms of depressive disorders may prove to be more apparent than real. Eventually, neurobiology may clarify the brain mechanisms underlying the experience of intoxication. Contemporary behavioral science has only just begun to try to characterize the nature of alcohol and drug intoxication by looking directly at the phenomenon while it occurs. We have learned that the pleasures of drinking are distorted into their antithesis as intoxication increases beyond an ever-fluctuating individual limit. Once that limit is reached, further drinking can rarely restore the subjective states associated with mild intoxication.

The evanescent, inconstant effects of alcohol intoxication are far more complex than one might predict from its simple molecular structure. And perhaps that inconstancy, the ever-certain uncertainty, is for some a more desirable state than absolutely predictable good cheer.

PART III

Alcohol and Health: Facts and Illusions

CHAPTER 13

The Role of Alcohol in Sickness and Health

In the thirteenth century, Arnauld de Villeneuve, a distinguished professor of medicine at the University of Montpellier, extolled distilled spirits as "aqua vitae," the "water of life," and wrote, "this name is remarkably suitable since it is really a water of immortality. It prolongs life, clears away ill humors, revives the heart and maintains youth." Arnauld de Villeneuve also asserted that "there is undoubtedly something to be said for intoxication. Inasmuch as the results which usually follow do certainly purge the body of noxious humors."

These claims were readily accepted by many European physicians during the thirteenth and fourteenth centuries. This was a period of undaunted optimism, when alchemists tried to transform base metals into gold and physicians sought a universal cure for all disease. The scientific era brought discoveries of specifics in the areas of medicine and metallurgy. Yet in some contemporary cultures, the medicinal properties often ascribed to alcohol echo that distant time. A contemporary French treatise on the global health benefits of wine, circa 1980, makes Arnauld de Villeneuve appear conservative. Even today, Chinese herbal remedies often contain up to 40 percent alcohol and the prescription advises the patient to drink as much as possible. However, these are exceptions to contemporary medical opinion about alcohol's effects on health.

Unqualified medical endorsement of alcohol as a pharmacological panacea has evolved over six hundred years into fervent condemnation of alcohol abuse for its deleterious effects upon biological function and behavior. Similar polarizations of medical opinion have periodically oscillated around many pharmacotherapies, particularly those which affect the brain. Arnauld de Villeneuve's opinions about the effects of alcohol on life expectancy, cardiovascular function, growth and development, are balanced today by somewhat more objective data. Yet there remain many unanswered questions about the role of alcohol use and abuse in sickness and health.

ALCOHOL AND THE HEART

Coronary artery disease is the leading cause of death of adult males in the United States. Does alcohol use or abuse contribute to this disorder? At present, the answers are often conflicting, since not only the amount and frequency of alcohol consumption but many other life-style variables may affect cardiac function. For example, individuals who abuse alcohol may also engage in other activities that favor the development of premature coronary artery disease. These include heavy smoking, poor dietary habits, obesity and lack of adequate exercise. Moreover, a sudden change in usual drinking and activity levels, strenuous exercise and heavy drinking over a weekend or on vacation have been associated with the so-called Holiday Heart Syndrome, characterized by irregular heart beats or dysrhythmias. These cardiac dysrhythmias may occur in individuals without clinical evidence of heart disease and can lead to fibrillation (abnormally rapid heart beat) and death.

The direct actions of alcohol on the heart are poorly understood. Relatively small amounts of alcohol (3 ounces) can induce a form of cardiac arrhythmia called ventricular tachycardia during an exercise test in people who have normal test results during sobriety. The mechanisms underlying these effects are unclear. However, since alcohol depresses cardiac contractility (or pumping ability) and changes conduction rates, these effects may contribute to the disturbances in cardiac rhythms.

Alternatively, there is accumulating evidence that moderate alcohol users may have a lower risk for heart attacks than abstainers. Several studies conducted in the continental United States and in Hawaii have shown that men who abstain from alcohol are more likely to have coronary artery disease than individuals who are moderate drinkers (one or two drinks per day). These studies were rigorous in controlling for the possible contribution of other factors such as hypertension, obesity and cigarette smoking. Even taking these factors into account, moderate drinkers had a lower risk for coronary artery disease than alcohol abstainers. Other recent studies suggest that moderate alcohol use may also have a protective effect against premature cardiovascular disease in women. "Premature" in this context means before age fifty. Moderate drinkers had a significantly reduced risk of heart attack relative to those who never drank. There are now over a dozen studies that suggest that the rate of coronary artery disease is reduced by about 30 percent among moderate drinkers. These relationships are not specific to the United States. Cardiac deaths were negatively related to moderate alcohol consumption in surveys carried out in Australia, Scandinavia and England.

The optimal amount of alcohol needed to reduce the risk for heart disease is unknown. Much more research is necessary before any firm answers can be obtained. However, the conclusion that "some" alcohol may be better than none appears to be justified on the basis of the existing research on heart disease.

Although moderate drinking may be benign, excessive drinking is not. Alcohol abuse can lead to severe and potentially lethal heart disease. One specific form of heart disease linked with alcohol abuse is "Alcoholic Cardiomyopathy." The heart muscle can be damaged by steady drinking of large quantities of alcohol as is characteristic of alcoholism. Such damage can result in severe abnormalities of the contractile ability of the heart, which may progress to serious heart failure. While the exact amount of alcohol required to damage cardiac function in people is unknown, the condition appears to develop only in very heavy drinkers and alcohol-dependent persons.

The amount of alcohol consumed, rather than the type of alcohol (distilled spirits, wine or beer), appears to be the most important factor affecting heart disease. With rare exceptions no specific constituents of beverage alcohol have been shown to damage the heart. In a few instances, alcohol-related heart damage has occurred following consumption of alcohol that also contained toxic substances. Heavy consumption of beer that contained trace quantities of heavy metals, such as cobalt added by manufacturers to stabilize the foam, led to cardiac damage. Once the toxic properties of these trace metals were identified, the beer manufacturing process was changed to eliminate that hazard.

Heart disease takes many forms, and the determinants of each type of heart disease are still poorly understood. Great advances have been made in the medical, pharmacological and surgical treatment of cardiac disease, and the types of pathological changes in cardiovascular structure have been clearly described. However, until the multiple origins of cardiac dysfunctions are clarified, the way in which alcohol may benefit or injure the heart will remain an issue of intense controversy.

ALCOHOL AND GASTROINTESTINAL DISORDERS

The role of alcohol in gastrointestinal function is far less ambiguous than the effects of alcohol on cardiac function. Even traditional anecdotes that alcohol aids digestion may be incorrect. Alcohol inhibits the absorption and transport of vitamins from the gastrointestinal tract and thus impairs nutrition. There is no scientific controversy about the generally injurious effects of excessive alcohol intake on the gastrointestinal tract.

As we discovered in following the odyssey of the alcohol molecule, some alcohol is absorbed from the stomach but most is absorbed primarily from the small intestine. In order to reach the small intestine, alcohol must traverse the course of the esophagus, the stomach and the duodenum. Alcohol abusers take large volumes of alcohol into their gastrointestinal tract, and there is good evidence that prolonged alcohol exposure is hazardous. Alcohol irritates the mucosal lining of the stomach, and in certain susceptible individuals, such chronic irritation may facilitate the development of gastric ulcers. Ulceration of the duodenum may also be related to heavy alcohol intake. While controversy remains concerning the cause of gastric and duodenal ulcers, there is no doubt that chronic drinking of high doses of alcohol aggravates these conditions.

Even in the absence of overt ulcer development, alcohol-induced irritation of the gastric mucosa can have serious health consequences. There are an alarmingly large number of documented case reports that alcohol, taken in combination with other gastric irritants such as aspirin, may produce profound gastric irritation and severe bleeding. In fact, most cases of "hemorrhagic gastritis" appear to be caused by alcohol taken in combination with some other substance that irritates the stomach.

Pancreatic function may also be impaired as a consequence of prolonged consumption of high doses of alcohol. The *pancreas* produces enzymes that are secreted into the intestinal tract to break down food products so that their nutrients can be absorbed. Pancreatitis is an imflammatory disorder of the pancreas, which may cause severe pain, debility and even death. While the cause of pancreatitis is not known, sustained consumption of large amounts of alcohol may increase individual vulnerability to this disorder. It is also recognized that once an individual develops pancreatitis, for whatever reason, drinking may cause an exacerbation or worsening of this condition.

The *liver* is the organ most vulnerable to damage from alcohol abuse. Several types of liver disease have been found in alcohol abusers who are otherwise healthy and have good nutritional intake. Cirrhosis, the most severe form of liver disease, is highly correlated with alcohol abuse, although it can also develop in nonalcoholic individuals secondary to viral infections of the liver. Cirrhosis of the liver together with its complications, is one of the leading causes of death among adult males in the United States. The relationship between cirrhosis and alcoholism in many societies has been so highly correlated that formulas have been devised to estimate the number of alcoholics in a population based upon the incidence and prevalence of cirrhosis-related fatalities. While this technique cannot provide an accurate index of the number of persons with alcoholism, it does illustrate the relationship between cirrhosis and alcohol abuse.

The liver is especially vulnerable to toxic effects of alcohol because it is the primary site for the breakdown of alcohol and its elimination from the body. The liver contains the necessary enzymes and biochemical machinery for converting alcohol to a product called acetate, which the body uses for energy. Alcohol is recognized by the liver as a substance that requires detoxification. But there are critical limits to the quantity and number of toxic substances the liver can handle before damage occurs to the liver itself. This is why the combined use of alcohol and other drugs that are detoxified by the liver may have more adverse consequences than use of any single drug alone.

The amount of alcohol usually consumed by social drinkers can easily be broken down by the liver without any permanent ill effects. However, transient and reversible changes do occur in the liver when only moderate doses of alcohol are metabolized. Even cocktail-party drinking sometimes results in an increased degree of fat concentration in the liver, a phenomenon described as "fatty metamorphosis." It is not known why fat tends to accumulate in the liver during the breakdown of alcohol, but this temporary change is not believed to have residual effects. There is also a change in the structure and composition of elements in the liver cell during alcohol metabolism, which has been detected with the electron microscope. The most profound alcohol-induced changes occur in the microscopic filamentous bodies in the liver cell that are called the endoplasmic reticulum. This system becomes enlarged and swollen and disrupts normal cell functions.

Social drinking may produce temporary changes in liver cell morphology and function, but chronic alcohol abuse eventually leads to serious derangements of liver function. One form of severe liver disease observed in alcohol abusers has been termed "alcoholic hepatitis." This form of hepatitis is similar in some ways to viral hepatitis but it usually has a more sudden onset. Patients with alcoholic hepatitis may become extremely ill within forty-eight to seventy-two hours following heavy drinking, and massive liver cell death may occur. Symptoms include high fever, severe abdominal pain, nausea and vomiting. Fortunately, most attacks of alcoholic hepatitis are relatively mild and the patient survives. However, alcoholic hepatitis may prove fatal if blood pressure falls rapidly and the patient goes into shock.

If heavy drinking continues, hepatitis attacks may occur more frequently, and this condition may progress to the most serious form of liver disease, cirrhosis. Cirrhosis involves liver cell death with subsequent "scarring" of the liver. Such scarring or growth of fibrous tissue in the liver may occur after repeated episodes of alcoholic (or viral) hepatitis or it may develop insidiously with little or no warning. In many instances alcohol abusers are not aware that they have cirrhosis of the liver until they develop jaundice or other signs of severely impaired liver function. Loss of appetite, abdominal pain, weakness and debility may increase in severity as cirrhosis progresses.

Once cirrhosis is detected, what is the prognosis? Continued alcohol abuse, especially in combination with a deficient diet, will eventually lead to death from liver failure. When this occurs, there is an accumulation of toxins such as ammonia (a by-product of normal nitrogen metabolism) that the liver is no longer able to detoxify. Patients may become confused, stuporous and eventually die in coma.

Cirrhosis can also kill by exsanguination. When progressive scarring of the liver occurs many other abnormalities besides jaundice also develop. One of the most serious is the enlargement of veins which surround the esophagus. Severe enlargement of esophageal veins ("esophageal varices") may lead to spontaneous venous bleeding, which can be so severe and profuse that death occurs in a very short time.

Although alcohol abstinence cannot guarantee that the disease process will be arrested, improved nutrition and cessation of drinking usually lead to improved liver function if the damage has not been too severe. It was once believed that a good diet could protect the liver from alcohol-induced disease, but this has not been confirmed by recent research. It has been possible to produce hepatitis and cirrhosis in well-nourished baboons by giving them a liquid diet in which alcohol accounted for 50 percent of the total calories. These data demonstrate that alcohol is the toxin directly responsible for these forms of liver pathology. The dose and time required to produce liver damage in baboons varied from 4.8 to 8.3 grams of alcohol per kilogram over nine months to four years. Similar individual variations in susceptibility occur in humans. Although it has been estimated that consumption of 210 grams of alcohol per day (about 3 grams per kilogram) for twenty years leads to a 50 percent chance of developing cirrhosis, it is not known with any degree of certainty how much is too much for healthy liver function.

ALCOHOL AND CANCER

Does alcohol "cause" cancer? No. Are certain risks for the development of cancer increased in alcoholics and alcohol abusers? Probably. Although cancer has been one of the leading killers of mankind since antiquity, surprisingly little was known about environmental and dietary factors which contribute to this disease until very recently. It is now recognized that many chemicals and toxic by-products of industry, which are present in our environment, may have a major role in carcinogenesis. But alcohol per se has never been shown to be carcinogenic in any test system. In rare instances, constituents other than alcohol in alcoholic beverages have been implicated in the causation of cancer, but the trace quantities of these substances are found not only in alcoholic beverages, but in many food products. The amounts of these compounds which have been shown to produce cancer are hundreds or thousands of times greater than the amounts consumed by social drinkers or even alcohol abusers. Thus, at present, there is no good evidence that alcohol itself is carcinogenic, or that any specific constituent in alcohol or alcoholic beverages increases the risk of cancer in humans.

If alcohol itself is not carcinogenic, are there factors that increase the risk of cancer with heavy alcohol use? Unfortunately, the lifestyle of alcohol abusers often tends to compromise good health. For example, heavy drinking is often associated with heavy smoking. Studies have shown that both social drinkers and alcoholics smoke significantly more cigarettes during drinking than at other times. Since it is well known that cigarette smoking may lead to cancer of the lung, alcohol abuse may indirectly increase the risk of lung cancer. Moreover, the incidence of certain types of cancer, principally cancer of the esophagus, throat, and mouth, are more frequent in alcoholics and alcohol abusers.

Is there any evidence that individuals who abstain from alcohol have a lower probability of developing cancer? Unfortunately, this does not appear to be the case. In fact, no form of cancer appears to be underrepresented in alcohol abstainers as contrasted to alcohol users. The cause of cancer is unknown, and since drinkers and abstainers alike frequently succumb to this disorder, it is highly unlikely that moderate alcohol use is a major contributing factor.

ALCOHOL AND THE BRAIN

During the decade of the seventies, a steadily increasing number of admissions to state mental hospitals in the United States were for alcohol-related brain disorders. Does this mean that more people sustained damage to their nervous system because of alcohol abuse? Probably not. As treatment for other serious mental disorders became more effective (primarily through new drugs), fewer patients with chronic schizophrenia or depressive illness required long-term hospitalization. Consequently, more beds were available for persons with other types of mental disorders. Patients with alcohol-related brain disorders were excluded from state hospitals during the fifties and sixties, but were able to gain admission during the seventies.

How much alcohol is actually required to damage the brain remains a matter of controversy. Some scientists have shown that moderate to high doses of alcohol can produce irreversible changes and death of brain cells in experimental animals. Alcohol can permanently damage the biochemical system responsible for conducting ions into and out of the nerve cell body. For example, alcohol can damage the structure of the outer membrane or wall of the nerve cell and impair the manufacture and transport of substances called nucleic acids. These nucleic acids are crucial for maintenance of the protein structures of nerve cells. Since certain types of nucleic acids have also been implicated in the process of memory storage, alcohol-related changes in memory function have been linked to adverse effects of alcohol on brain cell nucleic acids.

Other scientists argue that these damaging effects of alcohol on brain function, based on animal studies, cannot be validly generalized to the human nervous system because the amounts of alcohol necessary to damage brain cells in animal models far exceeds what most people could possibly drink. In fact, some studies that report adverse effects of alcohol on brain cell function have used such enormous doses of alcohol it could precipitate proteins! By analogy, the alcohol doses used would be equivalent to immersing nerve cells in a solution containing concentrations of salt and vinegar sufficient to pickle cucumbers. A nonchemical example of protein precipitation is hard-boiling an egg.

Alcohol and Memory

Although brain cell damage in animal models has not been produced with realistic physiological doses of alcohol, this does not mean that alcohol abuse is not a factor in some forms of disordered brain function. For example, transient memory dysfunctions may occur during alcohol intoxication. The spectrum of memory disorders associated with alcohol abuse range from brief episodes of forgetting to severe amnestic syndromes accompanied by structural changes in the brain. The fragmentary loss of recall for certain events during intoxication is sometimes called a dissociative effect of alcohol. Forgetting the negative emotional consequences of drinking (e.g., increased anxiety and depression) is one dissociative effect of intoxication. Forgetting rather commonplace occurrences (e.g., where money or alcohol was hidden during intoxication) is another example of a fragmentary lapse in recall. Curiously, the hidden money may be found only during a subsequent episode of intoxication.

Another common and more severe form of transient memory loss during drinking is called the "blackout." As the term so vividly implies, there is a complete amnesia for events during a period of severe intoxication. The blackout can last for several hours or several days depending on the length of the drinking spree. However, it is temporary and usually does not include the period just before or after drinking began.

It was once believed that only alcoholics suffered from blackouts when they were drinking heavily. However, it is now known that normal, healthy social drinkers, and even people who usually abstain from alcohol, may have blackouts if they drink enough alcohol on a given occasion. The blackout seems to be related to the amount of alcohol consumed and the resultant blood alcohol level. The higher the alcohol dose, the greater the probability and severity of a blackout. Although anyone who becomes severely intoxicated may have a blackout, they are most probable in alcoholics, who drink large amounts of alcohol.

Does the blackout reflect some kind of permanent damage in the central nervous system produced by alcohol? Probably not, but no one knows for sure. The brain contains billions of nerve cells or neurons, and there is great plasticity and redundancy in the central nervous system. However, once damaged, nerve cells cannot regenerate like many other cells in the body. Nerve cells cannot divide and form new cells in response to a chemical or physical insult. Thus, any toxic substance or condition that destroys nerve cells causes a permanent depletion of nerve cells from the brain.

The awesome specter of permanent nerve death has led to some apparently exaggerated speculations about the effects of small doses of alcohol on the brain. It has been difficult to demonstrate rapid nerve cell death in experimental animal models fed alcohol over long periods of time. However, alarmists speculate that thousands, perhaps millions, of nerve cells may be destroyed during moderate social drinking. Since the human brain contains a finite number of nerve cells, usually estimated at about ten billion, the loss of "millions" of cells during a drinking episode could exhaust the neural reserve in a relatively short period of time. Despite the assertions of some cynics, a totally denervated human being has yet to be discovered.

Although the mechanisms underlying a temporary memory loss associated with alcohol intoxication are unknown, it was once speculated that alcohol may have a specific disruptive effect on "short-term" memory. Short-term memory has been imprecisely defined but usually refers to recall of events over seconds or minutes. It has been argued that disruption of the process of coding information into short-term memory stores would result in an amnestic syndrome, since information received would never be transferred to long-term memory stores. Moreover, it was believed that individuals who were prone to blackouts during alcohol intoxication might show more serious disruption of short-term memory function than control subjects when tested during drinking. Even though these arguments were plausible and provocative, clinical studies failed to demonstrate that alcoholics with a history of blackouts performed more poorly on tasks designed to assess short-term memory function than alcoholics without a history of blackouts. One of the procedural problems that has confused the scientific literature on the effects of alcohol on memory (as well as many other performance-related tasks) is that subjects usually are not paid to perform well, but merely to participate in the testing. When the payment is directly related to the excellence of the performance, many commonly reported alcohol-induced performance deficits are difficult to discern. For example, when alcoholics were working for alcohol on a short-term memory task, they performed very well when heavily intoxicated and often performed better than sober staff attendants. Accurate evaluation of the behavioral effects of alcohol (and other drugs) requires that a subject attends to the task and is sufficiently motivated to perform at maximum capacity.

The belief that alcohol specifically disrupted short-term memory led General Motors to develop an automobile ignition interlock system to try to decrease drunken driving. The PhysTester device required subjects to remember the information in visual display for three to six seconds, then to punch in the same information on a keyboard wired into the ignition system. The correct information had to be entered before the ignition could be activated by a key. This device might have been effective with normal social drinkers who became heavily intoxicated at a cocktail party. However, alcohol addicts develop tolerance for alcohol and can perform short-term memory tasks and many other simple motor and cognitive tasks when very intoxicated. Unfortunately, it is the alcohol-tolerant individual who accounts for many of the drunken driving fatalities. Statistics show that the preponderance of blood alcohol levels in deceased drivers are above 150 mg/100 ml. The nontolerant individual would probably be unable to drive at all at these blood alcohol levels.

In addition to alcohol-related blackouts, it has long been believed that chronic alcohol abuse may produce persistent memory loss and damage the brain. Profound dementias and disorders restricted to memory function have been associated with alcoholism. Despite a wealth of clinical data on various forms of brain dysfunction related to alcoholism, it has never been possible to establish that alcohol alone accounted for the deficits observed. It now appears that nutritional deficiencies are the primary factor in many neurological disorders once attributed to alcoholism. Many common (Korsakoff Psychosis) and esoteric (Marchiafava-Bignami disease) diseases of brain function, which were once believed to be primarily caused by alcohol, have also been discovered in individuals who had no past history of alcohol abuse, or, in fact, any history of alcohol use.

The incidence of severe memory disorders in alcoholic patients is relatively low (1 to 3 percent). However, in those people with Wernicke-Korsakoff Syndrome, the profound memory dysfunction is unremitting, and postmortem studies reveal pathological changes in portions of the brain. This syndrome is characterized by an inability to recall recent events, although remote memory for early life experiences remains intact. Affected patients may not remember what they ate for lunch or the route to the dining room, but can recall a childhood incident in some detail. Deranged memory and thinking processes, accompanied by impaired control of the eye muscle and ataxia (a stumbling gait), appear to result from vitamin deficiencies, especially a thiamine deficiency. Alcoholics are at high risk for vitamin deficiencies because of poor nutrition and because alcohol inhibits the absorption and transport of vitamins from the gastro-intestinal tract.

We now recognize that alcohol abuse is often accompanied by other substance abuse disorders, such as polydrug abuse and heavy caffeine and nicotine use. The effects of polydrug use on brain function are unknown, and it is possible that combinations of drugs may have different effects than any single drug. Thus, alcohol-related disorders of the central nervous system are probably due to an interaction of heavy alcohol use with poor nutrition and concurrent abuse of other drugs.

In summary, there is good evidence that alcohol abuse may contribute to diseases of the nervous system, but there is no evidence that alcohol is the sole culprit. Moreover, there is no good evidence that moderate alcohol use produces any direct damage to the brain. Most memory disorders associated with alcohol abuse are reversible and remit within three to four weeks after cessation of drinking.

Alcohol and Sleep

The belief that alcohol improves sleep is probaby as pervasive today as in ancient times. Alcohol is still a favorite common remedy for occasional sleep disturbances. Problem drinkers and alcoholics often justify continued drinking as necessary to treat recurrent insomnia. The mythic god of sleep was not Bacchus, but Hypnos (or Morpheus), a winged cherub who sprinkled drops of opium from an opium horn and brought somnolence to man. Despite the lyrical and persuasive images of several generations of poets, neither opiates nor alcohol improve sleep. In fact, both alcohol and opiates profoundly disrupt normal sleep patterns. The search for a drug that induces and sustains sleep, without altering normal sleep rhythms, continues to challenge modern psychopharmacology.

Sleep is a complex and fascinating process that may hold one key to the mysteries of brain function. Sleep is not a simple uniform phenomenon but rather a series of changes in brain activity. The oscillating rhythmic pattern of sleep can be measured through surface electrodes placed on the head. The resulting electroencephalographic record, or EEG, shows that the depth of sleep correlates with identifiable patterns of neural activity. These patterns have been called sleep states and are usually designated by Roman numerals I through IV. Sleep researchers have been able to characterize normal sleep according to both the sequence and the amount of time spent in each stage of sleep. Comparison of normal EEG sleep profiles with EEG patterns from patients with sleep disorders has advanced basic understanding of the sleep process and made it possible to evaluate the effectiveness of sleep medications with quantitative as well as subjective measures.

Normal aging is usually accompanied by adverse changes in sleep defined by the proportion of changes in EEG sleep stages that occur during a night's sleep. For example, deep sleep, or Stage IV EEG patterns characterized by slow wave activity, decline progressively through life. Stage I, or light sleep, often associated with Rapid Eye Movements or REM sleep, declines sharply between the sixth and eighth decade. Fragmented sleep, characterized by many arousals and wakefulness begins to increase after age forty. Some scientists have postulated that these changes in sleep patterns with aging are intrinsically related to parallel changes in cognitive function in the elderly.

Alcohol changes sleep patterns to resemble the sleep of an elderly person. Even a single drink in the evening can decrease the proportion of REM sleep below the level considered normal (i.e., 20 to 25 percent of total sleep time). However, studies of normal subjects showed some adaptation to this alcohol effect over successive nights. Alcoholic subjects studied during a period of chronic drinking also showed suppression of REM sleep and sleep fragmentation defined by frequent awakenings, frequent changes in EEG sleep stages, and brief periods of sleep.

Studies of the effects of alcohol on slow wave Stage IV sleep have yielded conflicting results. Both increased and decreased Stage IV have been reported. These studies are not conclusive because adequate baseline measures were lacking and subjects were not adapted to the sleep laboratory. Moreover, alcoholic subjects may have been in alcohol withdrawal, a condition that itself uniquely disrupts sleep. Clinical studies of alcoholic individuals during alcohol withdrawal have consistently reported disordered sleep with severe insomnia. Thus, either the presence or absence of alcohol may disrupt sleep in persons physiologically dependent on alcohol. Moreover, sleep disturbances in alcoholics may persist for weeks during a period of sobriety. The sleep of abstinent alcoholics also resembles the sleep of elderly people more than the sleep of nonalcoholics of the same age. The process by which alcohol disrupts sleep patterns is unknown.

Only one of the sleep-related benefits commonly attributed to alcohol has been confirmed by sleep research. Alcohol does induce somnolence and people do fall asleep more quickly after drinking. However, it appears that the more rapid sleep onset cannot compensate for the subsequent sleep disturbances associated with alcohol intoxication. Even as little as one ounce of alcohol before retiring suppressed REM activity in normal subjects.

ALCOHOL AND REPRODUCTIVE FUNCTION

The adverse effects of alcohol on male reproductive function were described in an earlier chapter (11) on sex and aggression. As we have seen, even moderate amounts of alcohol depress testosterone, the primary male sexual hormone. During chronic alcohol abuse, testosterone levels remain abnormally low. In alcoholic men, the testicles may become shrunken and the ability to maintain penile erections greatly impaired. Complaints of sexual impotence and loss of sexual interest are frequent among alcohol abusers. Some men even develop enlarged breasts, a condition called gynecomastia. These disorders may be somewhat reversible depending on the cumulative dose and duration of excessive drinking. Alcoholic men who become abstinent may enjoy significant recovery of sexual function.

Considerably less is known about the effects of alcohol on sperm. Recent animal studies indicate that alcohol may impair both sperm production and viability. If these findings are generalizable to man, the result would be a decrease in fertility, since the likelihood that sperm in the male ejaculate would successfully impregnate the female ovum would be reduced.

Alcoholic women may also suffer severe derangements of reproductive function, and cessation of menstruation, or amenorrhea, is often reported. Some alcoholic women continue to menstruate, but more subtle disorders of the menstrual cycle may impair fertility. For example, if ovulation does not occur at midcycle, there is no ovum available for penetration by the male spermatozoon. Anovulation and amenorrhea in alcohol abusers ensures infertility.

Alcoholic women who do menstruate and ovulate also may be infertile because the luteal phase of the menstrual cycle is abnormal. The luteal phase is the period after ovulation when there is an increase in levels of progesterone, a steroid hormone produced by the ovary. It is during the luteal phase that several changes occur in the walls of the uterus, which permit it to receive the fertilized ovum. There are changes in the myometrium, or outer walls of the uterus. Muscle cells and connective tissue increase to provide better structural support for the embryo. The vascular supply of the endometrium, or inner wall of the uterus, also increases and there is a proliferation of secretory glands. This highly vascular endometrium develops into the placenta, the embryo's major life-support system, if pregnancy occurs. If the ovum is not fertilized, the endometrial lining sloughs off in menstruation. This cycle of preparation and decline is repeated every month throughout each woman's reproductive period.

However, if the luteal phase is too short, or if progesterone levels do not increase enough, the uterine environment may be inadequate for the growth and development of the fertilized ovum, and a failure of ovum implantation or a spontaneous abortion may occur. Even if impregnation and implantation of the ovum does occur, the process of development is very fragile, especially during the first trimester. Moderate social drinking has been associated with spontaneous abortions during this vulnerable period, and the effects of alcohol on fetal development are discussed more fully in the next section of this chapter.

This triad of menstrual cycle disruptions, amenorrhea, anovulation, and luteal phase inadequacy, which so often afflicts alcoholic women is poorly understood. As with impotence and depressed testosterone levels in alcoholic men, these disorders are often reversible if women become abstinent after three or four years of alcohol abuse. Menses may resume and normal pregnancies may occur. However, if alcoholism continues, irreversible structural changes in the reproductive system may occur and amenorrhea can persist for twenty years or more. The ovaries of alcoholic women on autopsy are often smaller than normal, suggesting either atrophy or absence of structural components necessary for production of the ovum. The intricate process by which hormones from the pituitary gland (luteinizing hormone and follicle stimulating hormone) and from the ovary itself (estrogen and progesterone) regulate the selection and development of the ovum is a fascinating chapter in the annals of reproductive physiology. Innovative research on reproductive function has led to remarkable progress in understanding the neuroendocrine regulation of the menstrual cycle and the implications for alleviating infertility.

The mechanisms by which alcohol disrupts these basic hormonal rhythms are not yet known. Perhaps alcohol is directly toxic to the ovary as it is to the testes and this accounts for the functional and structural defects seen clinically. Alternatively, alcohol may interfere with the hormonal signal from the hypothalamus that is required for the pituitary to release luteinizing hormone and follicle stimulating hormone, which in turn regulate ovarian function; or alcohol may interfere with the capacity of the pituitary to produce and/or release these hormones. Since the ovarian steroid hormones also regulate release of the pituitary hormones in a "feedback" relationship, alcohol could disrupt this crucial hormonal balance by interfering with any portion of the system. It may be that alcohol's toxic effects interfere with the normal functions of the hypothalamus, pituitary and ovary simultaneously. Only further research can answer these questions.

Several interacting problems have conspired to impede interpretation of these issues from clinical studies alone. Alcoholic women often have other disorders that could also disrupt reproductive function, so it has been difficult to accurately identify the contribution of alcohol itself. For example, liver disease is a frequent consequence of chronic alcoholism, and liver disease alone could disrupt the menstrual cycle. Alcoholic women are often malnourished because of marginal diets and alcohol-related impairment of absorption of essential vitamins and minerals. Malnutrition and profound weight loss can also disrupt menstrual cycle regularity. Amenorrhea is not uncommon among very lean female athletes who tend to abstain from drinking alcohol.

These factors have led scientists to examine the effects of alcohol on reproductive function in animals where nutrition and overall health can be controlled. Thus far, data from studies in female monkeys and in rodents strongly implicate alcohol as the most probable culprit in menstrual cycle disruptions. Chronic alcohol intoxication has resulted in amenorrhea in female rhesus monkeys, whose normal menstrual cycles are almost identical to those of human females. Upon autopsy, abnormally small ovaries and atrophy of the uterus were found in alcoholic monkeys. The ovarian pathology suggested that ovulation had not occurred because there were no corpus lutea, structures which are necessary for secretion of the hormone progesterone. Similar pathology has been found in rats fed a liquid diet containing alcohol for several weeks.

One interesting feature of the monkey studies was that females learned to self-administer alcohol through a surgically implanted venous catheter, and thus controlled their daily dose of alcohol. Some monkeys took high doses of alcohol, developed alcohol dependence, amenorrhea and reproductive system pathology. Other monkeys self-administered very little alcohol and maintained normal menstrual cycles. So the dose of alcohol appears to be critical for inducing pathology of the reproductive system. The duration of intoxication also appears to be an important factor.

Single high doses of alcohol do not have demonstrable effects on the reproductive hormones essential for female reproductive function. Single high doses of alcohol did not suppress luteinizing hormone or estradiol in young women or in female monkeys. This suggests that a single episode of intoxication is unlikely to disrupt menstrual cycle function, although individual susceptibility may vary.

One remarkable feature of most biological systems is their inherent resiliency. The reproductive system is no exception. Alcoholic men and women do have children during the time they are abusing alcohol. Thus for some, there is an adaptational process, a type of alcohol tolerance, that permits conception and reproduction. Today we are as far from understanding the mechanisms of alcohol adaptation as we are from understanding how alcohol disrupts the reproductive system. Clearly, alcohol adaptation is, in some basic sense, the obverse of alcohol-induced disruption. Advances in techniques to measure changes in reproductive hormones, combined with improved animal models for studying alcohol dependence, should eventually clarify the several unresolved questions about alcohol effects on reproductive function.

ALCOHOL AND FETAL DEVELOPMENT

Admonitions against drinking during pregnancy can be found in writings from antiquity and the Bible. Aristotle commented that "drunken women bring forth children like to themselves," but physicians paid little or no attention to the possible adverse consequences of alcohol use or misuse during pregnancy until the nineteenth century. In 1849, Dr. William B. Carpenter, Examiner in Physiology at the University of London, Professor of Jurisprudence at University College, was awarded a prize of one hundred guineas for his essay on "The Use and Abuse of Alcoholic Liquors in Health and Disease." In his essay, Carpenter concluded that "it is scarcely necessary to accumulate further proof in support of the assertion, that, of all the single causes of insanity, habitual intemperance is the most potent, and that it aggravates the operation of other causes." Dr. Carpenter also quoted the resident physician of the Crichton Lunatic Asylum in Dumfreis, a Dr. W. A. F. Brown:

> the drunkard not only injures and enfeebles his own nervous system, but entails mental disease upon his family. His daughters are nervous and hysterical; his sons are weak, wayward, eccentric, and sink insane under the pressure of excitement, of some unforeseen exigency, or of the ordinary calls of duty. At present, I have two patients who appear to inherit a tendency to unhealthy action of the brain, from mothers addicted to drinking; and another, an idiot, whose father was a drunkard.

More than one hundred years passed before other physicians published observations on the role of maternal alcohol abuse on growth and development of the newborn. During the 1950s and 1960s several isolated case reports by physicians in Germany and France were published in medical journals with modest circulation. These reports described abnormalities in behavior, physical appearance and growth of infants whose mothers abused alcohol. Children of alcoholic mothers were found to have delayed development of language, low I.Q. scores, abnormal hyperactivity, often accompanied by problems in school and difficulty in relationships with peers. Many of these observations formed the basis for what was later described as the "fetal alcohol syndrome" in case reports published by physicians in the United States. American investigators found children whose mothers were alcohol abusers were only about 65 percent normal birth length and 38 percent normal weight. These children were also reported to have distinctive abnormalities of the face and head. Many were mentally retarded with I.Q.'s ranging between 50 and 83.

These observations stimulated many clinical and experimental animal studies, which led to the general conclusion that alcohol abuse by pregnant women can result in significant physical and mental damage to their children. There remains some controversy about whether alcohol alone, or alcohol in combination with a number of other factors, is the major cause of harm to the newborn. Many women who abuse alcohol during pregnancy also use and abuse other substances that may damage the fetus. For example, many women who are heavy drinkers are also heavy smokers. Women who abuse alcohol during pregnancy also often have poor health habits, poor nutritional intake, and maintain a life-style that is not conducive to the best possible intrauterine growth and fetal development. Thus, some scientists have suggested that a more appropriate name for the "fetal alcohol syndrome" would be the "fetal alcohol–substance abuse–nutritional deficiency–stress related syndrome." Whatever the relative weight of these various factors, it is clear that alcohol abuse during pregnancy does not favor a healthy pregnancy or the birth of a healthy normal child.

Fortunately, most women do not abuse alcohol during pregnancy. But should women drink at all when they are pregnant? There is much disagreement on this point. Scientists who have studied the fetal alcohol syndrome point out that no safe level of consumption of alcohol has been determined. There is no assurance that moderate drinking will not harm the fetus; hence, they argue that women should abstain completely from drinking when they are pregnant. Other scientists point out that many women drank occasionally during pregnancy with no evidence of harm to their children. In fact, some scientists believe that by attributing special significance to the role of alcohol in fetal damage, unnecessary guilt may be induced in those mothers who drank moderately and gave birth to physically handicapped or mentally retarded children. Since the major cause of growth abnormalities and mental retardation in children is as yet unknown, it is unfair and cruel to indict mothers who occasionally used alcohol or drank moderately during pregnancy, especially if they were not aware of any potential relationship between alcohol use and birth defects. There is no evidence that maternal alcohol abstinence insures a normal healthy fetus or that moderate alcohol use can harm a fetus.

At present, the issue of alcohol use during pregnancy remains an unresolved question, and the amount of alcohol necessary to damage the fetus is unknown. The decision of whether or not to drink during pregnancy obviously must be made by each individual woman. Such decision-making should occur in a context where women are fully informed about the existing evidence concerning alcohol and pregnancy. An informed decision should consider the woman's own health and the health of her unborn child. This rational decision process should not be confused or confounded with efforts to modulate alcohol use or abuse by pregnant women with paternalistic or chauvinistic threats, intimidation or cajolery.

ALCOHOL AND HEALTH: ILLUSIONS AND IMPLICATIONS

Over the past six hundred years, pronouncements about alcohol use and health have vacillated between extolling alcohol as a medical panacea and indicting it as a cause of many diseases. In contrast, reliable data about the consequences of alcohol *abuse* have only been obtained within recent years. At present, there is good consensus that alcohol *abuse* may place an individual at risk for a number of serious health problems. Among these, alcohol-related liver disease is a major cause of death and disability in contemporary societies. While alcohol abuse does not cause cancer, abuse of alcohol does appear to increase the risk for development of certain types of cancer. Both transient and irreversible changes in memory function may follow alcohol intoxication and sustained abuse. Alcohol cannot cure insomnia and may disorder sleep.

Moderate alcohol use has not been linked to any significant health problem in humans. While this statement may seem inaccurate and even inflammatory to some (including some eminent scientists, journalists and politicians), this conclusion is based upon the best, most objective and dispassionate scientific data currently available. There remain important questions about the possible salutary effects of moderate alcohol use on certain health parameters. One of the most interesting questions is the possibility that moderate drinking may protect against the development of coronary artery disease. Much more careful research is necessary before this question can be answered satisfactorily.

A time-honored illusion held by man is that engaging in or abstaining from certain behaviors, such as eating only special foods, drinking esoteric beverages or even thinking appropriate thoughts,

CHAPTER 14

Alcoholism: A Search for Origins*

Alcohol abuse can take many forms, and the origins of alcohol problems remain obscure. No single factor has been found to explain or predict why some people develop alcohol problems and others do not. Alcoholism can develop in anyone, and neither youth nor age, affluence nor poverty, piety nor atheism provides an effective barrier against this insidious process. Alcoholism has almost always been with us. This protean disorder, which transcends time and cultures, appears to have only one common determinant: chronic, repetitive, excessive drinking. Little is known about how much or how often alcohol must be consumed in order to produce behavioral problems or physiological dependence upon alcohol. Severe alcohol-related problems can disrupt the intrinsic fabric of life experience well before the onset of alcohol withdrawal signs and symptoms that herald the development of alcohol addiction.

The search for "causes" of alcoholism continues in an effort to predict who is at high risk for developing a serious drinking problem. Identification of any significant predisposing factor could lead to more effective techniques for prevention. If, for example, a single biological variable or a single pattern of disruptive early life experiences was found to uniquely determine the risk for alcoholism, then

* Portions of this chapter have been adapted from "Etiological Theories of Alcoholism" by N. K. Mello, which was awarded the Mary Cullen Research Trust Etiology Monograph Prize in 1980. The Monograph appears in Volume III, *Advances in Substance Abuse, Behavioral and Biological Research*, Greenwich, Connecticut: JAI Press, Inc., 1983.

targeted efforts to modify these effects could be mobilized. However, it is evident that many factors within the individual and in the environment influence whether alcohol problems occur or not, and whether these remain constant through time.

The number of causative factors alleged to account for alcoholism is almost as varied as the individuals who suffer from alcohol problems. The emphasis of leading theorists necessarily reflects their scientific training but the possibilities are virtually unlimited. Some believe that alcoholism reflects characteristics that are unique to the individual. For example, a special personality structure, a particular type of behavior disorder or related form of psychopathology, or a genetic predisposition each might increase the risk for alcoholism. Others believe that more global social factors are of primary importance. Cultural influences, ethnic patterns of alcohol use, peer pressures, particular forms of social deprivation all may contribute to the risk for alcoholism.

The distinction between individual-specific and sociocultural is, of course, artificial but useful for identifying spheres of emphasis. Each fragment may contribute to the puzzle of alcoholism to a greater or lesser degree and may vary from individual to individual. Since the expression of alcoholism involves biological, biomedical, psychological and sociocultural factors, it is likely that all may contribute to alcoholism to some extent. Although fervent "true believers" in many disciplines might argue otherwise, at present there are no absolutes in alcoholism. Our understanding of the behavioral, biological and sociocultural bases of alcohol problems is a continually evolving process.

Before we consider some of the major ideas about why some people develop drinking problems and others do not, it is important to recognize that little is known about the origins of other drug abuse problems as well as most major behavior disorders. Alcoholism and alcohol abuse are not unique in posing difficulties for identifying critical antecedents. The recent history of psychiatric research reveals that it has not been possible to identify any single cause, any single internal or external deficiency, which can explain the process of most disordered behaviors. Actually, more is known about the etiology of alcoholism than about the origins of depression or schizophrenia, since at least one factor in the essential equation, i.e., alcohol, has been clearly identified.

ALCOHOLISM: NATURE OR NUTURE?

Alcoholism runs in families. One parent with an alcohol-related problem significantly increases his or her child's risk for developing alcohol abuse or alcoholism. If two parents have alcohol-related problems, the child's risk for developing similar difficulties is increased even further. But is this due to learning or to the genetic heritage transmitted from parent to child? Untangling the relative contribution of genetic variables and familial learning factors is very difficult. This issue, often referred to as the "nature or nurture" question, has complicated our understanding of many psychiatric disorders, including schizophrenia and depression.

In an effort to distinguish between the contribution of learning factors and genetic factors to the development of alcoholism, studies were done with adoptees who had been separated from their biological parents soon after birth and raised by nonrelatives, with no subsequent contact with the biological parents. Adoptees with no parental history of alcoholism were compared to adoptees with an alcoholic biological parent. The findings suggest that genetic factors may be important in determining relative risk for alcoholism.

These studies were carried out in the early 1970s in Denmark where there is a centralized national registry of adoptions, psychiatric hospitalizations and criminal records. The study involved 133 Danish men who had been separated from their biological parents within two or three weeks of birth and then adopted. When these men were approximately thirty years old, they were interviewed by psychiatrists who were unaware of their biological background. Fifty-five men had a biological parent with alcoholism and seventy-eight men did not. The occurrence of depression, anxiety, sociopathy, drug addiction and neurosis was low, and essentially the same in the two groups of men. Approximately 40 percent of both groups were heavy drinkers, but the rate of alcoholism among men with an alcoholic biological parent was almost four times that of men with no alcoholic biological parent.

The relatively low frequency of alcoholism among Danish women made it difficult to determine if genetic variables are equally important in females. However, comparison of forty-nine adopted women with alcoholic biological parents with forty-eight adopted daughters of nonalcoholics showed that 4 percent of both groups had alcoholism or serious drinking problems. This is greater than the estimated prevalence of alcoholism (between 0.1 and 1 percent) among Danish women. The investigators concluded that these findings were only suggestive because of the small sample size but alcoholism may have a partial genetic basis in women.

These studies of adoptees proved the strongest evidence that genetic factors increase the risk for alcoholism, independently of any learning factors. The adoption studies are consistent with previous studies of twins. It has long been argued that if genetic factors are important in the causation of alcoholism, two genetically similar individuals should have a similar propensity for developing alcohol problems. Twins may be *identical* (both individuals developed from the same egg following fertilization; homozygous) or *fraternal* (two different eggs were fertilized by two different sperm at the same time; heterozygous).

Homozygous or identical twins are more alike in all respects than heterozygous or fraternal twins. Identical twins share the same genetic material from both parents. Thus, if alcoholism were in part genetically determined, identical twins should be more alike with respect to presence or absence of alcohol problems than fraternal twins. A number of studies have shown that this is indeed the case. Concordance between alcohol problems in identical twins is significantly greater than concordance of alcoholism in fraternal twins.

However, it has been objected that behavioral as well as genetic factors may account for those findings. Identical twins tend to share more similar behaviors than fraternal twins. For example, identical twins often are dressed alike and they often share many of the social and cultural attributes of growth and development since they are always of the same sex. Thus, it has been argued that high concordance rates of drinking problems among identical twins may reflect a greater tendency to emulate each other's behavior than usually occurs with fraternal twins. However, studies of identical twins who have been reared apart show a high rate of concordance of alcohol problems. However, among fraternal twins who were reared apart, there was a relatively low concordance of alcohol problems.

Nature, in the form of genetic factors, does appear to influence the relative risk for developing alcohol problems. But what of "nurture"? What is the role of learning in alcoholism? The answers to these questions are complex and somewhat surprising. Parental role models do not appear to be as powerful a determinant of alcohol abuse as was once believed. Exposure to an alcoholic parent figure does not necessarily increase the risk for development of alcoholism in the children. One study showed that children raised by their alcoholic parents (usually the father) had the same rate of alcoholism (about 18 percent) as their brothers who had been adopted out and raised by nonalcoholic foster parents. Similar results were found in another study of individuals raised with either a biological or foster alcoholic parent. Only 20 percent of the nonalcoholics had a biological parent with alcoholism whereas 65 percent of the alcoholics had an alcoholic biological parent. Whatever adverse learning and experiential factors result from growing up with an alcoholic parent figure, alcoholism in the biological parent appears to predict the development of alcoholism in the children with far greater reliability than any other social and environmental factors studied.

Although many people develop alcohol problems independently of a family history of alcoholism, the evidence for the importance of genetic factors in the genesis of this complex illness is quite convincing. At present, the biological mechanisms underlying familial transmission of high risk for alcohol problems are unknown. However, since children of alcoholics are at high risk for the development of alcohol problems, it is important to try to identify other factors that may facilitate *or* protect these people from alcoholism in later life. It is not known if alcoholics with alcoholic parents differ from alcoholics without a familial history of alcoholism in terms of symptom patterns or prognosis.

Moreover, if genetic factors do contribute significantly to the development of alcoholism, independently of learning and experience, what type of biological difference is inherited? One possibility is that genetically determined differences in alcohol metabolism could affect individual reactions to alcohol and influence drinking behavior. For example, rapid metabolism of alcohol could increase drinking since intoxication levels would decline more rapidly. Alternatively, slow alcohol metabolism could decrease drinking since intoxication would persist and perhaps produce a relative "intolerance" to alcohol.

Although the processes involved in alcohol metabolism are well known and sensitive techniques for measuring rates of alcohol metabolism are available, clarification of its role in the development of alcohol problems has proved to be difficult and complex. Among the types of problems that complicate study of this issue is the control of the many factors, other than genetic factors, that influence rates of alcohol metabolism. Recall that following absorption, alcohol is uniformly distributed throughout the body water. Consequently, body weight and body structure also influence the resulting concentration of alcohol in the blood. For example, a person weighing 180 pounds will have a lower blood alcohol level than a person weighing 130 pounds after consuming the same amount of alcohol under the same conditions, because alcohol is distributed over a larger water volume. But in individuals of equal weight, a relatively obese individual may have a higher blood alcohol level than a muscular individual because lipids decrease the available water compartments for distribution of alcohol. Each of these variables must be considered in any attempt to establish the relationship between alcohol metabolism and the etiology of alcohol abuse or alcoholism.

In addition to differences in individual body structure, patterns of nutrition can also influence the rate of alcohol metabolism. A malnourished or starved individual will have a slower rate of alcohol metabolism than a well-nourished individual. Since patterns of eating and nutritional status may vary considerably between individuals as well as between cultural groups, this may limit the validity of such comparisons.

Finally, even if all the variables that may influence alcohol metabolism and the behavioral effects of alcohol could be accurately identified and controlled for, and individuals of identical body structure, nutritional status, and drinking history could be selected, one further potential difficulty remains. In order to establish unequivocally that inherited differences in alcohol metabolism affect the probability of alcohol abuse, it would be desirable to conduct *prospective studies* of the comparison groups. Ideally, children in each group would be followed over many years before drinking problems developed. Until it is possible to use variations in metabolic rate to *predict* the development of alcohol problems, it will be very difficult to distinguish between natural variations in alcohol metabolism and changes induced by chronic exposure to alcohol. As we have discussed earlier, rates of alcohol metabolism may increase slightly during sustained drinking. But after a period of sobriety, alcoholics and normal drinkers metabolize alcohol at about the same rate. Although genetic aspects of alcohol metabolism cannot satisfactorily account for a differential risk for alcoholism at present, recent rapid advances in studies of genetics hold great promise for the future.

ALCOHOLISM AND PERSONALITY

In addition to one's genetic heritage, "personality" is that unique constellation of attributes which differentiates one person from another. "Personality" is a complex construct that remains elusive despite many thoughtful psychological theories and ingenious efforts to measure and characterize it. This discussion of personality in relation to alcoholism will not address the theoretical and psychometric nuances of this construct but uses the term in its everyday commonplace sense.

The hypothesis that alcoholics differ from others along certain definable dimensions of personality is seductive. If correct, the people at risk might be identified early and helped to avoid alcohol problems. Moreover, uniform personality profiles among alcoholics might facilitate better psychotherapeutic interventions. These are not trivial considerations. The search for an "alcoholic personality" continues, despite the conflicting and inconsistent data that now exist. There is a vast literature on this topic, which can be interpreted to indicate that either very much or very little is actually known.

Yet, if there are consistent common personality features among alcohol abusers which reliably predict the development of alcohol problems, these have eluded the determined efforts of several generations of psychiatrists, psychologists and social scientists. Heterogeneity rather than homogeneity seems to best describe alcohol abusers as well as persons with many other behavior disorders. Although the notion that there is "an alcoholic personality" is an unfortunate fiction from many standpoints, it is valuable to review some of the leading hypotheses and try to appreciate their appeal and their limitations.

The "alcoholic personality" is sometimes defined in terms of specific traits. An infantile dependency, low frustration tolerance, general immaturity and orality are often ascribed to alcoholics. Sometimes the "alcoholic personality" is described in terms of dominant response patterns to stress, such as reactions of shame, guilt and anger. Among the more occult and insidious precursors of the "alcoholic personality" are severe deprivations during infancy and childhood and unresolved Oedipal conflicts.

Psychoanalytic writers often ascribe drinking problems to a failure to employ normal healthy mechanisms of defense against stress and anxiety. Some psychoanalysts believe that alcoholism may be the outcome of faulty "ego development." Some theorists even interpret the spectrum of impairments during severe intoxication as an attempt to "regress" or return to earlier or more primitive stages of personal development. Why? In order to resolve conflicts that occurred at these stages or to recapitulate problem-solving strategies that may have been effective early in life. Terms such as "regression in the service of the ego" have been invoked to explain drunken and disorderly behavior.

Consider the case for "oral dependency," another psychodynamic concept associated with alcoholism. A serious student of psychodynamic psychology would probably argue that all dependent persons have life difficulties because of unresolved oral dependent problems, which can be traced back to early interactions with parental figures. Naturally, this process can be ascertained only after the fact. Oral dependent persons do seem to have many troubles. They develop neuroses and psychoses. They suffer from depressive disorders. And they have problems with obesity. However, psychodynamic conceptualizations fail to explain why one person may drink too much whereas others may fear heights, become periodically expansive or despondent, hear voices of God or overeat. The alcoholic has often been described as an "oral dependent personality," primarily because alcohol is ingested through the mouth. Imagine the theoretical possibilities if alcohol were used socially in suppository form.

Still, the sonorous quality of these formulations imparts an authority almost independent of content. Labels and slogans, especially obscure ones, may convey an aura of knowledge that is useful in certain highly stylized, socially sanctioned forms of noncommunication. Renaming a phenomenon to clarify its origins and predict its eventual outcome is something of a tradition in many professions. For example, the doctor who informs the patient with chest pain that he or she has "angina pectoris" may help the patient feel more comfortable. The attorney who advises the client with a poor case to plead "nolo contendere" may also bring comfort. Assigning big words to important problems, perhaps with the hope that a really big word will reduce the size of the problem, is not malicious or deviant. When in doubt, most people would prefer to do (or say) something rather than just ponder the imponderable. The illusion of knowledge seems more comforting than the certainty of ignorance.

Other less theoretical personality dimensions that allegedly relate to problem drinking could be the effect of alcoholism as well as the cause and therefore appear to be of limited value. Do alcoholics consume alcohol because of low frustration tolerance, or does their tolerance for frustration change as a consequence of drinking? Do people abuse alcohol to reduce feelings of shame and guilt, or do they feel shame and guilt because of excessive drinking? In order to answer these questions, research strategies which have been called prospective studies are necessary. Prospective studies examine biological and behavioral processes *before* rather than after a specific problem has developed. An example of sophisticated studies of this type are research undertakings that are carried out to determine the role of dietary factors, smoking and environmental pollutants on health. While these studies are not perfect, their major strength is they proceed from a hypothesis testing base. Data obtained permit the hypothesis to be substantiated or refuted. With few exceptions, prospective studies have not been carried out to ascertain the role of personality factors in the causation of alcoholism and alcohol abuse.

At present, there is no strong evidence that alcoholism can be attributed to any special, unique, or circumscribed personality dimension. Indeed, there is little agreement on what is an "alcoholic personality trait" or if characteristics so labeled are the cause or effect of alcohol abuse. There is no valid way to measure hypothetical constructs such as personality traits, unconscious conflicts and motives which cannot be objectively verified. However, personality-based explanations persist in part because of their all-encompassing simplicity and fundamental untestability. Persons with a wide spectrum of personality attributes have either developed alcoholism or alcohol abuse or have been free from such problems. Moreover, there is compelling evidence that persons with severe emotional problems do not necessarily develop problems with alcohol. The scientific news in this area, such as it is, is both good and bad. The good news is that bad personality problems will not necessarily lead to problem drinking. The bad news is that problem drinking may develop in mature and psychologically healthy individuals.

Depression and Alcoholism

Many alcoholic patients report that they began drinking heavily because they were despondent. Alcoholic patients often explain continued drinking as a form of self-medication for recurrent depression and despondency. These clinical reports have an intuitive plausibility, since there are probably few individuals who have not sought solace in drinking when depressed.

It is important to distinguish between depressive illness, a pathological condition, and depressive reactions to severe stress. Some people respond to severe stress by seeking refuge in alcohol intoxication, although many others do not. Many alcoholic patients relate their drinking problems to a particular calamitous event or to bereavement. Yet life stress alone has not been shown to be a critical determinant of alcohol abuse.

However, depression, like alcoholism, is not unidimensional. Depressive disorders, like drinking problems, span a continuum from mild symptoms to severe life disruption. Some depressive disorders have a familial, probably genetic component. Depressive illness is not always characterized by despondency. The so-called bipolar depressive disorders are characterized by fluctuations between depression and mania. These patients may drink more during the manic than the despondent phase of their illness.

The interweaving of alcohol abuse and depression, and depression and alcohol abuse is far too complex to attribute primacy to either condition. Alcohol abuse and alcoholism may coexist with depression as well as other psychiatric disorders, but alcoholism no longer can be dismissed as "simply" a symptom of a major neurotic or psychotic condition. Controversy about the relationship between alcoholism and depression has profound implications for treatment. Some experts believe that 25 to 50 percent of all persons with alcohol problems have "a primary affective disorder," and therefore the underlying depression should be treated with antidepressant drugs. Although antidepressant drugs have been remarkably effective in relieving depressive illness, these drugs have not proved effective with all alcoholics. Indeed, this is not surprising, since not all alcoholics have major depressive disorders. Antidepressant pharmacotherapy has been most effective with a subgroup of alcoholic patients who also suffer from a primary depression. As diagnostic techniques are increasingly refined, accurate identification of these patients should improve. For other alcoholics, antidepressants may be both ineffective and dangerous. Heavy drinking and antidepressant medication can result in a lethal drug overdose as described in the chapter on alcohol and other drug use. Pharmacotherapy is discussed more fully in the chapter on the treatment of alcoholism.

Other expert clinicians believe that very few alcohol abusers or alcoholics have a primary depressive disorder. The disagreement continues because of the differences in groups of patients studied. For example, more depressed patients with alcoholism appear to be admitted to Veterans Administration hospitals than to proprietary hospitals where patients are responsible for payment of their own hospital bills. Moreover, the types of diagnostic instruments or assessments that are used for categorizing alcoholism and depression significantly determine the final diagnosis. Real variations among alcoholic patients and diagnostic instruments complicate the basic limitation of most studies which try to clarify the contribution of any psychiatric disorder to alcoholism. The fundamental weakness of most studies is that they are retrospective in nature. In other words, patients are seen after they have an alcohol-related problem and after they are depressed. It is difficult if not impossible to decide if alcoholism followed depression or if depression emerged as a consequence of alcoholism. One major finding to emerge from retrospective clinical studies is that in women, depression often appears to precede the development of severe alcohol problems. Yet depression occurs more frequently among women than men. This suggests that an apparent difference in depression among men and women with alcohol problems should be evaluated against differences in the general population before ascribing a unique causal significance.

The best way to determine whether or not depression usually precedes alcoholism or not is by prospective longitudinal studies. Because such studies involve following the same individuals over many years, these are difficult to conduct. However, recent findings from prospective studies suggest that alcoholism may develop first, and depression may be a consequence of problem drinking. College and

professional school students were followed by psychiatrists and social scientists for long periods of time, up to twenty years, to ascertain the relationship between emotionality and life-style changes. It
was found that development of depression before problem drinking
was far less common than severe depression following long periods of
alcoholism and alcohol abuse. These findings are consistent with the
accumulating evidence that alcohol is a poor antidepressant. As we
discussed earlier in chapter 12 on alcohol and mood, anxiety, tension
and depression may increase during drinking and increase progressively over a protracted period of intoxication.

If depression is more often the result of alcoholism than the converse, this offers hope for the despondent alcoholic. If persons with
alcohol problems would at least entertain the hypothesis that their
depression may be due to problem drinking rather than the cause of
problem drinking, their prospects for a better future might be enhanced. People who stop drinking excessively are often surprised to
find their depression is significantly relieved as well. It is ironic that
those who rely on alcohol and other drugs to relieve their periodic
despondency may actually aggravate the condition. Remission of
depression during a drug-free period is not unique to alcoholism; it
has been observed clinically in the treatment of polydrug abuse and
other drug problems. It is as though some joyless, self-denying, puri-
tanical programmer limited the extent to which the brain could sustain drug or alcohol intoxication and code it as pleasurable.
Alternatively, the momentary illusion of happiness during intoxication could be thought of as an essential vigilance system to protect
the brain from the many toxic effects of chronic inebriation. Such
analogies only underscore how little is now known about alcohol
and drug effects on the brain and affective states.

Greater certainty has been achieved in the realm of rejecting hypotheses. There does appear to be little evidence to support the notion that certain maladjustive patterns uniquely predict alcohol
abuse. There is no question that alcoholism and a variety of psycho-
pathologies may coexist. But depression does not necessarily precede
alcoholism and more often may follow serious drinking problems.

SOCIAL AND CULTURAL INFLUENCES ON ALCOHOLISM

What role, if any, does society play in increasing or decreasing the
occurrence of alcohol-related problems? As we move from the indi-

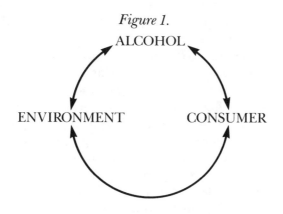

Figure 1.

vidual to the more global influences of culture and society, the difficulty in identifying alcohol-specific factors increases even further. Differences in alcohol use and abuse among groups within a culture may be greater than differences between cultures. A kaleidoscope of cultural attitudes and drinking practices has been assembled by social scientists and anthropologists — and these intricate patterns change as each viewer offers still another interpretation. Yet exploration of these myriad patterns is a fascinating process which yields still another perspective on alcohol abuse and its origins.

At the very simplest level, it is clear that the effects and availability of alcohol, the characteristics of the individual drinker, and social and cultural influences of the environment in which alcohol is consumed all interact to determine the use or abuse of alcohol. The deceptively uncomplicated schematic shown below labels the major factors and emphasizes their mutual interactions. We have already seen the many ways in which alcohol can affect the brain and behavior and we have begun to examine the diverse individual factors alleged to influence alcohol abuse. And now, with some trepidation, we focus on the social-cultural environments, which of course include all of the foregoing as well as family, religion, ethnicity, nationality and social status within that matrix of variables.

Preliterate and Transitional Cultures

It is a common assumption that preliterate societies are somehow "simpler," and that simpler life-styles, rituals and regulations make for fewer social problems. Although the "noble savage" image may no longer receive the uncritical acclaim it once commanded, the notion that society corrupts and that complex society corrupts completely retains a certain reductionistic appeal. Patterns of alcohol use and abuse have been studied in at least 139 preliterate cultures. Despite enormous variations in customs, there was a positive relationship between the frequency of drunkenness and the frequency of ceremonial drinking. This indicates that drinking in a sacred or ritual context does not necessarily inhibit drunkenness in all societies. Chronic, addictive drinking comparable to that seen in American society was extremely rare in preliterate societies, but this may reflect, in part, the limited availability of alcohol. Perceptions of drunkenness varied, but in many cultural groups, drunkenness is generally approved, expected, prepared for and anticipated with pleasure. It may even be socially required and is by no means universally disruptive.

Anthropological studies have suggested that problem drunkenness was less frequent in societies where people take special care of the physical and emotional needs of infants and children; use permissive rather than punitive methods of socialization; exert relatively little socialization pressure toward achievement and independence; tolerate dependent behavior in adulthood; engage in communal eating; and relate folk tales that tend to describe the world as essentially kind and friendly. These studies depend largely upon subjective impressions and case study analysis of the interrelationships of what are intuitively identified as significant variables, a method somewhat similar to psychoanalytic case studies in this society. Yet, the conclusions reached are very different. The finding that in preliterate societies, indulgence of dependence in infancy and childhood is associated with *less* drinking and drunkenness is not consistent with psychodynamic accounts of alcoholism.

Today, the preliterate society insulated from contemporary cultural influences has all but vanished. Studies of societies in transition from traditional customs to acculturation have provided an especially dismal chapter in the annals of alcoholism. The American Indian is one example of social, cultural and political influences on alcoholism. The frontier society valued the man who could "hold his liquor," and his image has been glamorized in most of the Grade-B Westerns enjoyed by generations of American children. All manner of villainy was usually attributed to the Indian, and folklore about his inability to tolerate distilled spirits has been preserved and embellished. The myth that the American Indian becomes extraordinarily drunk and malicious after drinking a small amount of alcohol was perpetuated because it served several social needs. Depictions of "the Indian" as unable to "hold his liquor" facilitated the macho myth of the white man's superiority. The myth also provided a rationale for believing that the Indian had an inherent degree of savagery or barbarism. White men might become drunk and rowdy, but the Indian allegedly engaged in "uncontrollable vile and brutal acts" when intoxicated. Hence, the solution was to control the Indian and especially to control the Indian's drinking behavior, a political decision that has persisted until the present day.

The sale of alcoholic beverages is legally restricted on many Indian reservations. Therefore, the Indian must procure and, in many instances, consume alcohol at long distances from the reservation. The consequence of having to drive over poor roads for about forty miles to acquire alcohol and then drive back the same distance while intoxicated is a high rate of auto fatalities associated with alcohol abuse. Thus regulations intended to restrict access to alcohol by Indians on reservations fail to curtail drinking and enhance the probability of alcohol-related highway fatalities.

Alcoholism is a major problem among American Indians and Eskimos, but this does not indicate a unique biological vulnerability. Real economic and social disadvantages combined with progressive changes in traditional mores and values are more probable contributors. It is clear that alcoholism is not equally distributed across all strata of American society. In addition to rural Indians and Eskimos, the largest proportion of problem drinkers are persons of lower economic status who live in urban areas, and a higher rate of severe drinking problems is also reported to exist among ghetto-reared black men. Longitudinal studies suggest that the same childhood patterns of early school problems, delinquency, drug use and broken homes predict adult drinking patterns for ghetto-reared blacks and whites. Although the economically disadvantaged appear to have more alcohol problems, it is important to recognize that none of these factors permit reliable prediction of the development of problem drinking.

Japan is another example of a culture in transition, with changing patterns of alcohol use and abuse. In a single generation, the usual image of the Japanese as quiet, moderate drinkers of rice wine (sake) has expanded to include highly visible hedonistic overindulgers in beer and distilled spirits, usually at corporate expense. Both patterns of alcohol use probably occur since drinking practices in Japan, as elsewhere, are varied. Still, a Western-style of recreational drinking has become more prevalent as successful corporations underwrite relaxation evenings for their executives. Recent social expectations that executives share recreational drinking and the "macho" overtones with which it is imbued may subtly encourage heavy drinking. The contemporary social value of drinking in Japan is reflected in a curious idiom. A nondrinker is called a "geko" (lower door), and the phrase used to refuse alcohol implies the refuser is "ill-mannered." The drinker is called a "jogo" (upper door) with the implication of superiority.

The increase in alcohol abuse problems among the Japanese today challenges a popular stereotype about Japanese alcohol intolerance. It has been well documented that many Orientals show an enhanced sensitivity to alcohol called the "flushing" response. Flushing (peripheral vasodilatation) is accompanied by increased heart rate, a fall in blood pressure and, in some instances, complaints of dizziness, abdominal discomfort and muscle weakness. These changes appear to be exaggerated versions of the usual peripheral physiological changes produced by alcohol, e.g., peripheral vasodilatation resulting in a sensation of skin flushing and warmth; muscular relaxation; and stimulation of gastric secretion and motility.

In one study, 83 percent of Oriental adults (Japanese, Taiwanese and Korean) showed visually discernible flushing following ingestion of a test dose of alcohol in contrast to 34 percent of Caucasian adult controls. Only two Caucasians showed alcohol-related peripheral vasodilatation comparable in magnitude to that observed in the Oriental adults, as determined by appropriate physiological measurements. In order to determine if alcohol-induced changes in peripheral blood flow might reflect learned responses rather than a genetically determined physiological reaction, alcohol-induced flushing responses were compared in Caucasian and Oriental infants. Oriental infants also showed a significantly enhanced rate of peripheral dilatation following alcohol, in contrast to Caucasian infants. Thus neither expectancy about alcohol effects nor cultural factors that affect drinking practices and determine dietary intake could account for the different effects of alcohol on peripheral blood flow in the Oriental and Caucasian infants.

Interviews with adults revealed that Orientals had more subjective complaints about the adverse effects of alcohol than did Caucasians. These complaints could be related to the differential effect of alcohol on peripheral blood flow, e.g., dizziness, irregularity of heartbeat, abdominal discomfort and muscle weakness. The extent to which these adverse physiological effects of alcohol in fact influence drinking behavior is unknown. When it was generally believed that alcoholism was less prevalent among Orientals than Caucasians, it was tempting to speculate that sensitivity to, or intolerance for, alcohol might play some protective role. Careful epidemiological studies of alcoholism in Japan remain to be done, but the increase in recreational drinking suggests that "flushing" is not a serious deterrent to alcohol use in Japan.

Considerably less is known about drinking practices in modern China. Brandy is said to be a preferred accompaniment to meals. Examination of a recent display of Chinese medicinal wines and other pharmaceuticals under labels exotic to Western eyes revealed that many contained 40 percent alcohol. Certain standard prescriptions for pains of childbirth, arthritis, rheumatism and muscle strain advised the prospective patient to "drink as much as possible." Continued use of such time-honored remedies suggests that alcohol remains a basic ingredient in Chinese pharmacotherapy.

Cultural Stereotypes of Alcoholism

Societies with well-established alcohol use patterns are often the target of cultural stereotypes. Alcohol use is almost universal and most cultural or ethnic groups include some alcohol abusers as well as some abstainers. Still, cultural, ethnic and national stereotypes of alcohol use and abuse evolve and persist, despite the conspicuous lack of data to confirm or disprove these notions. Moreover, formal theories about social and ethnic determinants of alcohol use and abuse have been constructed which rely heavily on these stereotypes. One frequently advanced premise is that in societies where moderate use of alcoholic beverages occurs regularly, abuse is uncommon, and vice versa. Italy is one of the major producers and consumers of wine, yet it is stated that Italians rarely abuse alcohol, at least as inferred from the incidence of public drunkenness. Some social scientists believe that the Italian term for drunkard, "umbriago," conveys enough disdain to exert some kind of socio-linguistic force for temperance. Conversely, France has a pattern of wine production and consumption quite similar to that of Italy, but public drunkenness is not uncommon. Do the French and the Italians really differ in the extent of alcohol abuse? No one really knows because systematic comparative studies with all of the safeguards and controls necessary for insuring adequate date collection and analysis have not been carried out.

A second common belief is that alcoholism is infrequent when drinking customs, values and sanctions are well established, known and agreed upon, and consistent with the rest of the culture. Alternatively, in groups with marked ambivalence toward alcohol and no agreed-upon ground rules, conflicts about drinking and related guilt feelings and uncertainties, may be associated with relatively high rates of alcoholism. Whether or not ambivalence about alcohol contributes to development of alcohol problems has not been determined. As with many such hypotheses it is difficult to support or refute the basic premise with adequate empirical observations. In fact, a dispassionate examination of most ideas about national, ethnic and cultural aspects of alcohol abuse reveal that these are speculations, not theories, and certainly not conclusions drawn from careful observation.

Consider for example the stereotypes about the Irish and the Jews. It is commonly believed that the Irish or people of Irish extraction are heavy drinkers and often alcoholic, whereas Jews (whatever their original nationality) tend to drink moderately and rarely abuse alcohol. Curiously, representatives of both groups regard these questionable perceptions as indications of an ethnic identity and social pride. Some who are Irish, or perhaps aspire to be Irish, ascertain that hard drinking parallels hard work and courage. Alcohol is an important ingredient in Irish hospitality, present at celebrations and funerals. Drinking cements friendships, assuages all varieties of illness, and is well tolerated by the traditional Irish Catholic Church. Some Jews who regard heavy drinking as macho foolishness stress that moderation and temperance reflect intelligence and social concern. *"Shikker iz a goy"* (drunken is a gentile) is a Yiddish song that stereotypes gentiles as destructive drunkards. The song is believed to have originated in the ghettos of Russia, and sociologists still debate the extent to which Jewish identity with sobriety arose in response to the marauding drunken peasants. Among the Orthodox, the degree of disdain for drunkenness is conveyed in the folk-saying *"A Yid a Shikker, zoll Geharget veren"* ("A Jew who's a drunkard, may he get killed"). Are these social attitudes (or their vestigial remains) in fact reflected in different drinking patterns among the Irish and the Jews? If so, which Irish and which Jews, and how do they differ from the rest of their group?

Attempts to assess cultural differences in alcohol problems face the same methodological difficulties that affect accurate assessment of rates of alcoholism generally. The social stigma still associated with alcohol abuse tends to result in underreporting and alternate diagnoses. Spontaneous changes in drinking patterns and reports of drinking problems, within only a few years, illustrate another dimension of the problem. For the most part, information about alcohol use and abuse comes from two sources: surveys and interviews. Surveys involve asking people to report their alcohol use and abuse patterns. Many consider this form of data rather weak, since most individuals are reluctant to divulge details of their personal behaviors. Also, there are certain conditions that favor over- or underreporting of drinking or drug use. Youths from subcultures with macho values may overreport drinking behavior and other drug use. In social groups where alcohol abuse is considered deviant, underreporting of both use and abuse tends to occur.

More reliable data come from interviews with alcohol abusers who provide information about their ethnic and social background and describe their patterns of use and abuse. Information volunteered by people with a verifiable alcohol problem is more likely to be relevant to the question of correlations between social factors and alcohol use. Much information about alcohol abuse and social factors has come from interviews with known alcohol abusers, but these patients may comprise a nonrepresentative or biased sample for two reasons. These alcoholics are detected or "found" cases and may or may not resemble a more general sample of cases which are as yet undetected. Also, most studies of alcoholic patients have been conducted in public institutions such as municipal hospitals or state-funded treatment facilities. Since these facilities primarily offer services to the indigent or financially comprised individual, data obtained in these settings must be interpreted with the knowledge that an economic bias exists.

Since more individuals of Irish than Jewish extraction utilize municipal and state treatment facilities, perhaps one reason for an apparent higher incidence of alcohol problems among Irish is that the Jewish alcoholics simply are not detected. There is recent evidence that indeed this may be the case. Third-party reimbursement for medical expenses, including alcohol-related medical expenses (such as Blue Cross–Blue Shield), indicates an increasing incidence of alcohol-related problems among Jews. Some theorists postulate that this is because Jews are becoming more acculturated into American society and hence drinking and drunkenness is becoming more prevalent. However, the most recent studies of Jewish alcoholics suggest that neither abandonment of Jewish identity and religion nor severe emotional problems were characteristic of these patients. Alcoholism occurs among both religious and nondevout Jews. An alternative hypothesis is that alcohol problems always existed among Jewish individuals but that such problems were never detected in medical surveys or entered on death certificates. An analogous example of such a phenomenon is the purportedly low incidence of suicide in predominantly Catholic countries, such as Italy and France. The actual rate of suicide may be higher than the official figures, because physicians are loath to write suicide on death certificates and inflict shame and degradation on the families of the deceased. Clearly, the process of case finding and reporting is a critical determinant of the reliability of findings about the relationship between social factors and alcohol use and abuse.

Within any culture, heterogeneity rather than homogeneity of drinking practices appears to predominate. Recognition that wide variations in alcohol use and abuse exist within any single cultural, national or ethnic group argues against premature labeling of deviant drinking practices as a generalized group phenomenon or assuming that certain cultural practices are likely to prevent alcohol abuse. No particular pattern of child rearing has been shown to facilitate or to protect against the development of alcohol abuse across cultures. No single cultural attitude has been shown to be universally effective in promoting or impeding alcohol abuse. Finally, there is accumulating evidence that traditional stereotypes may be less valid today as belief is supplanted by better case finding in the context of changing societies.The provocative question is why alcohol abuse occurs in many societies despite wide social and cultural differences. It is the universality of the problem across contemporary societies and across time that commands our attention. Much remains to be learned about the extent to which alcohol abuse can be modulated by social and cultural variables.

In conclusion, it is evident that no single biological or psychosocial variable can account for the origins of alcohol abuse and alcoholism. This complex behavior disorder is determined by so many interrelated factors that it probably is unrealistic to search for any one single cause. The great diversity of developmental and social experiences of individuals who develop alcohol problems suggest that efforts to unravel the origins of alcoholism may prove less productive than efforts to understand how alcoholism is maintained. Stated another way, all behavior, including excessive drinking, is presumably maintained by its consequences. Although these consequences are complex and varied, by examining the functional relationship between particular consequences and drinking behavior, it is theoretically possible to sketch a meaningful profile of alcoholism in each afflicted individual. The goal is to identify and understand the factors that determine an *ongoing* pattern of alcohol abuse, abstinence and relapse.

In the next chapter, you will meet some people with serious drinking problems and learn something about their drinking patterns and the effects of alcohol abuse on their lives. Although it is difficult to isolate and define the factors that periodically initiate and maintain excessive drinking episodes, these proximal determinants are critically important. Once identified, these "maintenance" factors could be analyzed and more effective forms of individualized treatment could be developed.

CHAPTER 15

Alcoholism: The Process in Profile

Polemics about definitions of alcohol abuse and alcoholism are a predominant theme in the clinical literature — dissonance with baroque overtones. No one questions that alcoholism exists, but competing claims for appropriate criteria are advanced with a distracting stridency. Before addressing the crucial questions "How Much" and "How Often," the chimerical Scylla and Charybdis guarding the Straits to Alcoholism defined, we will pause for a glimpse of the disease process.

Four people with drinking problems are described at a particular point in their lives. These profiles do not describe any single person, but are composite pictures of many people. Yet each one is real, abstracted from our clinical experience with many patients with alcohol-related problems. Some of the evanescent effects of alcohol on mood and behavior described in earlier chapters are illustrated in these case vignettes. The difficulties in ascribing current drinking problems to any single facet of developmental experience is also shown.

PROLOGUE: SOME PROBLEM DRINKERS

Mary Beth: The facade of urban liquor stores has achieved a certain dismal uniformity, but in suburbia, neon and wire-mesh security windows are often replaced by an American Colonial exterior. Pink and green sale bubbles disappear from the windows, and prices increase on discreetly labeled bottles. The clientele worries over the price of French wine while stocking up on scotch and gin. Customers usually appear sober and purposeful. The chronic inebriate is rarely in evidence — the affluent alcoholic can have liquor delivered. Soft lights and Muzak soothe, and the array of colored bottles is a sensory experience in itself.

Mary Beth often wondered if the proprietor of her local liquor store was aware of her drinking problem. She always tried to supplement her personal alcohol supply with the large-volume acquisitions for routine cocktail parties. But this morning she was especially anxious about her impending purchase. The holiday weekend was approaching; several cocktail parties were scheduled. Anticipation of the holiday weekend had prompted a better-than-average sale at the local liquor store. Mary Beth had persuaded her husband of the wisdom of a very large alcohol purchase, but how much would be enough? She folded the newspaper to outline the advertisement and began to read it again, trying to calculate the costs.

Mary Beth was continually surprised by her lack of ability to keep some balance between her personal liquor supply and the entertainment reserves. It was like the problem she had with her checking account. No matter how carefully she noted debits and credits, things seldom worked out as planned. More often than not she would have to borrow from the "public reserves" in order to replenish her private stock. This borrowing required another diversion of household money allocated for other purposes. Sometimes she "saved" household money for alcohol and hid it, then forgot where. She used to hide her alcohol where no one could find it, but often she couldn't find it when she wanted it.

Mary Beth examined her private alcohol supply (now consolidated in a single repository) and was comforted that it would not become critically low, at least for a while. She decided to have a small drink to celebrate a successful inventory and to prepare her for an afternoon excursion to the liquor store. Suddenly impatient, she quickly opened the aluminum seal on a "secret" half-gallon bottle of vodka and cut her finger.

Muttering to herself, she cleaned some of the blood from her finger and from the shiny rim of the cap and remembered reading somewhere that many home accidents are caused by alcohol. She poured a clear half-tumbler full of vodka. She dressed the glass with two ice cubes and added a miserly splash of tonic. After a small sip, she relaxed, set aside the newspaper, and began her usual morning ritual of listing her plans for the day. Writing a list forced her to organize — helped her to remember — made her feel more "in control." She made up her mind not to have another drink until she had accomplished at least the next three things on her list. She decided to push herself this morning, then she would deserve to enjoy the holiday liquor sale and maybe even a late lunch at the mall.

Item number three on her list was calling Laura. Mothers and daughters were supposed to become closer after the daughters had finished school and were on their own. But it seemed to Mary Beth that at least over the past year, she had become angrier with Laura and increasingly critical of her way of life.

She sucked the last bit of liquid from the tumbler and began chewing on the ice cubes. That good feeling was there again — so comforting but so brief.

Her marriage, she thought, was like her life; adequate but dull. Even empty, although she enjoyed most of the material, social and cultural luxuries she had ever desired. She thought about her parents frequently. Although she had not been excessively despondent after their deaths, now, many years later, she increasingly recalled details of their lives together. She began to have dreams that, at first subtly, and then more overtly, recapitulated long forgotten experiences with her mother and father. Fleeting images of laughter and closeness came almost unbidden.

Because of strong and almost dramatic attempts to establish her own identity and independence during early adolescence, she was surprised that she usually thought of her parents with such affection. The thread of reveries about her parents inevitably led to thoughts about her husband, Robert. She was certain that he loved her and wanted her to have a good life. Her marriage was really more than adequate; it was in many ways rich and fulfilled. But even as she reassured herself, she became overwhelmed with a sense of enormous emptiness and loss. Without knowing why, she wished that it had all been different and that she could have the power to change it all for something unknown, maybe worse, but at least different. Why pretend that her marriage wasn't achingly dull and boring when it was. The ice was melted now and she toyed with the empty glass, tracing moist rings on the counter. Thinking about this unfocused sense of emptiness, while recalling the many good qualities of her childhood and marriage, made her feel guilty and unworthy, conflicted and sad.

Then a strange metamorphosis in her feelings occurred. She could recognize and describe this now familiar change in feelings with accurate detachment. But she could not control the outcome. Her feelings of guilt and worthlessness and sadness gave way to intense anger at her family. Petty grievances with her husband and daughter were rehearsed with outraged indignation. This scenario often repeated itself after the first morning drink, the recapitulation of good fortune, fulfillment, longing, boredom, guilt, remorse. The terminal focal point of anger at her husband was often diffuse, but today she felt her feelings were justified. He had left for work early without even saying goodbye.

She licked her cut finger; it ached. The rapid flush, the warm feeling had waned. There really wasn't much to do this morning; there wasn't much to do most mornings. Lists on lists on lists — did it matter? She poured herself another half-tumbler full of vodka and added some ice and tonic.

The second drink was never really as good as the first, but at the same time, it seemed more urgently necessary. Mary Beth thought about the vodka, colorless, odorless — on the breath anyway — as she caressed the bottle.

The most wonderful thing was how quickly it worked. Maybe the kids who were using drugs had a similar experience, like the needle freaks she heard about. But she knew there was more to it than that. In fact, Mary Beth knew a number of tricks to make the good feeling arrive as quickly and as strongly as possible. One thing was to take the first drink before breakfast, never make the first drink too strong, and always mix the alcohol with a little carbonated water or tonic.

She read and reread the list of special values offered in the preholiday sale. Gin — $7.97. Vodka — $6.79. Bourbon — $8.17. She refilled her glass slowly. The newsprint smeared on her fingers. The ad became blurry as droplets of moisture fell from the glass.

Robert: Robert justified his fast, aggressive driving as the only way to avoid boredom during the monotonous commute. Actually he delighted in edging ahead, cutting in, running the light. He felt it got him into shape for his day. Robert was a successful, although not self-made, businessman of Irish-American extraction, and like many businessmen, he did a lot of business over a good drink.

Today's session was routine — the entire afternoon had been scheduled for the meeting. A table telephone was always available at the restaurant in case of an "emergency." Two phone calls were preprogrammed for Robert during the course of the luncheon to keep up appearances. In fact, everyone at the meeting received two, but usually not more than two, phone calls. To the extent that urgent phone calls conveyed importance, a mutual agreement about status had been achieved without formal negotiation. However, the number of drinks consumed during lunch was negotiable. Robert beamed at the waiter in the first flush of pre-business cordiality, then ordered a round of drinks. Cohen refused at first, then gave in — ostensibly to prove he was acculturated. Parsons seemed detached, indifferent to the requisite pre-lunch ritual.

Robert hoped that Cohen wouldn't drink too much. It made him nasty. Not nasty-nasty, but nasty in a shrewd, aggressive manner. Cohen was bright and doing business with him was fine as long as you dealt with his intelligent side. Cohen's belligerence was another matter. Robert could control his own temper except when he was drinking heavily, and he didn't want to be provoked or hassled this afternoon.

The second and third martini did little to change Robert's outward appearance. But as he drank more, he became less concerned about business matters and distracted by other things happening in his life. He really should do something about the strange, hostile contest that had been going on between himself, his wife and daughter. They were all so different, yet they were becoming more alike through some invisible link that was drawing them together and toward more frequent and savage combat. He couldn't understand it. They had their own lives, why couldn't they leave each other alone? He ruminated sadly while making appreciative noises in response to Cohen's sallies. Cohen was becoming expansive.

Parsons had withdrawn into his strong silent tycoon pose. Robert felt too distracted to mobilize Parsons and outmaneuver Cohen. He felt Cohen watching him, waiting, just waiting for a false move. He knew he should conclude the meeting as soon as possible. When the second prearranged phone call came from his secretary, he hurriedly excused himself to deal with an urgent problem at the office.

But Robert didn't return to the office. Instead, he went to a nearby bar. Checking out local taverns and lounges after a long business lunch had become something of a routine. Robert enjoyed visiting different kinds of bars and fancied himself as something of a sociologist. He would sit, quietly drinking, and survey the people about him in an almost mechanical manner. During this time, he would construct elaborate fantasies, enjoying the posture of the detached observer. As the afternoon passed and drinking continued, however, the vividness and coherence of his fantasies changed and his thinking became more and more convoluted.

There was a system in his drinking. His goal was to prolong the flow of fantasy and keep it as rich as possible. Drinking too little clogged his thinking, drinking too much made associations too chaotic and sometimes frightening. It was a narrow line, and Robert knew it was essential to pace his drinks.

Maintaining control of his fantasies while drinking could become almost an end in itself, although as the afternoon wore on, Robert found this harder and harder to achieve and began to take pleasure in abandoning control. Often, he was unaccountably distracted from his fantasies as once again concerns about his family, his work began to intrude insidiously. The alcohol had stopped working — he began to feel heavy, angry, morose. The bar was quiet now — tables were being cleared. He forced himself to watch the football replays on the seven o'clock news and prepared himself for the long drive home. Plenty of time to plan tonight's excuse for being late.

Laura resented her mother's usual midday call, which invariably came immediately before her lunch break. As often as not, the call provoked anger, guilt and self-recrimination that could last the rest of the afternoon. On the other hand, the absence of the daily call also made Laura uneasy and apprehensive.

Laura was still uneasy that evening as she set the table for dinner. When sober, her mother was always the paragon of self-righteous propriety. She would speak with high praise about her daughter and husband, and almost everyone else, remonstrating only about her own petty failures. And when sober, her father seemed to be the opposite — bullying and impatient.

Her parents seemed to realize what they were doing, but apparently were helpless to change. At some point, when Laura was still young, both had learned that alcohol could change the ways in which they felt about themselves and others. Intoxication seemed to free them to relate in more varied, less constrained ways. But the change was so transient. Mother's relaxation and openness would soon degenerate into surly anger and vindictiveness. Father's flexibility and ability to express warmth would degenerate into uncritical acceptance of everything, no matter how mediocre, and finally he would become totally enmeshed in almost autistic self-pity.

And what was the product of these two tortured and alcohol-manipulated souls? A daughter, who herself was more frequently turning to alcohol in order to deal with rage and frustration. Perhaps the thing that bound her parents together and she to them was some invisible thread having something to do with drinking. It was so paradoxical because at one time Laura would not even consider drinking when she was in trouble. She recalled how isolated she felt as a small child when her parents were both drunk. She vowed that she would never become like them. Yet, although she had never really been intoxicated, she felt she was slipping toward an insidious drinking problem of her own, particularly when drinking with Bill. She remembered her grandfather, proud of his heavy drinking, once boasting that good drinking ran in the family.

Laura realized she was obsessed with something that was not really a problem, or was it? Why, after reacting with shame and rage to her mother's phone call the day before, had she immediately gone out and had two drinks? Was she imitating her mother in order to get even with her? Did she envy her mother's ability to let go and become a bitch? Had she inherited something she didn't want? Had they all inherited something?

Bill: After graduation from high school, Bill could have attended a good college; in fact, the vigor that his guidance counselor applied to affirmative action programs would have made it possible for him to attend a first-rate Ivy League school. Instead, he enlisted in the army.

It was not so much that Bill felt he needed the structure and discipline of the army for himself. His sense of security came from knowing that all the others in the army were bound by the same code and discipline that regulated his life. Uncomfortable as this regimentation might be at times, he felt the system protected him and insured his chances for promotion. And the system did work well for Bill. Within three years after enlistment he had been recommended for and completed officer's candidate training. At the time of his third reenlistment he had earned a Bachelor of Science Degree and had achieved the rank of lieutenant. At thirty, Bill was a graduate of a good college and a captain. In some ways this was a fantastic accomplishment for a ghetto-reared black man. On the other hand, giving orders to surly youths, white and black, made him nervous. Taking orders from suspicious and covertly angry superiors was frightening. Although Bill was never formally excluded from the military's social fringe benefits, he never felt wholeheartedly accepted.

However, there was one activity that was inbued with a general sense of well-being — the Friday afternoon happy hour. Bill recalled the relaxation and camaraderie of heavy drinking in ghetto bars and taverns when paychecks were cashed late Friday afternoon. At these times some of the feelings of oppression appeared to be lifted, if only temporarily. The same feeling occurred when Bill became mildly intoxicated on Friday afternoon. But he soon recognized that one man's relief from oppression might be another's escape from boredom. The Friday afternoon happy hour seemed to become longer throughout the succeeding weeks and months, starting earlier and ending later. Then they escaped the old bounds of the Friday-only tradition; a happy hour was convened at any time for virtually any excuse — good news, bad news, or no news. At the time of Bill's third reenlistment, he was becoming heavily intoxicated at least three or four times a week.

Bill met Laura at a Wednesday afternoon happy hour convened by officers' wives to kick off a community fund campaign. Laura was invited to represent a social agency for rehabilitating delinquent youth. At first Bill feared his heavy drinking might offend Laura, but they formed an easy relationship and at the end of six months were living together. Although Bill continued to attend happy hour functions, Laura was not included and they shared little social interaction with other people. At first, Bill left the happy hour parties early, but as time passed, he began to continue drinking after he returned home for dinner with Laura.

Although his heavy drinking caused no impairment in his work or his relationship with Laura, Bill was aware that something was wrong. On several occasions he tried to solicit some comment and perhaps even a reprimand about his heavy drinking from Laura. Laura was uneasy about this and would try to direct their conversation toward some other subject. Occasionally she did acknowledge that Bill's drinking might be somewhat excessive but quickly contrasted the control that Bill had over his drinking to the loss of control that her mother had often shown. Laura bitterly told Bill that both of her parents were probably alcoholics and that his drinking and behavior when intoxicated were neither self-destructive nor damaging to others. If Bill persisted in being critical about his drinking, Laura would seductively suggest that his drinking might be more beneficial than harmful. She pointed out that their lovemaking was more satisfactory when Bill was intoxicated. And while Bill knew he might be more amorous when drunk, his own degree of pleasure at such times was blunted. What seemed most paradoxical to Bill was that although Laura condemned her parents for irresponsible drinking, she not only was permissive about his drinking, but subtly encouraged it.

When Bill was thirty-five he was promoted to lieutenant colonel and commanding officer of an army nuclear missile unit. His military performance was outstanding, but his personal relationship with Laura had gradually deteriorated. By then they were both drinking frequently and heavily. Their sexual relationship was poor. Fights were frequent. Laura became belligerent when drunk and Bill responded to her outbursts of verbal abuse by drinking more, regressing from passive indulgence to oblivion, stupor and sleep. The final separation occurred shortly after Laura's mother attempted suicide. Laura had accused Bill of ruining her life and causing the estrangement she felt between herself and her parents.

Bill readily accepted an assignment as army technical consultant with the Japanese government. During his first year in Tokyo, he had relatively little to do and he began to drink more and to drink more often. He had developed an enormous tolerance for alcohol and now found that consumption of a quart of vodka per day was not only routine but necessary. Happy hour drinking was increasingly less satisfying, and Bill began to drink alone. Soon after he became intoxicated, he would feel more relaxed and happier, but this would quickly fade into a morass of dejection and self-depreciation. He began to have frequent episodes of severe despondency while drinking and felt alternatively overwhelmed by great anger or deep depression. He began to have difficulty sleeping and morning nausea, vomiting and shakes could only be suppressed or averted by heavy morning drinking.

Bill was surprised by his own ability to function as well as he did after drinking almost a pint of vodka each morning. Most of his staff and colleagues overlooked his occasional unsteadiness or confusion and even the junior medical officer was only mildly concerned about the finding of some liver damage in Bill's annual physical examination. Bill was advised to cut down on his drinking, but there was no social or medical pressure to change his way of life. Gradually, Bill dismissed the idea of requesting reassignment and hoped he would be able to remain overseas indefinitely.

At age thirty-six, after eighteen years of military service, Bill was retired from the army with a medical disability. He was told that his service record in no way prompted this decision, but that his liver disease precluded his adequately carrying out military responsibilities.

COMMENT

Prognosis is the medical term for probable outcome of a disease, i.e., "the prospect as to recovery from a disease as indicated by the nature and symptoms of the case." Disregarding for the moment the particular symptoms, most would agree on a general rank order of severity of drinking problems. Based only on overall impressions, Bill's life was most seriously disrupted by alcoholism. Robert drank heavily, often alone, in a way that let him escape from real and imagined work pressures and into a realm of fantasy and introspection. Despite this he was able to keep his business and his marriage relatively intact. Mary Beth drank heavily but was able to keep up her home, shop, entertain, and maintain her marriage while recognizing this dichotomy. Laura seemed just on the threshold of problem drinking in response to frustrations, but most of her drinking was social, in the context of meals.

Now consider the actual symptoms.

- Drinking daily (Mary Beth, Robert and Bill)
- Drinking alone (Mary Beth, Robert, Bill and Laura)

- Drinking to intoxication (Mary Beth, Robert and Bill)
- Drinking in the morning (Mary Beth and Bill)
- Sequestering private alcohol reserves (Mary Beth)
- Depression associated with alcohol intoxication (Mary Beth, Robert and Bill).
- Withdrawal signs (shakes, nausea and vomiting) in the morning (Bill)
- Liver damage secondary to alcohol abuse (Bill)
- Memory dysfunction associated with alcohol abuse (Mary Beth)

Even without other background information (years of alcohol misuse, family drinking history, developmental history, medical history), can we hazard some predictions as to prognosis on the basis of these symptom clusters? Counting the number of symptoms is not necessarily a good predictor, since some of these symptoms have more serious medical implications than others. Liver damage, for example, is not reversible, and the capacity of the remaining healthy portions of the liver to provide adequate function depends entirely on the extent of the damage. Memory dysfunction may or may not be reversible depending on the adequacy of nutritional status and many other factors. Withdrawal signs are reversible but they indicate a physiological dependence on alcohol and are one of the definitive criteria for alcoholism.

Still, it is possible that each of these patients could resolve his or her drinking problems and resume a normal healthy life. For some, complete abstinence from alcohol might be necessary. Very rarely, others could enjoy occasional social drinking without recurrence of alcohol abuse. Alternatively, each of these patients might continue addictive drinking, with progressive deterioration of their health, work and family relationships. Death could result from the severest forms of alcoholism. No alternative was an absolute certainty.

EPILOGUE: TEN YEARS LATER

Case Report: Mary Beth

Drinking increased progressively over a four-year period; physical dependence on alcohol developed, as evidenced by recurrent episodes of tremulousness and hallucinosis; patient initiated and terminated individual psychotherapy, group psychotherapy, and membership in Alcoholics Anonymous; one suicide attempt; hospitalized three items for anemia and cardiac disease secondary to alcoholism; suddenly and spontaneously terminated all alcohol use and has been abstinent for over five years; divorced husband and remarried with reasonably good new marital adjustment; current state of health, good; current interpersonal adjustment, good.

Case Report: Robert

Divorced; currently involved in long-lasting relationship with a woman who has had multiple alcohol and drug abuse problems; business is flourishing and personal wealth has increased considerably; drinking behavior remains essentially unchanged; physical health good with exception of mild withdrawal signs and symptoms during protracted abstinence (several days to several weeks); has been involved in individual psychotherapy (three times per week) for the past four years; believes he has made significant progress in understanding his motivations and behavior but continues to drink heavily.

Case Report: Laura

Married and living in a small Midwestern community; working actively as social worker at a local hospital; not interacting with parents except for occasional phone call during past five years; terminated two-year course of psychotherapy (with concurrence of therapist); has experienced much less anxiety, guilt and despondency; occasional (once or twice yearly) episodes of binge drinking; not drinking daily, use of alcohol in most social settings is moderate and appropriate.

Death Certificate: Bill

Deceased, age forty-eight; cause of death — Laennec's cirrhosis, bleeding esophageal varices, portal hypertension, ascites, terminal hepatic coma and cardiac failure; time from onset of alcohol dependence to death was fourteen years.

CHAPTER 16

Alcoholism and Polydrug Use Examined

DEFINITIONS AND DISEASES

What is alcoholism? This question has occupied hundreds of people for countless hours. Many scientific societies, government agencies and international organizations have convened task force after task force to define alcoholism and develop criteria for its diagnosis. To date, no universal consensus has been achieved, and the discussion continues. Although most would agree that alcoholism involves repetitive, excessive drinking that results in injury to an individual's health, adequate social function or both, there is dissension about the criteria for each component of this definition.

One recurrent and central point of controversy is whether or not alcoholism is a disease. The answer is important because it determines how the alcoholic is treated by physicians, by health insurers, and by society at large. In order to understand some of the emotionalism surrounding this issue, it is useful to examine our assumptions about disease.

Cancer, multiple sclerosis, allergies, Huntington's chorea, diabetes, schizophrenia, poliomyelitis, arthritis, emphysema, pre-senile dementia, appendicitis, and disorders of the kidney, the heart, the gallbladder are all diseases. The pathogenesis and nature of these common afflictions varies greatly, but no one would argue that arthritis, for example, is not a disease. What, then, distinguishes arthritis or heart disease from alcoholism? Certainly not the potential seriousness of the medical consequences. Arthritis can be painful and incapacitating, but heart disease and alcoholism can also kill, so lethality is not the distinguishing criterion.

The distinction comes not from medicine, but from a seldom articulated moral and social perspective. Alcoholism belongs to that special category of self-imposed disorders once attributed to deficiencies of character, moral laxity, and self-indulgent hedonism. The condition of alcoholism is more likely to evoke condemnation than sympathy. "Drunken bum," "lush," "booze hound," "souse" are some contemporary derogatory descriptions of persons who indulge in the sin of inebriety. The long tradition of equating intemperance with sin was more fully discussed in earlier chapters on drinking in America. But echoes of these early attitudes still reverberate today as contemporary society weaves an erratic course between punitive disposition and grudging dispensation of insurance-compensated treatment for alcoholism.

As more diseases are shown to have a self-imposed component, it will be interesting to see if these pejorative attitudes toward alcoholism persist. Or if other afflictions will become the subject of equally scurrilous attack. For example, will obese smokers with heart disease be castigated by a society newly obsessed with a morality of health? Or perhaps the emerging subspecialty of sports medicine, which now enjoys general approval, will be disdained, as all forms of "do-it-yourself" disorders become designated as a menace to society and the public purse.

These are absurdities perhaps, but social attitudes toward alcoholism have often been irrational. On the subject of diseases, medicine is much more charitable in its formally articulated admission policies. A leading medical dictionary defines disease as a definite morbid process having a characteristic train of symptoms; it may affect the whole body or any of its parts, and its etiology, pathology and prognosis may be known or unknown. Physicians treated the medical consequences of alcoholism (e.g., liver disease, esophageal varices, peripheral neuropathies), but until quite recently tended not to treat alcoholism. Some psychiatrists tried to treat alcoholism but usually with the stipulation that the alcoholic must be sober during the interview. This is somewhat analogous to refusing to treat a diabetic or cardiac patient when he or she is in relapse.

It was not until the mid 1960s that enough physicians and scientists agreed that alcoholism is a disease to persuade the public and the Congress of this view. In 1967, a small federal program called the National Center for the Prevention and Control of Alcoholism was established as part of the National Institutes of Mental Health (NIMH), with authority to conduct research on alcoholism. In 1970, Congress passed new legislation establishing a larger bureaucracy, the National Institute on Alcohol Abuse and Alcoholism (NIAAA), which was independent of the NIMH and was concerned with both research and treatment. Once resources became available, scientists began to study alcoholism as a biomedical disorder, and what might justifiably be called the dark ages of alcoholism came to an end. In very practical terms, that's what it means to be designated as a bona fide disease instead of a moral aberration. Once alcoholism was legitimized as a respectable area of inquiry, a broad spectrum of biomedical, behavioral and social research began. The few scientists who had braved the dark ages were gradually joined by colleagues in many disciplines. Federal research and training grants related to alcoholism increased from twenty-six in 1966 to ninety-two in 1983. Most of the new knowledge about alcoholism described in this book has been acquired in these few years.

There are still some who object to the notion that alcoholism is a disease. Opposition based on more sophisticated variations on the moral laxity theme still occurs. Others argue that the medical model is "too restrictive" and does not take into account the many social and psychological variables that may contribute to alcoholism. This argument is based on a misunderstanding of the basic assumptions of the disease model of alcoholism.

One essential element of the disease model of alcoholism is that its expression depends on an interaction between the drinker, the agent of the disease (alcohol) and the environment. Thus alcoholism resembles infectious disease in which host-resistance factors and environmental variables may be more important than the existence or even the virulence of the infectious agent. Most diseases with specifiable agents depend on the interrelationship between these three factors, rather than any specific factor alone. But perhaps the element most essential to defining alcoholism as a disease is the concept of addiction.

ALCOHOL ADDICTION

The process by which concepts about alcoholism evolved from a quagmire of moral turpitude to the status of medical illness is in itself a fascinating story. The transformation of alcoholism from depravity to disease began with a clinical report by two eminent neurologists, Drs. Victor and Adams, in 1953. Their careful observations of 226 alcoholic men admitted to the Boston City Hospital culminated in the first systematic description of the alcohol withdrawal syndrome. Within six to eight hours after their last drink, alcoholics develop a series of signs (observable by others) and symptoms (reported by the patient), which usually become most severe within twenty-four hours, then gradually diminish over the next forty-eight to seventy-two hours. The major signs include tremor of the arms and hands, and sometimes the tongue and torso, sweating, a flushed face, a mild increase in heart rate (tachycardia), rapid involuntary movements of the eye (nystagmus), hyperactive reflexes, nausea and vomiting. Patients often report some disorientation, nervousness, insomnia, nightmares and occasionally hallucinations. These signs and symptoms may remit spontaneously, but a number of drugs (chlordiazepoxide, paraldehyde) have been shown to make the patient more comfortable.

The withdrawal syndrome that defines physiological dependence on alcohol differs both in severity and duration from the hangover after occasional overindulgence. The headache, stomach discomfort and mild anxiety that characterize the hangover usually disappear within a few hours. Some scientists believe that the hangover is a mini-withdrawal syndrome, but this has not been conclusively demonstrated.

Victor and Adams noted that the common alcohol withdrawal syndrome has two variants that occur far less frequently. One variant is alcoholic epilepsy or "Rum Fits" in which grand mal seizures occur in addition to the aforementioned signs and symptoms. A second variant is delirium tremens, which is the most severe form of alcohol withdrawal and occasionally results in death. Delirium tremens develops in about 5 percent of alcoholics in withdrawal and in about 30 percent of those with "Rum Fits." In popular parlance the term "delirium tremens" is often used incorrectly to refer to all complications during alcohol withdrawal. But in fact delirium tremens usually develops well after the other two withdrawal syndromes, within three to five days after drinking stopped. It is characterized by profound confusion and disorientation, vivid hallucinations, and fever as well as tremor, agitation, sweating and tachycardia. Mortality from delirium tremens has decreased from an estimated 15 percent to about 1 percent as medical management has improved.

Although much more systematic than prior reports, the observations of Victor and Adams in 1953 were not entirely new. An association between cessation of heavy drinking, tremor and sometimes delirium had been reported in the medical literature between 1978 and 1813. Centuries earlier, Hippocrates was credited with the observation that "If the patient be in the prime of life, and . . . if from drinking he has trembling hands, it may be well to announce beforehand, either delirium or convulsion." What was new was the neurologists' interpretation of this cluster of signs and symptoms. Victor and Adams believed that these syndromes were the direct effect of alcohol and could not be attributed to any associated condition such as poor nutrition, metabolic or infectious disease so often seen in alcoholics. To quote, "It is difficult to escape the conclusion that the clinical states under discussion depend for their production, not only upon the effects of prolonged exposure to alcohol, but temporally on abstinence from the drug."

This idea that after habitual alcohol intoxication, the cessation of drinking could, in and of itself, produce an alcohol withdrawal syndrome was startling and not entirely accepted until more than a decade later. Other drugs such as opiates and barbiturates had long been known to produce specific withdrawal syndromes. But somehow, alcohol was not thought of as a "drug" in the same sense. Some vague amalgam of social propriety and sanction urged a distinction between alcohol and addicting drugs. Like it or not, alcohol is, and always has been, a drug and an intoxicant with specifiable pharmacological properties. Moreover, it is addicting, as evidenced by the fact that withdrawal signs and symptoms occur after repeated severe or prolonged intoxication.

In the early 1950s many remained skeptical and continued to argue that infections and poor eating habits, not alcohol, were responsible for these syndromes. Two studies, almost a decade apart, demonstrated unequivocally that the syndrome was caused by alcohol withdrawal and not by other extraneous factors. In 1955, it was shown that healthy, well-nourished men with a past history of morphine addiction developed an alcohol withdrawal syndrome after drinking nine to sixteen ounces of alcohol each day for between forty-eight and eighty-seven days. Still, skeptics could object that perhaps the prior condition of morphine addiction contributed in some way. It was not until 1964 that a comparable study was conducted in healthy alcoholic volunteers who had been detained for public drunkenness at the Bridgewater Correctional Facility in Massachusetts. These men ate well and drank up to forty ounces of alcohol a day for twenty-four days. When they stopped drinking, they developed tremor, sweating, nystagmus, tachycardia, hyperreflexia and vomiting within twelve to twenty-four hours. The alcohol withdrawal syndrome described clinically in 1953 was thus finally experimentally verified by clinical research on alcoholic men in 1964. No one objected that it might be something unique to Massachusetts.

Still, the idea that alcohol was an addictive drug encountered opposition at many levels. By the early 1970s, several laboratories had succeeded in producing an animal model of alcoholism. Monkeys, rats, mice, dogs all showed alcohol withdrawal signs once alcohol, given orally, intravenously and intragastrically by a variety of ingenious techniques, was removed. These early findings, now confirmed repeatedly, showed without serious doubt that alcohol was addicting. Through prolonged, high dose exposure, the animal or person became physically dependent on alcohol, i.e., withdrawal of alcohol produced a characteristic abstinence syndrome.

Clearly depravity was not at issue in these clinical and basic studies. Psychological, social and cultural factors could not reasonably be invoked to "explain" alcohol addiction in the rodent or the monkey. Moreover, behavioral studies showed that these animals would work at an operant task to give themselves alcohol. Not only did alcohol produce the primary pharmacological criterion of addiction, physical dependence, but the primary behavioral criterion, continued self-administration and self-intoxication as well.

The demonstration that alcoholism was a form of addiction was pivotal for the disease concept of alcoholism. Now scientists could begin to ask serious questions about when and how and why alcoholism affected the brain so as to produce physical dependence.

Thus, in the quest to define alcoholism, it is important to recognize that the alcohol withdrawal syndrome is unambiguous evidence of alcoholism. However, it does denote the end stage in a long and often insidious process. Since alcohol withdrawal is the last point on a continuum from sporadic drinking problems to frequent heavy drinking to an addictive drinking pattern, it cannot aid in early detection of drinking problems.

Alcohol Tolerance

Increased tolerance for alcohol is a second pharmacological criterion of addiction. Tolerance means that the drinker must drink progressively more to achieve the changes in mood and behavior once produced by small doses of alcohol. This attenuated response to small amounts of alcohol is one factor that favors heavy drinking. However, tolerance is difficult to quantify and is more ambiguous than physical dependence, defined by withdrawal signs and symptoms.

Tolerance for alcohol differs from tolerance that develops to many other drugs in that the absolute amount of alcohol that can be consumed does not change. The alcoholic rarely develops blood alcohol levels above 450 mg/dl. Respiratory depression and death can occur at blood alcohol concentrations above 500 to 600 mg/dl. In contrast, barbiturate addicts can tolerate twenty to thirty times the usual hypnotic dose and opiate addicts can take 240 mg of morphine, over twenty-four times the usual therapeutic dose for relief of pain. Alcohol addicts also develop tolerance to many toxic alcohols such as methanol, and can drink these in amounts that would be fatal for others.

The fact that the alcoholic can tolerate blood alcohol levels in the 200 to 400 mg/dl range dramatically differentiates him or her from occasional social drinkers. Alcoholics may show minimal behavioral signs of intoxication at these blood alcohol levels, whereas social drinkers may be severely intoxicated at blood alcohol levels above 100 mg/dl. Since alcohol-induced changes in the rate of alcohol metabolism are not sufficient to account for tolerance, some adaptive process must occur in the brain. The nature of the brain adaptation is unknown, and as we have seen in the chapter on alcohol and the brain, even the brain mechanisms underlying intoxication are unclear.

Another interesting aspect of alcohol tolerance is that alcoholics, during sobriety, may metabolize some other drugs more rapidly. For example, alcoholics may require larger doses of anesthesia in surgery than nonalcoholics. This phenomenon, called cross-tolerance, has also been shown for alcohol and barbiturates, hypnotics and sedatives, but not for opiates. Cross-tolerance during sobriety also suggests that the alcoholic's brain has undergone some adaptive changes which persist. Intoxicated alcoholics may metabolize other drugs more slowly, since the detoxification capabilities of the liver are not infinite.

CAN ALCOHOLISM BE QUANTIFIED?

We now return to the questions "How much?" and "How often?," the whirlpool into which the most definitional forays have been drawn, usually to flounder perilously at the vortex rim. Quantification is seductive, and many have tried to develop a meaningful quantity-frequency index to define addictive drinking. Given a specified quantity of alcohol, the extent to which drinking is concentrated or spaced reflects very different patterns of drinking. Fourteen ounces consumed over seven days has far different implications than fourteen ounces consumed on the weekend. And is two or three ounces of alcohol a day too much?

Society tends to define alcohol abuse in rather personal terms, e.g., anyone who drinks more than I do. Since standards for acceptable drinking vary so greatly from state to state and country to country, it is impossible to achieve consensus about what deviant drinking is. Among religious Mormons, any alcohol is too much. The Moslem religion has always disdained alcohol, a prohibition enforced more vigorously under Ayatolla Khomeini, who also banned music. On the other hand, cocktails and wine with dinner is expected among some groups, and viewed with suspicion by others. Rural drinking patterns may differ from urban drinking patterns. One or two six packs of beer while watching the game is common for some, offensive to others.

Granted that it may be difficult to arrive at a socially sanctioned alcohol quantity-frequency index, perhaps it is possible to determine how much alcohol is necessary for alcoholism to develop. This seemingly simple question has also been difficult to answer. One approach has been to try to reconstruct the alcohol addiction process by asking alcoholic men how much they customarily drank and when they first developed alcohol withdrawal signs and symptoms. One hundred and twenty-nine alcoholic men volunteered for clinical research on alcoholism at St. Elizabeth's Hospital in Washington, D.C., over a four-year period (1969–1972). These men were asked to recall the details of their drinking history. Such self-report data are often unreliable because people forget or tend to over- or underreport their drinking. Despite these limitations, it appears that the amount of alcohol consumed did not predict how soon physical dependence on alcohol developed. Usually a minimum of at least two years of heavy drinking preceded the first episode of alcohol withdrawal but for most (72 percent of the sample studied) at least five years passed before an alcohol withdrawal syndrome developed. How much was enough? As little as one pint of whiskey, one quart of beer or four-fifths of wine per day. Most whiskey drinkers reported drinking four-fifths to one quart per day.

There is a folk belief that low alcohol concentration beverages like beer and wine "protect" against alcoholism. However, the critical ingredient is alcohol, and addiction can occur if enough alcohol is consumed in any beverage form. Intoxication is equivalent after drinking wine, beer and whiskey if the volumes are adjusted to yield equal amounts of alcohol.

When alcoholic men were studied on a clinical research ward over the course of a single heavy drinking episode, it appeared that the pattern of drinking was a more important factor in accounting for the severity of the withdrawal syndrome than the absolute amount of alcohol consumed. Men given a standard amount of alcohol every four hours over twelve days had far less severe withdrawal signs and symptoms than men who drank equivalent amounts whenever they wished. Steady drinking was rare among these men. Some drank heavily for several hours or days, then stopped, then resumed drinking in an erratic oscillating pattern. Some mild withdrawal signs (tremor, sweating) often occurred as blood alcohol levels fell (e.g., from 300 to 100 mg/dl). This suggests that a zero blood alcohol level is not required for a withdrawal syndrome to occur. A rapid decline in the level of alcohol circulating in the blood may be sufficient.

Also surprising was how well these alcoholic men tolerated withdrawal signs and symptoms. In the popular imagination, alcohol withdrawal (not to be confused with delirium tremens) is an awesome specter to be avoided at all costs. To test the validity of this notion, in the early 1970s, alcoholic volunteers were studied on a clinical research ward where they could determine when and how often they drank. Alcoholic men allowed to drink for thirty days or more tended to stop drinking every two or three days, even though this meant that withdrawal symptoms occurred. They had to work at a very simple task to earn alcohol and they tended to earn enough for a three- to four-day drinking eipsode, then become sober for a day or two. Although they could easily perform the task during drinking, they chose not to and continued to alternate between periods of heavy intoxification and sobriety. Such findings argue against the notion that alcoholics drink to avoid withdrawal. That explanation is further limited by the fact that heavy drinking necessarily preceded physical dependence and the withdrawal syndrome. Moreover, it is difficult to see how avoidance of withdrawal could account for resumption of drinking after a period of sobriety.

ALCOHOLISM REEXAMINED

Studies of alcoholics before, during and after alcohol intoxication have changed many of our preconceived notions about this disorder. A more rational consensus about what alcoholism is and what it is not is gradually evolving. One of the simplest definitions is perhaps one of the best —

Alcohol abuse is drinking that produces harm to an individual's health, his or her interpersonal interactions or both. Alcoholism is alcohol dependence manifested by increasing tolerance for behavioral changes produced by alcohol through time and by the appearance of withdrawal signs and symptoms following abrupt cessation of alcohol use. There are great variations in the expression of alcohol abuse and alcohol dependence with respect to both temporal sequences and degree of severity. The amount of alcohol that interferes with health and social functioning also may vary between individuals.

In 1980, the American Psychiatric Association revised its criteria for the diagnosis of alcoholism and attempted to provide clearer guidelines to help physicians recognize alcohol problems in their patients. The APA diagnostic criteria for alcohol abuse and alcoholism emphasized a pathological drinking pattern and evidence of social or occupational impairments attributable to alcohol abuse. The APA criteria are:

A PATTERN OF PATHOLOGICAL ALCOHOL USE DEFINED AS:

- *need for daily use of alcohol for adequate functioning*
- *inability to cut down or stop drinking*
- *repeated efforts to control or reduce excess drinking by "going on the wagon" (periods of temporary abstinence) or restricting drinking to certain times of the day*
- *binges (remaining intoxicated throughout the day for at least two days)*
- *occasional consumption of a fifth of spirits (or its equivalent in beer or wine)*
- *amnesic periods for events occurring while intoxicated (black-outs)*
- *continuation of drinking despite a serious physical disorder that the individual knows is exacerbated by alcohol use*
- *drinking of nonbeverage alcohol (such as antifreeze and shaving lotion)*

IMPAIRMENT IN SOCIAL OR OCCUPATIONAL FUNCTIONING DUE TO ALCOHOL USE:

- *violence while intoxicated*
- *absence from work*
- *loss of job*
- *legal difficulties (e.g., arrest for intoxicated behavior, traffic accidents while intoxicated)*
- *arguments or difficulties with family or friends because of excessive alcohol use*

Alcohol abuse is inferred if these disturbances have persisted for "at least one month." Alcoholism is distinguished from alcohol abuse by tolerance and physical dependence. The APA defines these as follows:

Tolerance: "need for markedly increased amounts of alcohol to achieve the desired effects, or markedly diminished effect with regular use of the same amount."

Withdrawal: "development of Alcohol Withdrawal (e.g., morning shakes and malaise relieved by drinking) after cessation of, or reduction in drinking."

Conspicuously absent from the APA criteria, as well as the guidelines published by the National Council on Alcoholism, is any mention of "craving." The notion that alcohol exerts an almost mystical force over the drinker that compels "drinking to oblivion" and entices the alcoholic to the bottle during sobriety has been enshrined in the alcohol literature for decades. From a logical standpoint, "craving" and "loss of control" are not very useful constructs since each is defined by the behavior it is invoked to explain. From an empirical standpoint, observations of alcoholics during intoxication have not confirmed "craving." As we have seen, alcoholics can turn off drinking for hours and days, despite mild withdrawal signs and symptoms. Although much remains to be learned about the determinants of alcohol abuse, it is unlikely that fundamentally untestable notions like "craving" will survive another decade of objective research. Possession by demons has long been expunged from serious consideration in Western medicine. Concepts like "craving" may soon follow into the realm of anecdote and superstition.

ALCOHOLISM: HOW MANY AND WHO

The most recent federal surveys estimate that approximately ten million Americans are alcoholic or have serious drinking problems. This means that 7 percent of the adult population or 10 percent of those who drink have alcohol problems. In 1975, the economic cost of alcoholism in this country was estimated at almost forty-three billion dollars. That figure, adjusted for inflation, would approach seventy-five billion in 1984. This estimate was based on men between the ages of twenty-one and fifty-nine and did not include women or alcoholic derelicts (i.e., those on skid row). Nor could the "costs" be attributed exclusively to alcohol abuse. However, the proportion of dollars assigned to each of the categories analyzed is surprising. Losses in production of goods and services accounted for 46 percent of the total estimated economic cost. Next came dollars spent for health care related to alcohol problems — a staggering 30 percent. Thus, in terms of economic cost as well as personal casualties, alcoholism achieves the status of a major public health problem, rivaled only by heart disease and cancer. An analysis of the nation's *total* expenditures for health (which included construction of research and medical facilities, training, as well as federal and private goods and services) showed that alcohol abuse–related expenditures accounted for about 12 percent.

Motor vehicle accidents and violent crimes related to alcoholism accounted for 12 and 6 percent respectively of the total estimated economic cost of alcoholism. Social welfare systems (unemployment and workmen's compensation, public assistance, special welfare and social service) and fire losses accounted for the remaining 7 percent of estimated economic costs. As more accurate and representative surveys are conducted, it is likely that these figures will be adjusted upwards.

Against this disturbing picture of billions of dollars lost because of alcohol-related problems, and the inestimable human costs to the ten million problem drinkers and their families, it is interesting to compare the federal dollars directed toward alleviation of the major diseases through research. In 1980, over nine hundred ninety-nine million dollars were allocated for research on cancer, over five hundred million dollars for research on heart disease and twenty-two million dollars for research on alcoholism. Would alcoholism be conquered if just one F-16 jet fighter were sold to benefit research? Unlikely . . . but there is a correlation between dollars expended, scientific talent and attention, and *eventually*, significant progress in an area of inquiry. Eventually, because major discoveries in biomedical science are usually the result of years of painstaking research.

Numbers and Truth

Numbers are the standard by which the seriousness of a problem is measured, the cudgel by which the political process is pushed and shoved, and the ultimate criterion for some. Yet to the extent that numbers are sanctified as absolute truth, they may be dangerously misleading. There are many problems involved in trying to count people with alcoholism or any other disorder. Inaccurate figures due to underreporting and nondetection are more common than overreporting. Before adequate surveys were available, the number of alcoholics was anyone's guess, and guesses escalated from five to ten, to fifteen to twenty million. Today's estimate of ten million problem drinkers is based on a composite of seven federally funded surveys and two independent surveys. It is probably a reasonable estimate, given all the problems inherent in the process.

It has always been difficult to count the number of people with alcohol problems. The social stigma associated with alcoholism leads to underreporting, and alcoholics often deny that they have a drinking problem. This difficulty is further compounded by the lack of consensus about what a drinking problem is, what the criteria for inclusion in that category are, and what differential weighting should be applied to what factors. The epidemiology of disease is an inexact but important science.

One approach to measuring drinking problems is to select a representative sample of households and survey the occupants with questionnaires and interviews. In 1965, a national survey of 2746 households examined the drinking practices of adults (over twenty-one). It was found that two-thirds of the population drinks, at least occasionally. About 12 percent were categorized as heavy drinkers — defined as almost daily drinking and five or more drinks on some occasions or weekly drinking of five or more drinks.

The remaining one-third of the population abstained from alcohol. Of these, most never drank and a few used to drink but had stopped. A survey taken in the late 1970s also showed that about one-third of the adult American population abstains from alcohol.

When the 1965 survey was repeated three years later, it was found that the categories of heavy drinkers and abstainers were not static and there was considerable fluctuation in individual drinking patterns. Although the relative proportions remained about the same, 31 percent had moved into or out of the problem drinker category and 15 percent had moved into or out of the heavy drinker category. This survey evidence of remission of problem drinking is encouraging but further complicates accurate counting.

Since the identity of abstainers and persons with heavy or moderate drinking patterns may change, but the total numbers in any category may remain the same, it might be expected that overall alcohol consumption might remain about the same. Indeed, translation of alcohol beverage sales into per capita alcohol consumption showed no important change during the decade of the 1970s. Sales analysis data indicate that between 1970 and 1976 alcohol consumption, divided by the number of people over fourteen, averaged between 2.61 and 2.69 gallons of absolute alcohol per person per year. Comparable levels of per capita alcohol consumption (i.e., above 2.50 gallons per year) were reported on only three previous occasions, in 1860, in 1911–1915 and in 1969. Of course alcohol sales cannot necessarily be equated with alcohol consumption since people may keep a supply of alcohol for several years.

The most recent and probably the most exact assessment of the prevalence of alcoholism in America was reported in 1984 by scientists from the National Institute of Mental Health who collaborated with sociologists and epidemiologists at major medical centers throughout the country. The study was the first and most comprehensive attempt to determine the prevalence of mental disorders in the United States by directly interviewing respondents in their own household with standardized, highly reliable questionnaries. Dr. Daniel X. Freedman, past president of the American Psychiatric Association and editor of the American Medical Association's Archives of General Psychiatry, has emphasized that the survey "is far more than a simple census. It is a landmark in development of American contributions to the psychiatric knowledge base."

Over ten thousand persons were interviewed in the survey. The procedures used for selecting persons to be interviewed were similar to those employed by pollsters for assessing the probable outcome of national elections. The sample reflected the composition of the American adult population with respect to age, sex, ethnic and occupational status. The interviewers who conducted the surveys were not physicians or psychologists but were extremely well trained lay persons who used a standardized interview schedule. This schedule was based upon the most recent and rigorous criteria developed by the American Psychiatric Association for defining various forms of mental illness, including alcoholism.

Two types of prevalence data for mental illness were reported in these studies. The first was called "six-month" prevalence and indicated the actual number of reports of mental illness during the six-month period before initiation of the study. The other type of data was termed "life-time prevalence" and included any type of mental illness that occurred at any time during the interviewee's life prior to the survey.

It was anticipated that mental illness characterized by symptoms such as anxiety or depression would be the most common disorders. *However, alcohol abuse and alcohol dependence ranked as the most common form of mental disorder in six-month prevalence rates.* Moreover, six-month prevalence rates showed that alcohol abuse and dependence was the first ranked form of mental illness in males whose age groups were from eighteen–twenty-four, twenty-five–forty-four and forty-five–sixty-four. For males sixty-five and older, severe cognitive impairment, loss of memory, abstracting and thinking facility was ranked first in the six-month prevalence data. It is highly likely that many of the cognitive impairment disorders were related to an earlier history of alcohol abuse or dependence. In women, rates of alcohol abuse and dependence for six-month and lifetime prevalence were significantly lower than in men, although alcohol abuse and dependence were the fourth most common psychiatric disorders for females between ages of eighteen to twenty-four years of age. The survey also revealed an astoundingly high rate for lifetime prevalence of alcohol abuse and dependence (19 to almost 29 percent of all male respondents).

At the time of publication of this book, the epidemiologic survey sponsored by the National Institute of Mental Health was near conclusion. A full report of the twenty thousand individuals surveyed should be available during the latter part of 1985. However, preliminary analysis of some of the data obtained suggests that alcohol abuse and alcohol dependence will remain America's most common mental health problem in men. In 1985, the best estimate for current or lifetime prevalence of alcoholism among Americans would be approximately 10 million, a figure which is remarkably consistent with older survey data.

Another approach to estimating the prevalence of alcoholism is to tabulate alcohol-related arrests, deaths and readmissions to treatment facilities, and estimate alcoholism prevalence on this basis.

An alternative approach is to count the number of cirrhosis deaths due to alcoholism for a given period. Such information is available in the 1975 report of vital statistics for the United States. Between 1950 and 1975, the percentage of cirrhosis deaths associated with alcoholism increased from 22.8 to 40.9 percent. It is impossible to know if this reflects a real increase or more accurate reporting.

Given the many real problems in accurately estimating the number of problem drinkers or alcoholics at any one time, it is unlikely that exact figures will ever be generated. However, insofar as very different approaches yield comparable numbers, our confidence in these estimates is increased.

Who Are the Problem Drinkers?

Although it is entirely reasonable to assume that men and women as well as the affluent and the underprivileged may differ in the prevalences of drinking problems, the same difficulties that plague accurate total counting also affect analysis of the subcategories. The most common approach has been to study the composition of groups of known alcoholics and problem drinkers. Despite the methodological limitations, there do appear to be some consistent trends.

Since 1968, Alcoholics Anonymous (AA) has surveyed its membership in the United States and Canada every three years. In 1980, reported membership was 476,000 and 24,950 individuals (about 5 percent) participated in the survey. *Three* changes in the composition of AA membership were especially striking. First, the number of women increased from 22 percent in 1968 to 31 percent in 1980. The percentage of younger alcoholics (i.e., under 30) increased from 11.3 percent in 1977 to 14.7 percent in 1980. Polydrug abuse [defined as addiction to some drug(s) in addition to alcohol] increased from 18 to 24 percent over the same period. Of the total number of AA members sampled, female polydrug users accounted for 34 percent and male polydrug users for 20 percent.

Other studies of identified alcoholics have shown similar trends. Among middle-class alcoholics, who were first admissions to a proprietary hospital group that specializes in alcoholism treatment, 23 percent were women in 1977–1978. This patient sample was considerably smaller than the AA sample, only 3,411 people in all. However, it did provide some demographic information about a group seldom subject to public scrutiny, the relatively affluent middle-class alcoholic.

In a sample of 14,000 lower socioeconomic class alcoholics seen at federally supported alcohol treatment centers between 1972 and 1974, women accounted for only 15 percent of the total. There were too few women to permit analysis of their response to treatment using random sampling procedures.

Information about drinking behavior and alcohol problems among women who are not diagnosed alcoholics is fragmentary. There are some indications that women who work are more often heavy drinkers and problem drinkers than women who do not work. Speculations about women alcoholics have often portrayed her as a housewife, drinking in secret, but this may not be the most prevalent pattern. One study compared the drinking patterns of married mothers, with traditional jobs, with unmarried (single, separated or divorced, widowed) childless women who were heads of households and held nontraditional jobs. Although the sample was very small, significantly more of the 102 nontraditional women reported heavy drinking and problem drinking than did the 463 traditional women. These differences in drinking patterns occurred among both low and high socioeconomic class women. On the basis of such findings, some theorists have eagerly embraced concepts such as "sex role conflict" and "fear of success" as probable explanations. Heavy drinking is alleged to reflect both "masculine striving" and a retreat into "femininity and dependency" with about equal frequency. As appealing as such trendy and simplistic "answers" may be to some, they offer little more actual information than yesterday's stereotypes about the drab and repressed alcoholic housewife. Only 25 percent of the AA sample were housewives; professionals (18 percent), sales and business (11 percent) and clerical workers (15 percent) accounted for the vast majority.

A number of interesting questions remain to be answered about alcoholism in women. There may be important differences between women and men in the natural history and expression of drinking problems and in its biomedical consequences. Women with alcohol problems may differ from women in the general population on several specifiable dimensions. Alcohol affects reproductive function adversely in women as it does in men and this is discussed in the chapter on alcohol and health. Problems specific to women, such as premenstrual tension, may periodically exacerbate an addictive drinking pattern. Now that alcoholism in women has finally been recognized as a real problem, which affects an estimated 2.5 million women, a clearer picture of the male-female similarities and differences should emerge.

Economic Factors: The public inebriate or "skid-row" alcoholic has long been the most visible problem drinker and for many, the symbol of the alcoholic. Although alcohol problems do appear to be overrepresented among the disadvantaged, it would be incorrect to conclude that alcoholism is only a disorder of the lower socioeconomic classes. More information is available about poor problem

drinkers because they are treated in public facilities. The more affluent are treated in private facilities, and until very recently, the
admission diagnosis was almost never alcoholism. The rate of alcoholism among middle and upper middle socioeconomic classes is unknown. But the rapid growth of private proprietary hospital groups
designed to treat alcoholics who can pay over five thousand dollars
for treatment suggests that these patients exist in greater numbers
than was once believed.

In a middle-class sample of 3,411 first admissions for alcoholism
treatment, over 62 percent described themselves as professionals, executives, managers or skilled laborers. Sales, business and clerical
workers accounted for over 15 percent of the sample, and unskilled
labor for 10 percent.

The 1980 Alcoholics Anonymous survey indicates that approximately 17 percent of its male members were professionals, 19 percent were in sales and business, 30 percent were laborers, and only 7
percent were unemployed. Retired persons (12 percent), office and
clerical workers (3 percent) and others (12 percent) accounted for
the rest.

The 1965 national survey found that more men and women of a
high socioeconomic level drank more than comparable age groups at
a lower socioeconomic level. However, the number of heavy drinkers
did not differ in the two socioeconomic groups. Abstainers are more
prevalent among people of lower socioeconomic status.

These several sources of information indicate that alcoholism affects many people at many socioeconomic levels. Alcoholism is not
necessarily a disease of the needy or the affluent.

ALCOHOLISM AND AGE

We tend to think of alcoholism as a disorder of the middle years;
however, both the young and the elderly may have serious drinking
problems. Alcoholism can develop at any age, although relative risk
is greater at some life stages than others.

Survey data indicate that among men, the highest proportion of
heavy drinkers are concentrated in late adolescence, the early thir-
ties, the late forties and the early sixties. Among women, the heaviest
drinkers were in the early twenties and late forties. Since heavy
drinking usually precedes problem drinking and alcoholism, it is

reasonable to assume that these may be periods of enhanced risk.

Alcohol and Youth: Alcohol and drug abuse among young people is an area of legitimate concern. Yet in part because of the emotionalism surrounding the issue, it has been more difficult to accurately gauge the extent of the problem in youth than in any other age group. The decade of the 1970s witnessed an interesting interplay between the counters, the Congress, the federal bureaucracy and the media over the issue of teenage drinking problems. Some analysts have argued that teenage drinking problems were, in part, the creation of a then infant federal agency (NIAAA) to compel congressional attention and compete more effectively with established agencies for funds. Since self-interest, altruism, and public responsibility often merge in amorphous ways, it is difficult to refute or substantiate the allegation that teenage drinking was orchestrated by "moral entrepreneurs" in the concerned bureaucracy. However, the analysis is worth reviewing because it illustrates one dimension of a poorly understood symbiotic relationship between bureaucratic spokesmen, politicians and the media.

During the late 1960s and early 1970s, there was considerable public concern over adolescent drug abuse. In 1971, the new National Institute on Alcohol Abuse and Alcoholism (NIAAA) began to assert that alcohol was abused more than other drugs and commissioned a survey on teenage alcohol abuse. By 1974, the NIAAA director told *Time* magazine that "the switch is on ... youths are moving from a wide range of other drugs to the most devastating drug — the one most widely misused of all — alcohol." Thus the NIAAA leadership sought to convince Congress of alcohol's share of the youth market, and, curiously, to establish that "drugs are us." The portrait of a teenage alcoholic was dramatized on television in 1975. In 1976, the U.S. Senate Appropriations Committee deplored the rising problem of teenage alcoholism, and instructed NIAAA to develop comprehensive education and information programs on this issue.

An examination of historic sources revealed no similar public alarm over adolescent alcohol abuse. A sociologist, R. L. Chauncey, suggests that "two possibilities emerge. Either the extent of the problems associated with teenage drinking had grown dramatically in recent years and the diligence of the newly created NIAAA served to publicize this burgeoning problem heretofore clouded by misinformation and blithe ignorance; or the NIAAA, in an effort to sustain itself, has seized on an emotionally charged topic certain to generate demands for a variety of educational and treatment programs." Chauncey argues cogently that the latter was the case, and that creation of new social problems and stirring the controversy surrounding old ones is a technique bureaucracies use to capture public attention and increase their appropriations.

Reputable scientists were commissioned by NIAAA to review the literature on teenage drinking. Although their findings were not prominantly featured in testimony to Congress or media statements, they were included in a 1978 NIAAA Report to Congress, only to vanish again by 1981. The Blane and Hewitt Report concluded that drinking among both college and high school students tended to increase between the early 1940s and mid 1960s. But the prevalence of alcohol use among teenagers remained relatively stable between 1965 and 1975. These findings did not suggest a growing teenage drinking problem.

Not surprisingly, several surveys have shown that alcohol is the most widely used drug among American youth, excluding tobacco. A national survey of 4,918 tenth- to twelfth-grade students conducted in 1978 showed that 32.6 percent abstained or drank very infrequently and 32.1 percent usually drank once a week, often two to five or more drinks per occasion. Less than 2 percent reported drinking daily. Drunkenness at least six times during the preceding year was reported by 31 percent of the sample.

Many have argued that occasional drunkenness poses more problems for the adolescent and society than frequent drinking of one or two beers. The figures on alcohol-related auto accidents involving drivers under twenty are a grim index of drunkenness. In 1978, the National Safety Council reported 5.6 million traffic accidents and 11,500 crashes with at least one fatality. Whether or not more stringent penalities for driving while intoxicated, such as California has recently enacted, will be effective in reducing the toll of drunk-driving tragedies at all ages remains to be seen. To date, there is little evidence that raising the legal drinking age has substantially reduced adolescent alcohol use. Neither leniency nor draconian regulations appear to significantly impact on adolescent drinking practices.

In 1982, there were 43,721 highway fatalities in the United States. Although this number was large, the death rate in 1982 decreased by 5,580 from a total of 49,301 in 1981. The decrease in highway fatalities between 1981 and 1982 was the largest ever recorded in the past forty years except for 1974, the year of the oil embargo. This trend has been paralleled by intensified efforts to promote legislation with harsher penalities for intoxicated drivers. Individuals and citizens action groups such as MAAD (Mothers Against Drunken Driving), RID (Remove Intoxicated Drivers) and CSD (Citizens for Safe Driving) have vigorously promoted such changes in existing legislation.

In April of 1982, President Reagan established a Presidential Commission on Drunken Driving, with the approval of the majority of members of Congress. In December of 1983, the commission concluded its deliberations and recommended a series of educational and legislative measures to reduce and prevent alcohol-related highway fatalities.

Most controversial was the commission's recommendation that the minimum age for purchase and public possession of any alcoholic beverage be increased from eighteen to twenty-one. The commission also urged more uniform and stringent law enforcement policies, including mandatory sanctions for the first DWI (Driving While Intoxicated) offense. A DWI arrest would result in suspension of the offender's driving license for not less than ninety days, and one hundred hours of community service or a minimum of forty-eight consecutive hours in jail. For repeat offenders, it was recommended that these sanctions be increased to include license suspension and a thirty-day jail sentence for those who continued to drive with a suspended or revoked license. The commission also recommended the elimination of plea bargaining in DWI cases.

Along with stricter laws against drunken driving, the commission also urged creation of an extensive public information and education program throughout the nation to dramatize the problems associated with drunken driving and to encourage safe driving practices. The use of seat belts, child restraints, and adherence to the 55 mph speed limit were recommended.

In order to facilitate implementation of these recommendations, the commission recommended the creation of self-funding mechanisms at state and local levels. They also urged involvement of the private sector, especially the motor vehicle insurance and the alcohol beverage industries. The commission emphasized the need for annual training sessions on the drunken driver problem for police, lawyers, prosecutors and judges. Finally, the commission strongly recommended that new task forces on drunken driving be created in states where they did not exist and that citizens action groups concerned with drunken driving issues be strengthened and expanded.

The majority of members of Congress were sympathetic to the commission's recommendations. Subsequently, Congress enacted legislation that restricts provision of federal highway trust funds to states that do not enact a minimum drinking age law of twenty-one years by 1986. This new law was signed by President Reagan in June, 1984. Although these restrictions of highway trust fund allocations would be limited to only 10 percent of normal appropriations, the real dollar loss sustained by states that do not enact a mandatory twenty-one-year drinking age law could be very great. It seems highly likely that this federal initiative will produce uniform twenty-one-year drinking age statutes in all fifty states within the next two or three years. But it remains to be determined how such uniform drinking age regulations throughout the nation will impact upon alcohol related highway fatalities.

The Elderly: There appear to be fewer drinking problems among the elderly than the middle-aged, but it is estimated that between 2 and 10 percent of men over sixty are alcohol abusers. About 10 percent of people enrolled in alcoholism treatment programs are over sixty. As the general population ages, it is likely that these estimates will increase. However, both the epidemiology and natural history of alcohol problems among today's elderly are poorly understood. The extent to which alcoholism reflects lifelong drinking problems or a response to leisure, to despondency associated with retirement, and the many medical and social concomitants of aging is difficult to determine. Loneliness and increased isolation often accompany aging, but do not invariably lead to problem drinking. Still, it is not surprising that recently widowed men, shown to be at high risk for potentially fatal medical disorders, are also at high risk for developing drinking problems. Such recent entries into the addictive drinking process may continue drinking until death. In contrast, the career alcoholic is likely to succumb to liver failure or other related diseases before reaching old age. Others may moderate their drinking as their capacity to handle alcohol diminishes because of liver impairment.

Aging is associated with changes in the body's capacity to tolerate alcohol and other drugs. These differences can be attributed primarily to changes in body structure. For example, older people may become more intoxicated at lower doses of alcohol than during their youth. But the rate at which people metabolize alcohol does not change as a function of age. Studies sponsored by the National Institute on Aging compared alcohol metabolism rates in healthy people between the ages of twenty-one and eighty-one. Although elimination rates were equivalent, older people reached higher peak blood alcohol levels. This is because older people are likely to have a relatively higher concentration of body lipid, a lower concentration of muscle mass and a smaller volume of body water in which alcohol is diluted. How this dimension of aging may influence alcohol use patterns is unknown.

Changes in the accessibility of alcohol is another factor that may affect drinking patterns among the elderly. Reduced mobility and a lower income may limit alcohol availability for many elderly people. Nursing homes and residences rarely provide alcohol, even wine or beer, for their clients. Yet several studies have shown that a cocktail hour may have very beneficial effects on morale and socialization in nursing home environments. It is an unfortunate aspect of institutionalization that healthy elderly people are often denied the simple pleasures of occasional social drinking. The roles of childhood and old age often merge and "protection" from intoxicants may become one part of that cycle.

Polydrug Use and Alcoholism

The simultaneous or concurrent use of multiple substances is hardly a new phenomenon. There is ample historical precedent for abuse of alcohol and opiates. In Victorian times, opiate addiction was not uncommon. Laudanum (tincture of opium) was readily available and often drunk with wine or water. Many literary figures of the period used opium and some preferred it to wine. De Quincey argued that "whereas wine disorders the mental faculties, opium, on the contrary (if taken in a proper manner), introduces among them, the most exquisite order, legislation, and harmony." Coleridge became addicted first to opiates and then to alcohol, thus anticipating a common pattern among contemporary opiate addicts maintained on methadone. Indeed, medical complications of alcoholism are a major problem among methadone-maintained heroin addicts today.

Addiction to alcohol and opiates is only one of a variety of possible addictive patterns. Alcohol is sometimes abused with barbiturates, stimulants, marijuana, depressants and, of course, tobacco. Almost every available medically prescribed and illicit drug can potentially be used with alcohol, though some combinations are more common than others. The National Institute on Drug Abuse (NIDA) surveyed preferences of multiple drug users in 1980 and found that some used alcohol plus only one other drug. In order of frequency, drugs used with alcohol were marijuana, minor tranquilizers, heroin, amphetamines, and barbiturates. Alcohol was frequently substituted when marijuana and minor tranquilizers were unavailable. Alcohol was most commonly used to alter the effects of opiates other than heroin, amphetamines, barbiturates, minor tranquilizers and marijuana.

The combined effects of alcohol and other drugs are complicated and potentially hazardous. A combination of alcohol and sedatives has an additive depressant effect, whereas the combination of alcohol and stimulants may modulate the effect of stimulants. The degree of impairment and the type of subjective effect of any drug combination is affected by the relative dose of each drug, whether or not the user is tolerant to the drugs, as well as the pharmacological characteristics of each drug. When more than two drugs are combined, the complexity of the reaction increases. There have been relatively few clinical assessments of combined drug effects using objective measures. Much of what is believed about the effects of specific drug combinations is an amalgam of street lore, self-reports by polydrug users, and inferences from pharmacological facts.

Counting and categorizing multiple substance abusers is even more difficult than trying to count persons with drinking problems and alcoholism. All the methodological problems are further compounded by the addition of a smattering of illicit and medically prescribed or over-the-counter drugs.

The casualties attributed to the combined use of alcohol and another drug appear to be increasing. Between 1973 and 1976, the cases reported by emergency wards, crisis centers and coroners' offices increased from an average of 1303 per month to 1929 per month. These figures are based on the number of alcohol-drug combination mentions collated from medical examiners, emergency wards and crisis centers in twenty-four standard metropolitan statistical areas. The Drug Abuse Warning Network, or DAWN, was designed to detect and monitor drug abuse trends. The number of drug overdose deaths related to alcohol plus other drugs is unknown. More details of the health consequences of polydrug use involving alcohol are described in the next chapter.

IMPLICATIONS OF MULTIPLE DRUG USE

Traditionally it has been assumed that alcoholism is a problem separate and distinct from opiate addiction or barbiturate addiction or stimulant abuse. Moreover, addicted persons were believed to have a consistent preference for their chosen drug. Evidence to the contrary has always been available, but attitudes and belief systems often have an autonomy that transcends contradictory information. From a legal standpoint, the distinction between alcoholism and the abuse of illicit drugs such as cocaine and heroin remains valid. But from a biomedical perspective, it becomes increasingly important to explore the similarities as well as the differences between the addictive processes. The single drug imperative has long dictated separate approaches to research and treatment of alcoholism and other drug abuse problems.

Indeed alcoholics resist and resent any association with drug addicts. The scorn and derision society accords to alcoholism is matched only by the disdain of alcoholics for drug addicts. This attitude is not reserved for "hard" drugs and extends to marijuana as well. In 1972 only five of twenty-five alcoholic patients would agree to try marijuana as part of a treatment trial. Most considered it to be more dangerous than alcohol. Antipathy towards marijuana by alcoholics foretold the failure of an otherwise innovative treatment effort, to reward alcoholics for taking disulfiram (Antabuse) with marijuana.

At the federal level, alcoholism and drug abuse are bureaucratically designated as separate entities. Any tentative suggestion of the possible advantage of combined alcohol and drug treatment programs incites the wrath of the alcoholism constituency and the NIAAA. One analyst comments:

> *it is intriguing and almost humorous to observe the resistances created by the suggestion that alcohol and drug-problem agencies join forces in their activities (usually on the basis of the contention that alcohol is a form of drug). Such suggestions sometimes create frantic responses such as intense arguments that alcohol problems are completely unique and that "everyone will lose" if the two sets of efforts are put together. These resistances clearly reflect vested interests which might be lost through a merger, as well as the greater stigmatization associated with illegal drug abuse: problem drinkers and their caretakers might have their reputations damaged by becoming associated with "drug fiends."*

Reorganization of the bureaucracy to combine NIAAA and NIDA is periodically considered, and was on the agenda of the Reagan administration. Whether the decision is to combine or to maintain the status quo, the arguments should be a revealing distillate of contemporary attitudes toward alcoholism and drug abuse. Pragmatically, the separate agencies insure increased visibility for both problems, and perhaps ultimately greater resources for research and treatment than might otherwise be forthcoming. Organizationally, a single agency might be more streamlined, but given the emotionalism over the issues, this might or might not translate into greater effectiveness. Politically, it is likely that the pro-alcoholism voices will triumph.

Will it matter? That is very difficult to predict. And perhaps the inherent unpredictability in the system is the most compelling argument for no change. Faced with the prospect of perhaps much better or perhaps much worse, many would elect the status quo.

Insofar as research and thinking about the addictive disorders can be dissociated from the supporting federal structure, renewed recognition of the prevalence of polydrug abuse has led to an important conceptual shift. Given that drug addicts abuse alcohol and alcoholics abuse drugs, it makes sense to look for similarities in drug use patterns and their behavior consequences. The possibility that various forms of drug abuse may have common characteristics that transcend the pharmacological distinctions between drugs is intriguing. Substance abuse is the new term that challenges, both conceptually and empirically, the long tradition of one drug, one problem.

Between 1977 and 1980, NIDA and the National Academy of Sciences sponsored a Committee on Substance Abuse and Habitual Behavior charged with examining possible commonalities among various forms of substance abuse. Substance abuse was defined to include not only alcoholism, opiate addiction and stimulant abuse, but also tobacco smoking, excessive eating and gambling. (Sabbath-breaking and profanity didn't qualify for today's vice inventory as they did in 1813.)

The Substance Abuse Committee's deliberations served to legitimize this area of inquiry. The stimulus of this provocative concept will help scientists to take a different, less parochial look at the addictive disorders. The result should be some significant progress in our understanding of drug abuse problems. This in turn should help society to deal with the yet unknown but inevitable substance abuse problems of tomorrow.

CHAPTER 17

Health Consequences of Alcohol and Drug Combinations

Alcohol intoxication during the use of other psychoactive drugs or prescribed medications has effects that range from benign to lethal. The type of drug, the amount used and the tolerance of the user all combine to influence the final outcome. Some drug mixtures, such as high doses of alcohol and barbiturates, are predictably life-threatening whereas others, such as alcohol and aspirin or alcohol and antihistamines, seem innocent enough but may temporarily incapacitate the unwary drinker. Many drugs are highly toxic when used alone and alcohol may enhance their inherent toxicity.

There are countless possible combinations of alcohol and other drugs. Over seven thousand medically prescribed drugs are described in the most recent *Physicians' Desk Reference*. The Drug Enforcement Administration lists thirty-nine narcotics, seventy-eight depressants, forty-nine stimulants and five known hallucinogens, each with a slightly different formula, which are commonly abused. Since purveyors of illicit drugs may mix several compounds to produce the final product, the recreational drug abuser is at highest risk for serious toxic drug reactions or fatal drug overdose. But the moderate social drinker may also be vulnerable during a prescribed course of antianxiety drugs, sleeping pills, pain medications and even certain types of antimicrobial agents.

Pharmaceutical manufacturers list every possible known side effect of their product in the package insert, but the prescribing physician seldom recites the entire list to the patient. To do so might lessen the patient's confidence in the medication and increase the probability of the side effects occurring — a sort of reverse placebo effect. The patient, caught between the overinclusiveness of insurance-conscious pharmaceutical manufacturers and the conservative physician, is best advised to ask his doctor if the medication interacts with other medications or with alcohol.

This chapter describes some common types of alcohol and drug interactions and the resulting medical problems. We have considered medicines that are seldom abused, such as antidepressants and aspirin, separately from medicines and illicit drugs that are often used recreationally and abused such as barbiturates, opiates and stimulants. The distinction between useful medicine and abused drug is often fragile, more determined by contemporary legal convention or evanescent street drug fads than the pharmacological class or chemical structure of a drug. Acknowledging these qualifications, it is still useful to differentiate medicines with low abuse potential from medicines with high abuse potential.

The available information is usually limited to the effects of alcohol in combination with a single other drug. When two, three or more drugs are used together with alcohol, the difficulty of predicting the effects increases proportionately. The human toll of drug combination fatalities is suggested by grisly statistics from the National Institute on Drug Abuse. A recent survey has shown that 34 percent of the drug-related deaths in nine cities were associated with drug combinations or the combination of alcohol and another drug or drugs.

ALCOHOL AND MEDICINES

A number of medically prescribed and over-the-counter drugs interact with alcohol. Even medicines that have minimal psychoactive effects may produce toxic reactions when taken in combination with alcohol. Most drugs with behavioral or physiological actions similar to those of alcohol are likely to produce an even greater effect when used with alcohol. Pharmacologists call similar drug actions "synergistic" and the summation of similar effects, "additive." An additive effect occurs when alcohol is used with another drug that also depresses brain activity or induces sleep. Many over-the-counter sleeping medications are in this category.

The interaction between alcohol and medicines is also influenced by the physical condition of the user. In elderly people, any adverse effects of alcohol and drug combinations may be greater than in young people. Changes in body structure and physiological function associated with aging may impair drug metabolism mechanisms and increase the probability of toxic side effects from conventional medications. Chronic illness may also compromise the body's capacity to detoxify alcohol and drugs. The elderly are more vulnerable to toxic drug reactions in three respects: they are more likely to take medicines; more likely to react adversely to them; more likely to react adversely to alcohol and drug combinations. Even therapeutic doses of sedative-hypnotic or sleep-inducing drugs may result in confusional states that may be misdiagnosed as senility. The interaction between alcohol and most sedative-hypnotic drugs is additive.

Prescription medicines that are seldom abused are considered separately from abused or illicit drugs. These medications are organized according to their major therapeutic function. Since drugs with similar medicinal uses may differ greatly in chemical structure, the effect when combined with alcohol is not always predictable from the general classification. This list of alcohol and drug interactions is not exhaustive; there may be other toxic combinations that have not yet been recognized.

Minor Pain Remedies

The unrelenting barrage of commercials for pain remedies suggests that headaches and minor pains of muscles and joints are very common afflictions which respond well to a variety of medications. Few medications are totally without unwanted side effects, and the minor painkillers are no exception.

Salicylates, such as aspirin, tend to be gastric irritants, partly because of the effects of salicylic acid in the blood as well as in the stomach. When alcohol and aspirin are taken together, the lining of the stomach, i.e., the gastric mucosa, may become inflamed, ulcerated and bleed. Since aspirin also interferes with blood clotting mechanisms, this can increase the danger of gastric hemorrhage.

Acetominophin, or Tylenol, is another type of minor pain medication; it is not a gastric irritant, but in large doses (10 to 15 g) it can have toxic effects on the liver. Doses of 25 g can be fatal. Since most Tylenol capsules are only 325 milligrams, huge quantities would have to be ingested to reach dangerous dose levels. Since alcohol also may have adverse effects on the liver, the combined use of alcohol and Tylenol may increase the possibility of liver damage.

Antihistamines

Next to pain remedies, the antihistamines are one of the most commonly used and readily available drugs. They relieve sneezing, runny nose and itching of the eyes, nose and throat often associated with colds and hay fever. Sedation and gastrointestinal disorders are among their major side effects. The combination of drugs such as Benadryl or Chlor-trimeton with alcohol has an additive sedative effect. Controlled studies have shown various performance impairments in persons using both alcohol and antihistamines. For example, the ability to drive or operate complex machinery is compromised by mixing antihistamines and alcohol. Thus, drinking drivers on antihistamines are at even greater risk.

Minor Tranquilizers

Minor tranquilizers are among America's most commonly prescribed drugs. It has been estimated that over eight thousand tons of one class of tranquilizers, the benzodiazepines, were consumed in 1977. Fifty-four million prescriptions for diazepam (Valium) and thirteen million prescriptions for chlordiazepoxide (Librium) were written that year. Today, the benzodiazepines are more commonly used than another class of tranquilizers, meprobamates (Miltown, Equanil and their several relatives), which were first synthesized in the early 1950s. One tranquilizer of the benzodiazepine group (flurazepam or Dalmane) was prescribed more (53 percent) than all other hypnotics in 1977.

The minor tranquilizers are useful in reducing feelings of anxiety and apprehension and as muscle relaxants. These drugs have hypnotic properties and are often prescribed as sleeping pills. Most minor tranquilizers also are anticonvulsants and are used in the management of some types of seizure disorders in children, and for emergency treatment of epileptic seizures. If used in high doses over long periods of time, meprobamates and benzodiazepines may produce physical dependence. Abrupt discontinuation of tranquilizers may lead to a withdrawal syndrome very similar to alcohol withdrawal. Tremors, sweating, ataxia, or a stumbling gait, and sometimes confusion, insomnia and hallucinations may occur. Yet one benzodiazepine, Librium, has been especially effective in the treatment of alcohol withdrawal.

The effects of alcohol and minor tranquilizers are similar, but these have not become generally preferred recreational drugs. Diazepam abuse has been reported among heroin addicts. Both monkeys and men with a history of sedative abuse will take diazepam when other drugs are not available, but its abuse potential appears less than that of barbiturates or alcohol.

Performance on a number of tasks is impaired by the combined use of alcohol and minor tranquilizers. Reaction time may be significantly slowed and ability to stop accurately in a driving task was impaired. Deficits in visual function (tracking and search) and information processing were found after relatively low doses of diazepam (5 to 10 mg) and low doses of alcohol sufficient to produce blood alcohol levels of 70 mg/dl. This pattern of impairments predicts that driving skills will be compromised by alcohol and tranquilizer combinations, and is borne out by statistics. Analysis of blood samples from hospitalized drivers showed that almost 11 percent had used alcohol plus diazepam before the accident and over 30 percent had used alcohol. Thus the combined use of alcohol and minor tranquilizers can increase vulnerability for auto accidents. Since alcohol and tranquilizers both are widely available, combined use is highly probable, and users should be aware of the potential hazards.

It has been suggested that alcohol enhances the absorption of diazepam (Valium) to produce an enhanced combined effect, but results from controlled studies have been inconsistent. At high alcohol concentrations some enhancement has been shown, whereas at low alcohol concentrations (10 percent) similar to those in many cocktails, diazepam absorption was slightly slowed.

Since the minor tranquilizers have many behavioral effects that are similar to moderate alcohol intoxication, it is not surprising that the combined effects of these drugs and alcohol are additive. Although the precise mechanism responsible for the additive effect is unknown, it is likely that it is mediated through the brain.

A Tranquilizer Receptor in Brain

The common use of minor tranquilizers in this society has often been cited as evidence of a pervasive anxiety unique to this period in time. Yet scientists have recently discovered that a receptor specific for benzodiazepines exists in the human brain and in the brains of all vertebrates. A receptor is a portion of the nerve cell membrane which has high affinity for and selectively attaches or binds to benzodiazepines. The adaptive and evolutionary significance of a brain receptor for the benzodiazepines, first synthesized in the 1930s, is of course unknown and poses many provocative questions.

In the late 1970s it was found that radioactive labeled benzodiazepine molecules bind stereospecifically to receptors in certain brain sites. Most molecules are three-dimensional, that is, they have length, width and depth. Some molecules have mirror-image configurations, and stereospecific binding refers to the fact that certain receptor sites will accept only the left or only the right mirror image. The highest concentration of benzodiazepine receptors is in the cerebral cortex, the cerebellum and the amygdala. These brain areas are primarily concerned with different functions. The cerebral cortex is believed to be the site of thinking and judgments, perception and sensation; the cerebellum with motor coordination and balance; and the amygdala with emotion. The relative receptor binding capability, or receptor affinity, corresponds to the potency of benzodiazepines as determined through clinical evaluations. The more potent the drug, the greater its receptor binding capacity.

The benzodiazepine receptor is the second drug-specific receptor to be identified in the brain. The first was the opiate receptor discovered in the early 1970s. Identification of drug-specific receptors has opened a new and exciting approach to the study of brain function. One fascinating implication of the discovery of opiate and benzodiazepine specific receptors is that natural substances may exist in the brain that have effects similar to the drugs. And indeed, discovery of the opiate receptor was paralleled by the identification of naturally occurring chemicals, called endorphans. Medicinal chemists have been able to construct synthetic analogues of the natural endorphans and these have analgesic properties similar to opiates. It appears that the natural endorphans may also mimic the psychoactive as well as the analgesic effects of opiates, since synthetic endorphan analogues are self-administered by monkeys as readily as morphine.

As yet a corresponding brain chemical with specific affinity for the benzodiazepine receptors has not been discovered, but several possible candidates are under investigation. We can speculate that the effects of relaxation techniques and meditation may be mediated through such an endogenous substance. If so, should one drink after meditating? To date, the effects of alcohol on performance after a relaxation therapy session have not been studied.

Antidepressants

Most people become depressed at some time in their lives. For many, depression is a reaction to bereavement, to a serious loss of social and economic security, to illness, to relocation, to loneliness, to defeat. Such reactive depressions are usually of limited duration, and supportive therapy can help to assuage feelings of sadness and despair. Since antidepressant medication requires about three weeks to reach effective levels, it is seldom used for reactive depressions.

For others, depressive episodes may occur unpredictably and last for weeks or months. This type of long-term depressive illness may or may not be triggered by loss or change. Feelings of despondency may be accompanied by other symptoms such as protracted insomnia, loss or increase in appetite, distractability and ruminative thinking. Depression may be associated with lethargy and fatigue or with frenetic activity and agitation.

Another form of depressive illness is characterized by an alternation between periods of severe despondency and sudden elation and excitement, often accompanied by expansiveness, garrulousness, grandiosity and impulsiveness. This syndrome is often called a manic-depressive disorder.

The pharmacological alleviation of depressive illness has been one of the major triumphs of biomedical research in the latter half of this century. Each of the three classes of drugs which are effective in treating depressive illness was originally developed and used for a different purpose. *Lithium*, often used for the treatment of manic-depressive disorders, was first used to treat gout in the middle 1800s, but its effectiveness was questionable. The chance observation that lithium made guinea pigs lethargic led to a clinical trial with manic patients in the late 1940s. The *Monoamine Oxidase Inhibitors* were first developed for the treatment of tuberculosis, and were derived from isoniazid and related drugs. Observation of striking mood improvements in tuberculosis patients led to clinical trials with depressed patients.

The *tricyclic antidepressants* (Elavil, Tofranil, Sinequan and others) are widely prescribed for the treatment of depressive illness, since these have fewer unwanted side effects than the monoamine oxidase inhibitors. The precursor of today's tricyclic antidepressants was first synthesized in the late 1800s and tried unsuccessfully as an antihistamine. One of its derivatives, imipramine, was first shown to be clinically effective in depressed patients in the late 1950s.

There are several hypotheses about the underlying changes in brain function associated with depressive disorders that make depression amenable to pharmacological treatment. However, the biological basis of depression and the critical mechanism of action of these chemically diverse drugs is, as yet, unknown.

The interaction between alcohol and the tricyclic antidepressants depends on the relative balance between sedative and stimulant activity of the particular compound. Those with greater sedative properties, such as amitriptyline (Elavil) may impair performance at blood alcohol levels of 80 mg/dl; or after drinking about three cocktails. Alternatively, desipramine (Norpramin, Pertofrane) may counteract alcohol effects somewhat because of its stimulant properties. However, the type and severity of the effect depends on the type of antidepressant and the amount of alcohol consumed. The monoamine oxidase inhibitors may act synergistically with alcohol. When beer or Chianti wine containing tyramine is consumed with monoamine oxidase inhibitors, a hypertensive crisis may ensue.

The *Physicians' Desk Reference* states that "untoward reactions to simultaneous use of lithium and other psychoactive drugs have not been reported." However, the difference between a therapeutic dose and a toxic (potentially life-threatening) dose of lithium is very small. Lithium intoxication is characterized by drowsiness, muscle weakness and poor coordination, vomiting and diarrhea. More severe toxic reactions include confusion, somnolence, stupor, coma and epileptiform seizures. It is reasonable to predict that the addition of alcohol intoxication could only result in adverse consequences for the drinker.

ALCOHOL AND ABUSED DRUGS

Illicit drugs of abuse such as heroin, marijuana and most hallucinogens are those which have no current medical use and therefore are assigned to Schedule I, the most restrictive category of the Federal Drug Enforcement Administration (DEA). Yet most contemporary drugs of abuse were used medicinally in an earlier era. This historical perspective is important to better appreciate current drug designations and the changes in scheduling that will probably occur in this decade. For example, in 1982, legislation was introduced to make heroin available for relief of pain in terminally ill cancer patients. Heroin is a potent opiate analgesic or painkiller like morphine, from which is it chemically derived. Marijuana is effective in reducing nausea and it is increasingly available for cancer patients for whom nausea is a disturbing complication of chemotherapy. Many synthetic compounds that mimic the active component of marijuana, delta-9-THC, are being developed for this purpose. Most other abused drugs also have some medicinal use. Cocaine is a powerful vasoconstrictor and is used as a local anesthetic. Barbiturates are useful as sedatives and also have anesthetic and anticonvulsant properties. The recreational or nonmedical use of drugs (e.g., narcotic analgesics, cocaine and marijuana) is illegal, although some states have recently "decriminalized" marijuana. However, illegality alone is not a sufficient criterion to define drug "abuse," as alcoholism and excessive smoking so eloquently illustrate.

Alcohol and Opiates

Opiates, like alcohol, have come to us from antiquity. The opium poppy or the poppy of sleep, *Papaver somniferum,* is one of more than one hundred species. It flowers briefly and then the seedpods swell with a milky latex. Crude opium is harvested by scratching the seedpod and collecting the exudate. Opium must be collected before the seeds ripen, within ten days after the petals drop, or it will be chemically destroyed. The familiar ripe poppy seeds and derivative oils do not have any psychoactive properties.

We can infer that opium smoking (and eating) in ancient civilizations was done for pleasure. The Sumerians described the poppy as the plant of joy over four thousand years before Christ. References to Nepenthes (presumably opium) appear in the writings of Homer over nine hundred years B.C. — when mixed with wine, the drug caused blissful forgetfulness. An association between joy and opium has persisted through the centuries in many cultures. Today, in modern Singapore, the faces of Taoist temple gods are smeared with crude opium to "put them in a happy mood for granting favors" (Hoefer et al., 1980, p. 205). However, as with alcohol, observations of heroin addicts during chronic heroin intoxication challenge these impressions of constant joy. Heroin addicts also tend to become increasingly irritable, anxious and depressed as heroin use continues.

In ancient times, opium was used medicinally for treatment of dysentery, and its derivatives are still used for this purpose today. Tincture of opium, or laudanum (a mixture of opium and alcohol), was developed before 1550 and generally used as a medicinal tonic throughout Europe and later in the American colonies. It was cheap and readily available, and apparently a staple in the damp and swampy fens of England, much as aspirin is today. In the United States, patent medicines containing opium, morphine and heroin as well as cocaine were sold in general stores and through mail-order catalogues until early in this century. Federal restrictions on the concentration of opiates in nonprescription medicines began with the passage of legislation in 1914, known as the Harrison Narcotics Act. Taxation and licensing involved in the manufacture, distribution and prescribing of opiates ended an era of uncontrolled opiate use.

Today the major medicinal use of opiates (morphine, hydromorphone, codeine and related compounds) is for the control of pain. The opiates are still the most potent analgesic drugs available. The discovery of opiate receptors (portions of the nerve cell membrane which selectively bind morphine and other opiate drugs) led to identification of naturally occurring opiatelike substances in the brain, known as the endorphans, which appear to mediate the body's own response to pain. For example, beta-endorphan levels increase significantly in women during labor and childbirth. Interpretation of these findings are qualified by the fact that the absolute beta-endorphan levels in these women were lower than those shown to be effective analgesics in experimental test systems. However, the analgesic properties of the endorphans and their synthetic analogues have been shown repeatedly.

Eventually, tolerance develops to the analgesic effects of opiates as well as to their euphorogenic and sedative effects. Opiate addicts must use more drug through time to attain the same subjective effects. Like alcohol, opiates induce physical dependence. Abrupt discontinuation of opiate use is followed by a characteristic pattern of withdrawal signs and symptoms. Coldlike symptoms, tearing, runny nose, sweating and yawning appear within about twelve hours after the last dose. The severity of symptoms is usually greatest at forty-eight to seventy-two hours and include insomnia, irritability, appetite loss, tremor, dilated pupils and often nausea, vomiting and diarrhea as well as an accentuation of the coldlike symptoms. The withdrawal syndrome gradually diminishes over seven to ten days. Scientists continue to search for an analgesic that will be as effective as the opiate drugs, but will not lead to physical dependence or abuse.

In addition to their analgesic and behavioral effects, opiates are also effective for the treatment of diarrhea, since they reduce gastric motility and lead to constipation. Opiates remain one of the most medically effective drugs for suppression of coughing.

In opiate-naïve individuals, these drugs also produce a number of unpleasant side effects, such as nausea and vomiting, dizziness, dysphoria, and sedation, even at low doses. These effects are familiar to many who have taken a tiny codeine tablet to relieve pain after dental surgery. At high doses, opiates can depress respiration and lead to coma and death. Opium was used as a poison in ancient times, and opiates (primarily heroin and methadone) are involved in more overdose and accidental deaths today than any other type of drug.

Although yesterday's tinctures of opium usually contained alcohol, it is wrong to assume that this combination is harmless. In fact, opiates and alcohol both are sedatives and may increase each other's depressant effects. The combination of morphine and alcohol may be lethal at high doses or in nontolerant people. For example, morphine at doses commonly used to relieve pain (15–30 mg) taken with high doses of alcohol (which produce blood alcohol levels of 180 to 200 mg/dl) could be fatal. Even among drug-tolerant opiate addicts who are also alcoholic, the combined use of alcohol and opiates poses a serious risk. Methadone is a long-acting opiate used to treat heroin addiction. The mortality rate of former heroin addicts, maintained on methadone, who are addicted to alcohol is ten times that of other methadone-maintenance patients. This reflects alcohol plus methadone overdose deaths as well as the serious medical complications associated with both addictions. Alcohol-induced liver disease plus viral hepatitis secondary to nonsterile hypodermic needles used to inject heroin is only one illustration. The prevalence of alcoholism among methadone-maintenance patients has been difficult to determine; estimates range from 12 to 40 percent. Although some impairment of motor performance has been observed after low doses of alcohol plus opiates, this seems an almost trivial consideration in view of the highly toxic and potentially fatal consequences of combined use at moderate to high doses.

Alcohol and Barbiturates

The barbiturates are a product of modern medicinal chemistry. Barbital was first introduced in 1903. Today a number of barbiturates are available, differentiated primarily in terms of their duration of action. For example, phenobarbital's effects may last for twenty-four to ninety-six hours, whereas methohexital may last for only ten to fifteen minutes if given intravenously.

Barbiturates are one of a class of compounds known as sedative-hypnotics. The term hypnotic means sleep-inducing, derived from the name of the Greek god of sleep, Hypnos. Until the advent of the benzodiazepines, one of the primary medical uses of the barbiturates was as sleeping medications. Barbiturates are sometimes used for pre-anesthetic sedation and for anesthesia, often in combination with inhalational anesthetics. But unlike opiates, barbiturates are poor analgesics. Barbiturates are also used for the emergency treatment of convulsions, including epilepsy. Phenobarbital is also used for long-term management of various types of epilepsies except temporal lobe epilepsy.

Barbiturate abuse by alcoholics is not uncommon. Despite their chemical dissimilarity, the subjective and physiological effects of alcohol and barbiturates are very similar, as is the resulting state of intoxication. Barbiturates may be used chronically or in intoxication sprees. As with alcohol, the barbiturate abuser may develop considerable tolerance to the intoxicating and sedative effects of barbiturates, but the dose at which barbiturates may cause death is not much higher in addicts than in nontolerant individuals. The similarities between alcohol and barbiturates are such that barbiturates can be substituted for alcohol during withdrawal and effectively reduce alcohol abstinence signs and symptoms. Long-acting barbiturates with anticonvulsant actions were once widely used for this purpose.

Barbiturates, like alcohol and opiates, induce physical dependence through chronic use. The typical barbiturate withdrawal syndrome is more severe than the common alcohol withdrawal syndrome. In fact, abrupt withdrawal from sustained high doses of barbiturates can be fatal. Tremor, nausea, vomiting, faintness upon standing (because of low blood pressure), weakness, anxiety and irritability may progress to include convulsions within two to three days. Patients with withdrawal seizures may develop delirium, agitation, hallucinations and profuse sweating similar to that seen in the rarest and most severe form of alcohol withdrawal, delirium tremens. Protracted barbiturate withdrawal may also be fatal.

Overdose deaths from intentional or accidental ingestion of large quantities of barbiturates occur frequently; they are second only to opiates in single-drug-related fatalities. Alcohol and barbiturates potentiate each other — coma and respiratory arrest may occur at doses of each drug in combination that would not be fatal for either drug taken alone. Forensic scientists estimate that barbiturates may be fatal at doses 50 percent lower than the usual lethal dose if taken in combination with alcohol. Respiratory depression, coma and death can occur at blood alcohol levels achieved during social drinking (100 mg/dl) if combined with low doses of barbiturates.

Alcohol and Other Sedative-Hypnotics

There are a number of other sedative-hypnotic drugs which are not barbiturates but may also be life-threatening when used in combination with alcohol. In general, the effects of these drugs resemble the hypnotic and sedative actions of barbiturates. Their chemcial structures are diverse but they have a similar capacity to induce coma and death at high doses.

Chloral hydrate is one of the oldest hypnotics, first synthesized in 1832. It is still prescribed for treatment of insomnia and for preoperative sedation. In combination with alcohol it acts synergistically to produce effects dramatized by its popular description as knockout drops or a "Mickey Finn."

Paraldehyde is another hypnotic that acts additively with alcohol. It has long been used to treat alcohol withdrawal and is preferred by many alcoholic patients. Despite its unpleasant taste and odor, addiction to paraldehyde has been reported and the withdrawal syndrome is similar to delirium tremens.

Methaqualone (Quaalude, Sorpor) is prescribed as a sleeping medication and for daytime sedation. It is abused, and when three to five times the usual therapeutic dose is combined with alcohol, stupor, coma, respiratory despression and death may result.

Glutethimide (Doriden) is another sleeping medication that is additive with alcohol and impairs motor performance at low doses. Death may result from the combination of alcohol and high doses.

Although this list is only illustrative, it is evident that persons who use any form of sedative-hypnotic should be cautious about the use of other depressant drugs and alcohol.

Alcohol and Inhalants

Before the introduction of anesthetics, the surgeon's assistants had to be strong men to hold down the struggling patient. Once inhalants or volatile anesthetics became available, the practice of surgery was greatly changed, and today the surgeon's skill accords him the pinnacle in the medical hierarchy.

Although *ether* was discovered in the early 1500s, its anesthetic properties were first demonstrated in 1846 in Boston. The amphitheater where major surgery was first performed under ether anesthesia still exists at the Massachusetts General Hospital and is called the Ether Dome in commemoration of that historic event. *Nitrous oxide* was discovered in 1776, but was not tried as an anesthetic until the middle 1800s. Its intoxicating properties were shown in public demonstrations of "laughing gas," and it is still used to reduce the pain of dental surgery. *Chloroform* also was introduced in the middle 1800s, and its sponsor was knighted by Queen Victoria. Today, other volatile anesthetics, such as halothane and methoxyflurane, have largely replaced their highly flammable predecessors for the alleviation of surgical pain. As with alcohol intoxication, the precise way in which the inhalants alter consciousness is poorly understood. One expert writes ". . . at the present time, a fundamental, predictive, and experimentally proven theory that can explain both the narcotic state and cellular changes produced by anesthesia is still lacking" (Cohen, 1975, p. 58).

The use of inhalants for psychoactive effects predates their use as general anesthetics. Legend reveals that the Oracle of Delphi inhaled carbon dioxide fumes emanating from a fissure in the temple floor to induce a trance before her prophetic pronouncements. The volatile anesthetics were used as intoxicants during the 1800s and are still subject to some abuse. Modern variations on inhalant abuse include sniffing of industrial solvents, gasoline, model airplane glue, cleaning fluids, antifreeze and many more. Relative accessibility and low expense seem to have popularized these intoxicants among some adolescents.

However, inhalant abuse is far from innocuous. Deaths from cardiac arrhythmias, laryngospasm, respiratory depression and even suffocation have been reported. Chronic exposure to some industrial solvents may also have serious toxic effects. Liver and kidney damage have been associated with carbon tetrachloride exposure, anemia and leukemia with chronic benzene exposure. Toluene exposure has been associated with brain damage and specifically cerebellar dysfunction. These are only a few examples in a very long list of documented inhalant-related toxicities.

The intoxicated state produced by inhalants is similar to alcohol intoxication. However, hallucinatory effects are sometimes described. The combined effects of alcohol and inhalants are additive, and alcohol is sometimes used to enhance inhalation-induced intoxication. Impaired judgment and motor coordination results from inhalant intoxication, and these effects are increased by the addition of alcohol. Since alcohol and inhalants have depressant effects on respiration and brain function, the combination of both drugs in high doses could be fatal.

Alcohol and Stimulants

Substances which may increase alertness and decrease fatigue have been classified as stimulants. These include a broad spectrum of drugs ranging from mundane and familiar items like coffee (caffeine), tea (theophylline) and tobacco (nicotine) to the most exotic and expensive of the illicit drugs — cocaine. In between are a number of medically useful stimulants that are prescribed for appetite suppression and for the treatment of hyperkinesis (hyperactivity) in children. It is paradoxical that a stimulant drug such as methylphenedate (Ritalin) would effectively calm a hyperactive child, but such are the mysteries of clinical psychopharmacology. Today's medically prescribed stimulants (dextroamphetamine, methamphetamine) are structurally similar to amphetamine, synthesized in the late 1800s and first used clinically in the 1930s to treat narcolepsy, a rare sleep disorder.

Occasional moderate doses of stimulant drugs increase feelings of alertness, energy and well-being and suppress appetite for food and drowsiness. Rapid euphoric effects probably account for most stimulant abuse. The nature of these effects is succinctly conveyed by the street term for intravenous amphetamines — "speed."

In the 1980s, cocaine abuse increased dramatically in the United States, and probably exceeds most other forms of stimulant abuse, although as with alcoholism, precise case counting is difficult. Yet cocaine is not in any sense a "new" drug. Like alcohol and opiates, cocaine has been used for centuries, and its exact origins are obscure. Cocaine is an alkaloid found in the leaves of the coca plant, indigenous to the Andes Mountains of South America. Coca leaves were highly valued by the Incas and used by the ruling classes and in religious rituals. After destruction of the Inca empire, the Spaniards came to recognize that coca chewing reduced fatigue and hunger, and encouraged the practice among Indian slaves. Perhaps coca leaves deteriorated in shipping, because coca chewing never rivaled the popularity of tobacco, tea and coffee in Europe. Its lasting introduction into Western civilization can be attributed to its medical usefulness as a local anesthetic and vasoconstrictor in ophthalmology and otolaryngology and to its popularization by Sigmund Freud. Freud took cocaine and described his reactions in a series of reports that have been translated and assembled by the Yale pharmacologist and scholar Robert Byck. Freud extolled cocaine's mood-elevating and fatigue-reducing effects and used it to treat the opiate addiction of a friend — only to induce cocaine abuse instead.

In the mid 1800s an enterprising vintner added cocaine to wine and "Vin Mariani" was endorsed with enthusiasm by composers and poets as well as Pope Leo XIII and President Grant. Even Coca-Cola, in its pre-1903 incarnation, contained active coca and was advertised as "the brain tonic and intellectual soda fountain beverage." In 1914, the pervasive nonmedical use of cocaine was prohibited by the Harrison Narcotics Act. Although cocaine is not a narcotic, the prohibitionistic climate of the time advanced relentlessly on the extant psychoactive substances, soon to include alcohol in its regulatory net. A curious amalgam of fear and prejudice toward foreign aliens contributed to the zeal for drug control. Public attitudes towards cocaine changed very rapidly. Cocaine had been endorsed by the Hay Fever Association as the official remedy for seasonal allergies. Suddenly, it was reviled as a source of insanity and crime. A recent resurgence in cocaine abuse has generated a new spate of rhetoric and fear. Systematic studies of cocaine's behavioral and physiological effects ceased with Freud, only to be resumed in the 1970s under the enlightened sponsorship of the National Institute on Drug Abuse.

The abuse of stimulant drugs can have serious medical and behavioral consequences. Repeated use of high doses of stimulants can be dangerously toxic, and overdose deaths have been reported. Seizures and respiratory failure may occur after high doses of cocaine; chest pain and loss of consciousness after high doses of amphetamine. Chronic cocaine free-base smoking is associated with impairments of pulmonary function similar to that seen after heavy tobacco smoking. Among the symptoms associated with amphetamine toxicity are suspiciousness, stereotyped and repetitious behavior, and intense preoccupation with the process of thinking and essential "meanings." Amphetamine-induced toxic psychosis is characterized by hallucinations, paranoid delusions, loose associations and picking at the skin. In monkeys, unrestricted access to stimulants results in similar skin mutilations, weight loss and eventually death. Toxic symptoms may occur to a greater or lesser degree in most stimulant abusers. It is disquieting to consider the impact of excessive stimulant use on long-distance motorists and those who share the road! After an extended period of stimulant abuse, people become depressed, lethargic and hungry. However, a withdrawal syndrome similar to the alcohol, opiate and barbiturate abstinence syndrome has not been reported.

Anecdotal accounts indicate that alcohol and stimulants are often used together. Alcoholics may use stimulants to counteract the sedative effects of alcohol. Stimulant abusers may use alcohol to modulate the high. Smoking coca paste, a recent variant of cocaine abuse, is often accompanied by alcohol consumption to control the cocaine-induced agitation and anxiety.

Information about stimulant-plus-alcohol fatalities is fragmentary at best. Amphetamine, methamphetamine and cocaine each are associated with reports of single-drug fatalities, but reports of combined stimulant-alcohol deaths are rare. Moderate blood alcohol levels (100–120 mg/dl) were found in 7 percent of drug-related deaths in which cocaine was the only other substance detected.

These findings suggest that an overdose of stimulants may be lethal, but the contribution of alcohol is unclear.

The behavioral effects of alcohol and stimulant combinations are unpredictable. Low doses of stimulants usually improve performance on skilled tasks and may enhance the stimulating effects of very low doses of alcohol. But, stimulants do not effectively antagonize alcohol intoxication. The impairments in judgment associated with either alcohol or stimulant intoxication are likely to be enhanced by their combined use. A combination of poor judgment and impaired performance due to intoxication on stimulants and alcohol can lead to a variety of hazards. An association between alcohol abuse and violence is well known. Recent statistics show that 60 percent of the cocaine-related fatalities were violent deaths. Stab wounds or gunshots dispatched the majority of homicide or suicide victims who also used cocaine.

ALCOHOL AND HALLUCINOGENS

LSD, mescaline, psilocybin, peyote, phencyclidine and DMT as well as ololiuqui derived from certain types of morning glory seeds and nutmeg (in very high doses) are commonly described as hallucinogens. The term is misleading, since many other drugs may also produce hallucinations at toxic dose levels and during drug withdrawal. In medicine, hallucinations are defined as perceptions which have no objective external source. The complex changes in sensations and perceptions associated with hallucinogen use are not necessarily hallucinations, but rather illusions, i.e., distortions or elaborations of an existing reality. Objects may appear larger or smaller, colors may appear more vivid and intense. These illusions may involve a blurring of usually distinct sensory modalities in which colors acquire musical qualities and sounds become associated with visual imagery. This blending of sensations has been called synesthesia. On a hypothetical scale of perceptual changes beginning with dreams, such illusory phenomena fall in about the middle and hallucinations at the extreme end.

The way in which drugs or psychotic disorders produce hallucinatory experiences is not understood. It was once believed that some hallucinations might be an overflow of dream activity into the waking state. Early studies of the correlations between brain activity reflected in EEG tracings and dream reports suggested that dreams occur most frequently during light sleep when there are distinctive eye movement patterns, a sleep stage referred to as Rapid Eye Movement (REM) sleep. Since alcohol and many other drugs suppress REM sleep during intoxication, the notion that drug withdrawal was associated with REM rebound and a spilling over of dream activity into waking hallucinations seemed plausible and provocative. Later studies showed that hallucinations could occur during chronic alcohol intoxication and that drug-related REM suppression did not invariably result in a REM rebound after drug use was discontinued. Moreover, it was soon recognized that dreams also occur during sleep stages other than REM, and it became more difficult to use EEG measures to trace the relationship between dreams and hallucinations. These exceptions still await integration into another theory. It may be many years before the biology of sleep and dreams, or illusions and hallucinations, is well understood.

Although the perceptual changes induced by hallucinogenic drugs have been widely publicized and often glamorized, these are only one aspect of the spectrum of drug activity. Physiological effects of LSD include weakness, dizziness, increased blood pressure, heart rate and temperature, tremor and nausea, and widely dilated pupils. These physiological and perceptual changes are associated with rapidly shifting fluctuations in emotion, ranging from elation to panic. If panic predominates, the user is described as having a "bad trip." If elation persists, then a sense of universal harmony and expanded awareness is reported by some. The ritual and religious use of the peyote cactus and psilocybin-containing mushrooms by the Mexican Indians had continued for many centuries before the discovery of LSD's psychedelic properties during the 1950s.

LSD was synthesized in 1938 from a substance found in the fungus of rye, and its psychoactive effects were discovered accidentally. The plethora of LSD "uses" during the 1950s seems something of an aberration today. LSD was tried as a therapy for disorders as disparate as alcoholism, pain associated with terminal cancer and psychosis. As with most instant panaceas, an uncritical enthusiasm soon waned and attention focused on the adverse side effects. The therapeutic value of LSD was never satisfactorily demonstrated. Prolonged psychotic, paranoid and depressive reactions to LSD were reported in susceptible individuals who may or may not have become ill eventually, even without LSD. Other more subtle disorders of thinking were more difficult to attribute unequivocally to LSD use, but few responsible persons would advise use of hallucinogens to improve abstract reasoning capacities.

The popularization of LSD by advocates of transcendent experience in the 1950s and 1960s, and its symbolic association with one segment of the American Youth Culture, is part of our very recent history. As with cocaine and opiates at the turn of the century, psychedelics became associated with murder, crime, insanity and an "undesirable" element in society. The genuine hazards associated with its use were amplified into a crescendo of concern. State and federal laws prohibiting the possession and recreational use of hallucinogens were enacted by the middle 1960s. The transition from possible panacea to evident menace occurred with unprecedented rapidity.

Enter Phencyclidine (PCP)

The marked decline in the use of LSD-like hallucinogens by the late 1960s was soon followed by a new drug problem, the illicit use of phencyclidine (PCP), also called angel dust. Between 1976 and 1977, the percent of twelve- to seventeen-year-olds who reported ever using PCP increased from 3 to 5.8 percent. Young adults (eighteen to twenty-five) who reported PCP use increased from 9.5 to 13.9 percent.

Phencyclidine is a dissociative anesthetic, widely used in veterinary medicine. Use of PCP as an anesthetic in man was discontinued in the late 1950s after post-operative delirium and psychotic reactions were observed in the patients. Unlike its predecessor hallucinogens, PCP has not been touted as a stepping stone to a mystical transcendent experience. Indeed PCP's known effects seem so aversive that its voluntary use is rather puzzling. Phencyclidine may be taken accidentally, since PCP is often misrepresented as LSD or mescaline, amphetamine or cocaine, or even THC (the psychoactive ingredient in marijuana).

PCP intoxication produces feelings of numbness in the extremities, incoordination, slurred speech, nystagmus, diffuse sweating and catatonic rigidity. Users may feel confused, disorganized and dissociated, and report changes in their body image. Some may use PCP intravenously to induce repeated episodes of unconsciousness.

At moderate to high doses, coma, stupor, fever, and convulsions are only a few of the medical complications observed, and overdose deaths from respiratory depression are not uncomon. Protracted psychotic reactions may occur after a single dose of PCP. Patients often appear schizophrenic; the clinical picture may include bizarre violent behavior, suicidal attempts, paranoid delusions, auditory hallucinations, confusion and incoherent speech. Some patients may resemble catatonic schizophrenics, with muscular rigidity and mutism. The cumulative effects of chronic PCP use on brain function have not been studied systematically. Some chronic users report persistent problems with recent memory, thinking and speech.

These impressions of phencyclidine toxicity are derived primarily from evaluations by emergency ward physicians and psychiatrists who treat PCP victims. Chronic phencyclidine users present a somewhat more salutary view of the drug. Increased sensitivity to stimuli, stimulation, dissociation, improved mood and feelings of intoxication were regarded as positive by one group. "Weird but interesting" was another typical description. Most also reported anxiety, disorientation, perceptual disturbances and many felt hyperexcitable, confused, irritable, paranoid and had difficulties talking.

On balance, the risks of severe toxic reactions would seem to outweigh the "positive" effects of PCP intoxication, but its use persists. PCP is the prototypic example advanced to support our hypothesis that a change in state of consciousness is the goal of the drug abuser and almost any change will do.

Alcohol and Hallucinogens

Since phencyclidine is a depressant drug, its use with alcohol should increase the depressant effects of both. The toxicity of phencyclidine may be enhanced by the addition of alcohol. It is obvious that the impairment of coordination and confusion associated with phencyclidine intoxication alone would argue strongly against driving, and the addition of alcohol would further increase the potential hazards.

The LSD-like hallucinogens are not depressants and little is known about their interactions with alcohol. Anecdotal accounts by polydrug users suggest that alcohol is not used to change the effects of hallucinogens and is not used as a substitute when hallucinogens are not available. Since the effects of hallucinogens include distractibility and distortions of visual perception, which probably would be increased by the addition of alcohol intoxication, driving or operating machinery during hallucinogen intoxication would be dangerous. The addition of any other intoxicant would be likely to further increase the potential hazards.

ALCOHOL AND MARIJUANA

Marijuana, like alcohol and opiates, was first used many centuries ago. It was valued as a medicine and an intoxicant in ancient China and India, and also as a source of hemp and paper. Its fibers were woven into rope and baskets and cloth, and crushed into pulp to make tablets for China's burgeoning bureaucracy. Hemp fiber fragments from Chinese tombs were made an estimated one hundred years before Christ. The extent of its use as an intoxicant in ancient times is unknown. It is believed that early Hindus revered marijuana for its inspirational, medicinal and aphrodisiacal powers. Even today Indian holy men use bhang to facilitate transcendental meditation — a practice shared by the Lord Buddha, who allegedly subsisted on a single cannabis seed each day during the period of his enlightenment and asceticism.

In America, marijuana has had a somewhat mercurial history. Britain ordered the Virginia colonists to grow hemp as early as 1611, but despite various proclamations and inducements, tobacco remained their primary crop. In New England, cannabis was an early and favored crop in the Massachusetts Bay Colony. Hemp was needed for cloth and rope for the developing shipbuilding industry. The first rope factory, known as a rope walk, was established in Salem in 1635. Riggings made from hemp launched New England's sailing vessels. During the Revolution, widespread shortages made hemp even more valuable as an item of barter and as raw material for clothes. There is little evidence to indicate that the hardworking hemp farmer learned of the intoxicating potential of his crop. As hemp farming declined during the Civil War, in response to decreased demand for and rising competition from iron cable, the fashionable European custom of hashish use began sporadically in America. At least thirty patent medicines in the post–Civil War era contained cannabis. It was commonly prescribed for a variety of aches and pains, as well as for loss of appetite and, curiously, for withdrawal from alcohol.

The transition in attitudes toward marijuana from indifference to alarm and revulsion began in the early years of the twentieth century and reflected many interacting social processes, including the anti-alcohol movement that culminated in national prohibition. Only eighteen years later, *four* years after Prohibition was repealed, marijuana use was outlawed by passage of the Marijuana Tax Act in 1937. In 1941, cannabis was removed from the U.S. Pharmacopoeia and National Formulary. The subsequent events are recent history. Marijuana regulation policies continue to dominate the nation's concern over drug misuse.

Marijuana is a hardy, almost ubiquitous annual weed. In 1753 the botanist Linnaeus named the plant *Cannabis sativa,* and today its psychoactive products are called ganja, hashish, bhang, dagga, kif and marijuana depending on the method of preparation and the country of origin. The resinous exudate of flowering leaves of the male and female plant contain the intoxicating substance. The active ingredient in the resin was isolated and chemically analyzed only as recently as 1964. It was named THC, an acronym for delta-9-tetrahydrocannabinol. THC can now be produced synthetically and is medically useful for reducing anorexia and nausea in cancer victims treated with chemotherapy as well as in the management of glaucoma.

Marijuana is a mild intoxicant — the effects of a cigarette are similar to those of a single cocktail. As with alcohol, both expectancy and experience are important determinants of the intoxication experience — inexperienced smokers may have little or no effect from a single "joint." Chronic marijuana users describe marijuana intoxication as a "high," and this term appears to include an increased sense of well-being and relaxation, an enhancement of visual and auditory perceptions as well as heightened sensations of touch, taste and smell. Marijuana also stimulates appetite, and users decribe a hunger for sweets as "the munchies." As with alcohol intoxication, motor coordination and the capacity for goal-directed behavior may be impaired and smoking may be followed by drowsiness. Time perception is also altered: the past, present and future may become somewhat confused, and time seems to pass very slowly.

At higher doses, this altered time sense may increase, thinking may become disorganized and confused, and sometimes hallucinations, delusions, feelings of depersonalization and paranoia may occur. Under these circumstances smokers may become very anxious and even panicked, but marijuana toxicity seldom requires acute admission to an emergency ward for either a psychiatric or medical emergency.

These somewhat evanescent changes in thinking and feeling states are accompanied by increased heart rate, increased systolic blood pressure and the appearance of bloodshot eyes due to reddening of the conjunctivae. Smoking of marijuana involves holding the inhaled smoke in the lungs for a time, and this has been shown to have adverse effects on lung function. Chronic smoking also has a deleterious effect on disorders such as bronchitis and asthma.

The effects of persistent, chronic use of marijuana are poorly understood. There is no evidence that marijuana produces irreversible changes in brain function, but there has been considerable concern about subtle changes in personality such as apathy and disinterest in conventional achievement goals, lethargy, unproductivity and indolence. This pattern of behavior has been called the "Amotivational Syndrome" and is often ascribed to a combination of the direct effects of marijuana and preoccupation with acquiring more marijuana. There is continuing controversy both about the validity of the concept of an "Amotivational Syndrome" and the extent to which marijuana can be justifiably implicated as "the cause." Among college-age students, these characteristics seem synonymous with clinical depression, and systematic diagnostic criteria to distinguish these symptoms from depression have not been developed. Large population surveys of American marijuana users and studies of cultures where heavy marijuana use is common (e.g., Greece and Jamaica) have not reported a higher than expected incidence of psychiatric problems or impairment of work performance and overall productivity. Clinical studies of young American men given the opportunity to earn money and marijuana cigarettes in a hospital research ward environment also found no obvious decrement in work capacity. Men smoked between four and six marijuana cigarettes each day and worked continually at a simple task for periods of six and a half to fourteen hours each day to earn *both* money and marijuana. Similar findings have been reported in several clinical studies and suggest that in a psychologically healthy individual, marijuana smoking per se does not necessarily decrease capacity for sustained work or impair interest in earning money.

Since alcohol and marijuana have several similar and overlapping effects, concurrent use of both drugs increases their intoxicating effects. Alcohol has been reported to enhance the subjective "high" following marijuana smoking. The combined effects of alcohol and marijuana on performance are "additive" in the sense that impairments following use of both drugs are greater than observed with either drug alone. Motor skills and attention factors related to driving are also more impaired by alcohol plus marijuana, and given the current death toll from alcohol related highway fatalities, this is an issue of continuing concern. Physiological effects of marijuana, such as increased heart rate and reddening of the conjunctiva, are also increased by the addition of alcohol.

Discussions of the probable consequences of decriminalization or legalization of marijuana usually consider the question of whether marijuana would be substituted for alcohol or more alcohol would be used with more marijuana to produce greater intoxication. This question was examined on a clinical research ward where young men could earn and use marijuana only, alcohol only, and then alcohol and marijuana in combination. It was somewhat surprising to find that fourteen of the sixteen subjects studied drank significantly less alcohol when marijuana was concurrently available. Moreover, although alcohol and marijuana were often used together, there was no evidence of unusual intoxication or toxic drug interactions. Marijuana smoking tended to increase through time, independently of concurrent alcohol availability.

The Quest for Sobriety

Although most of the interactions between alcohol and other drugs are undesirable or toxic, it is possible to imagine a beneficial combination. Perhaps the most useful drug-alcohol interaction would be one that reversed alcohol intoxication and restored sobriety. Another useful drug would be one that prevented alcohol intoxication or modulated its behavioral effects. Thus far, these pharmacological tools have eluded the medicinal chemist's craft, but serendipity has provided some promising possibilities borrowed from the realm of opiate antagonists, drugs that prevent or reverse the effects of opiates. This apparently successful tactic marks the latest event in a very long chain of improbable antidotes for intoxication.

Speculation and Sobriety

The search for a sobering agent spans many centuries. The remedies for drunkenness are entwined with the folklore and herbal wisdom of various cultures. It was the prevention of intoxication rather than its reversal that most intrigued the pragmatic ancients.

A jewel and a flower were believed to prevent intoxication in ancient Greece. Violets were used to decorate banquet tables and allegedly prevented drunkenness. Although there are about four hundred species of the genus *Viola*, the fragrant *Viola odorata* is believed to have been preferred by the violet-loving Athenians. The purple amethyst also was imbued with the power of the violet. *Amethystos*, the Greek word for violet and amethyst, means "not drunken." Today, the term "amethystic agent" is a welcome lyrical alternative to the more prosaic "sobering agent" as the search continues.

Milk is another time-honored prophylactic for intoxication. Its effectiveness is only marginal, but the belief does have a physiological basis. Any food in the stomach delays alcohol absorption, but milk, as well as cheese, yoghurt, eggs and meat, mobilize proteolytic enzymes, which digest these proteins very slowly. Thus, milk remains in the stomach longer than many other foods and delays alcohol absorption accordingly.

Another popular but specious strategy for maintaining sobriety is to drink the same type of alcoholic beverage all evening. It is not clear why mixing drinks is thought to increase intoxication. It is illogical to postulate the converse, i.e., that drinking only scotch will decrease intoxication. It is the alcohol in whatever beverage that intoxicates and equivalent doses of alcohol in beer, wine and scotch produce equivalent levels of intoxication.

When prevention fails, a series of somewhat questionable restoratives is offered. Cold showers, hot coffee, exercise, forced vomiting and eating are endorsed by some. None have been shown to have any measurable effect on blood alcohol levels or intoxication. In 1981, the effect of coffee (caffeine) on various performance measures and blood alcohol levels was examined once again in a sophisticated pharmacological study. High doses of caffeine (up to 500 mg) did not reduce the performance decrements associated with alcohol intoxication in a blood alcohol range often achieved by social drinkers. Moreover, subjects could not distinguish between the effects of caffeine and decaffeinated coffee. Comparable studies of cold showers have not been done. But given the kinetics of alcohol metabolism, any activity that engages the inebriate for about two hours should favor increased sobriety.

Science and Sobriety

The scientific search for an amethystic agent has been no less eclectic in scope. Vitamins, carbohydrates, hormones, amino acids, amphetamines and other stimulants, antihistamines, cholinergics, antianxiety and antidepressant drugs, as well as potassium and calcium have all been seriously considered and evaluated. The presumed mode of action believed to lead to sobriety differed as greatly as did the contenders. Some, such as carbohydrates and certain hormones, seemed promising because these might increase the rate of ethanol elimination from the body. Others, such as stimulant drugs, might counteract alcohol's depression of brain activity. In 1975, a comprehensive review on the subject concluded that "it can be stated unequivocally, that with the exception of dialysis, no effective method exists for rapidly eliminating ethanol from the body, nor are there any known chemicals or techniques that have been shown to reverse the CNS depressant effects of large intoxicating doses of ethanol in man."

Yet, as the echoes of that statement faded, a series of scattered clinical reports suggested there was a new amethystic agent that was effective for alcohol. The new amethyst was *naloxone*, a narcotic antagonist that can prevent and reverse the effects of narcotic drugs. This drug was described in the literature on opiate analgesics in 1967, but since there is no cross-tolerance between alcohol and opiates, there was no reason to expect that such a drug would antagonize alcohol. The ideological schism that separates the fields of alcoholism and drug abuse further reduced the probability of any meaningful exchange in research or treatment. But through serendipity and increased recognition of the nature of the opiate receptors in the brain, research on neuropharmacology has advanced in new directions.

Opiate antagonist drugs can both prevent and reverse opiate intoxication. If given before opiate use, the opiate antagonist blocks the subjective and physiological effects of heroin or morphine or dilaudid. Some antagonists, such as *naltrexone*, block the action of heroin for twenty-four hours or longer and are useful in the treatment of some well-motivated opiate-dependent patients. One 50 mg dose of naltrexone each day significantly suppresses heroin self-administration by heroin addicts. The narcotic antagonist is thought to occupy the opiate receptors in the brain so that opiate drugs cannot bind to those receptors and therefore cannot affect the user. No such pre-treatment antagonist exists for alcohol. (Antabuse [disulfiram] is *not* an alcohol antagonist and acts by a very different mechanism. It does not prevent alcohol's effects but does make the drinker very sick and may cause death if enough alcohol is consumed.)

Opiate antagonists can also reverse heroin intoxication and precipitate heroin withdrawal. *Naloxone* is an opiate antagonist with a relatively short duration of action, about four hours. Naloxone is often given to prospective admissions to methadone maintenance programs to insure that they are in fact opiate-dependent and not just looking for a free source of opiate drugs. Within minutes after naloxone administration, signs and symptoms of opiate withdrawal appear in opiate addicts. Although the exact mechanism of antagonist action is unknown, it is thought that the antagonist displaces the opiate from the opioid receptors in brain.

Clinical reports suggest that naloxone may also be useful in reversing alcohol intoxication in severely sedated individuals. Overdose patients brought to an emergency ward are usually unable to describe what they have taken and often are not accompanied by a reliable informant. Analysis of blood samples to determine the type of drug or drugs may take too long when a patient has depressed respiratory function and is comatose.

Low doses of intravenous naloxone restored normal breathing within minutes in some comatose patients later found to have very high levels of alcohol in their blood (250–580 mg/dl), and no evidence of narcotic drug use. The coma may be slightly reduced by naloxone injections, but several hours are usually required for complete restoration of consciousness. Since naloxone is a safe drug with highly specific opiate antagonist properties, and an emerging record of efficacy with alcohol and mixed alcohol-barbiturate intoxication, it may prove to be an important amethystic agent in emergency ward medicine.

Despite reports of dramatic reversals of alcohol-related coma, naloxone may not be an effective alcohol antagonist at lower levels of

CHAPTER 18

Alcoholism Treatment and Prevention: Rhetoric, Faith and Science

Alcoholism claims many victims. The toll in early deaths, disrupted lives and lost productivity is incalculable. Amidst the panoply of anguish, relatively little attention is paid to the fact that alcoholics do recover. And many have — some spontaneously, some through treatment, some through voluntary helping organizations like Alcoholics Anonymous. There is a potentially effective treatment for everyone who suffers from alcoholism today. The treatment may be different for different individuals and may even combine several therapeutic approaches.

Some recovered alcoholics have difficulty accepting that there is no single, unvarying treatment approach guaranteed to work for everyone. Well-intentioned former alcoholics often promote a pathway to abstinence, reminiscent of the true road to salvation extolled by fundamentalist religious reformers. Many recovered alcoholics tend to be ferociously intolerant of any treatment approach different from what was helpful to them, and rhetoric and an evangelical zeal often supersede dispassionate evaluation of treatment outcome. However, we contend that rigid adherence to any single treatment approach is usually not in the best interest of the patient and is often antithetical to a humane medical and scientific approach to alcoholism. The best treatment strategy is one that explores the options with the patient, tries the most promising techniques, evaluates the outcome, and if necessary tries again.

SOME BASIC CONCEPTS

Alcoholism treatment usually proceeds in two stages. The first phase, detoxification, focuses on the acute medical and emotional problems associated with prolonged heavy drinking. Patients are treated for symptoms of withdrawal from alcohol as well as for any related illness or malnutrition. The second phase of treatment involves management of the drinking problem and any other on-going psychiatric or medical disorders. Before describing the process of detoxification and some contemporary approaches to alcoholism treatment, it is important to consider three factors that critically influence the temporal course of treatment and its eventual outcome. These are the nature of alcoholism, the types of treatment offered, and the phenomenon of spontaneous remission of problem drinking.

1. Alcoholism Is a Chronic Illness

Alcoholism is often an insidious process that develops gradually at first and then accelerates. The alcoholic may deny that drinking is a problem for many years, despite accumulating evidence to the contrary. Unless disruptions of family, job, friendships or health force the issue, the alcoholic may continue addictive drinking until death. Confrontation does not necessarily change the drinking pattern. Some alcoholics choose alcohol above all else. Others recover, only to relapse again and oscillate between alcoholism and abstinence for many years. Many alcoholics respond well to treatment and recover completely.

Alcoholism is a chronic disorder that requires long-term care. It is more analogous to other chronic disorders such as diabetes and heart disease than acute infectious diseases. As do diabetes and chronic heart disease, it requires continuing medical attention rather than a one-time treatment such as the highly specific, targeted antibiotic treatment that can rapidly cure an infection. Initially, treatment of a chronic disorder may be conservative, but later, as the disease progresses, heroic efforts may be necessary. In the treatment of alcoholism, as with other chronic diseases, the immediate goal is to achieve remission or reduction of the severity of the disorder. Complete cures are often impossible, but treatment does alleviate suffering and prolong life. Since alcoholism causes chronic social and physical debilitation and shortens the life span, the first and most important goal of treatment is to curtail these insidious but not necessarily inevitable aspects of the disease process.

2. Multimodality Treatment Is Usually Most Effective

The strengths and weaknesses of alcoholism treatment programs often depend on the number of options available to the patients. A multimodal, pluralistic approach to treatment is preferable to reliance on a single form of therapy. There are few, if any, chronic diseases that respond favorably to a single mode of treatment. Patients with diabetes may require a spectrum of treatment procedures throughout the course of a chronic lifetime affliction. Treatment plans may range from the relatively benign management of dietary intake to the regular injection of insulin plus intermittent use of other drugs to manage intercurrent illnesses associated with diabetes. There are few specialists who treat diabetes who would argue that diet alone or insulin alone or any other specific intervention is the only answer for lifetime management of this chronic disorder. Similar considerations are necessary to develop the most effective treatment programs for alcoholism. For example, membership in Alcoholics Anonymous may be of great value for some, but may not be of help to others and may conceivably even exacerbate drinking problems in a few alcoholics. All forms of alcoholism treatment have advantages and limitations, depending on the genesis of the alcohol problem, the evolution of the problem over time, the severity and chronicity of the problem at the time of treatment, and the patient's perception of the value and need for different treatment procedures. The most successful treatment programs are those which encompass a variety of types of therapy and where treatment approaches are innovative and flexible rather than dogmatic or rigid.

3. Spontaneous Recovery from Alcoholism Can Occur

Many persons recover or achieve remission from an illness without specific therapeutic intervention. An apocryphal story alleges that one of the best-kept secrets of practitioners of internal medicine is that most people who are ill will get well without the physician's intervention. Indeed, the major maxim of medical care is to "at least do no harm" while treating the patient. These remarks should not be interpreted as reflecting a nihilistic attitude toward medicine or as an attempt to dissuade people from seeking medical attention when they are ill. Rather, the intent is to emphasize that remissions do occur in most chronic diseases, and it is not surprising that spontaneous remissions occur during the course of chronic alcoholism. While no one knows the exact rate of spontaneous remission of severe alcohol-related problems, evidence derived from longitudinal national surveys suggests that one out of every four or five persons with serious drinking problems will spontaneously improve within a three-year period.

Clearly, alcohol problems do not vanish as a function of some magical process. Very often remission is associated with a serious attempt by the problem drinker to modify drinking behavior in concert with help from friends or family. Because spontaneous remission does occur, a better understanding of how this process takes place is essential for judging the efficacy of any single treatment or combination of treatments. Ultimately successful treatment programs must achieve remission rates that are statistically significantly greater than what would be expected to occur by spontaneous remission alone.

In summary, alcoholism is a chronic disease. Complete cures are rare, but remission, with reduced disability and prolonged life, is quite possible. Alcoholism rarely responds to a single, unimodal form of treatment. An eclectic rather than a narrow treatment approach offers the best promise for recovery. Persons with alcohol problems do sometimes get better spontaneously, though usually with the help of many subtle and overt attempts at behavior modification by the problem drinker and those who are close to him or her. Given these three general considerations about the major factors that impact upon alcoholism treatment, let us now examine some specific components of treatment procedures, with the recognition that, ideally, each component should be part of a harmonious collaborative therapeutic endeavor.

DETOXIFICATION

The first step along the treatment pathway for many alcoholics is the process of detoxification. Detoxification may range from relatively simple and uncomplicated "sobering up" to intensive medical treatment of alcohol withdrawal states. Alcohol withdrawal may be so severe as to require intensive medical treatment in order to pre-vent serious debilitation and even death. Most alcoholics have expe-rienced repeated periods of detoxification in their homes with little or no assistance, except from friends and family. A plethora of home remedies have been used for sobering up and reducing the discom-forts of alcohol withdrawal but there is little evidence that any of these have any proven scientific value (see Chapter 17). Perhaps the most common treatment for alcohol withdrawal in ordinary, nonin-stitutional settings is the consumption of alcohol itself. Tapering off drinking or gradual weaning from heavy to light alcohol intake is often effective in reducing the severity of the alcohol withdrawal syndrome. Gradual reduction of alcohol intake has even been em-ployed for detoxification in some institutional and hospital settings. But usually the process of alcohol detoxification in institutions in-volves abrupt cessation of alcohol intake, in the context of social and medical management programs designed to ease the process of with-drawal.

Detoxification Centers

For many years, alcoholics who were homeless and had no personal resources were detoxified while temporarily incarcerated in a court-house or jail. As a consequence of documentation of the inhumanity of this process, as well as the strain placed upon an overburdened criminal justice system, many communities established detoxifica-tion centers to replace "drunk tanks" administered by the police and the courts. Enlightened community officials regarded this as a signif-icant step forward in the transition of the perception of alcoholism from a moral affliction to a disease. But as the detoxification centers began to develop a caseload record, others came to view alcohol de-toxification centers more cynically as only another revolving-door enterprise, or just another way to dispose of alcoholics. Despite high rates of recidivism and the limitations of the detoxification centers, a great gain in human dignity has been achieved for those alcoholics who endure chronic social and economic deprivation. At least the stigma of the jail and the usually deplorable conditions of the "drunk tank" no longer add to their misery.

Detoxification centers are undoubtedly more humane than jails as resources for short-term restitution of sobriety and treatment of withdrawal signs and symptoms. But more affluent problem drinkers often do not want to use these detoxification centers. What happens to the middle-class blue- or white-collar worker who has been drinking heavily for three consecutive weeks, and has not violated any laws, driven recklessly, engaged in barroom altercations or insulted a policeman, when he decides to stop drinking and is becoming increasingly anxious and shaky?

Whereas most state, municipal and federally supported detoxification centers care for individuals with very limited economic resources, the majority of employed Americans seek assistance from proprietary and nonprofit treatment centers where services are paid for by health insurance coverage or personal savings. Many of these detoxification facilities are part of comprehensive treatment units that provide multimodality care for alcoholism following detoxification. Anyone who wishes assistance for alcohol detoxification may be admitted to most of these detoxification facilities by direct application (a phone call or a personal visit) or by referral from a general hospital emergency ward facility or a personal physician.

The Detoxification Process

Most detoxification procedures take three to five days, although this may vary depending upon the duration and quantity of heavy drinking. Detoxification programs are usually carried out under direct medical supervision. Most patients progress through alcohol detoxification without serious medical problems. The degree of psychological discomfort may vary from mild anxiety to severe nervousness and depression. About one out of ten patients develops a moderate to severe medical problem that requires careful diagnosis and prompt and efficacious care. For example, many patients develop an increase in heart rate (called tachycardia) during alcohol withdrawal. Although in the majority of cases this problem alone requires no specific medical intervention, some patients who develop tachycardia during alcohol withdrawal also develop cardiac arrhythmia (irregular heartbeat). Some arrhythmias are relatively benign, but others may be life-threatening and necessitate expert treatment. Similarly, the majority of patients undergoing detoxification experience mild to moderate tremor of their hands, tongue and occasionally the whole body. Such tremor usually disappears spontaneously within twenty-four to forty-eight hours after cessation of drinking, but in a minority of cases severe generalized seizures may occur, with whole body convulsions and loss of consciousness. Such seizure disorders require diligent medical management. Heavy sweating during alcohol withdrawal is also common but very heavy sweating, associated with vomiting and poor fluid intake, may produce a state of severe dehydration which requires judicious treatment with intravenous fluids. In summary, the alcohol withdrawal syndromes usually are not life-threatening, but in a minority of instances, patients in alcohol withdrawal may require expert medical care.

It is often impossible to determine which patients will have a relatively easy course of alcohol withdrawal and which will develop major medical complications. Surprisingly, the duration of chronic intoxication and the amount of alcohol consumed do not reliably predict the relative severity of the alcohol withdrawal syndrome. Some patients have very severe withdrawal symptoms after relatively low alcohol intake for short periods of time. A past history of severe withdrawal symptoms often predicts an ominous course of withdrawal. Given these circumstances, it is obvious that alcohol detoxification programs should be equipped to provide comprehensive medical care to insure maximal safety for their patients.

There is a consensus that good general medical management is the most important feature of adequate and safe detoxification. This means that patients are under careful surveillance for possible illness caused directly by alcohol withdrawal or by general debilitation during withdrawal. Good medical surveillance involves routine, periodic monitoring of heart rate, temperature and blood pressure. It also involves careful attention to nutritional status. Patients who can take food and fluids by mouth usually do not require intravenous feeding. Other patients may be severely dehydrated and cannot retain fluids because of vomiting. When intravenous fluid therapy is undertaken, it is essential to monitor blood ion concentrations (i.e., sodium, chloride, potassium, magnesium, calcium). An excess or a deficiency of essential ions can be caused by fluid loss or fluid retention associated with overzealous intravenous fluid therapy. Abnormal ion concentrations can be life-threatening.

Drug Treatment of Alcohol Withdrawal

Many patients undergoing alcohol detoxification feel restless, apprehensive, fearful, despondent and have trouble sleeping. The use of drugs to reduce the severity of these symptoms is an important adjunct to the medical treatment of the alcohol withdrawal syndrome. The most commonly used class of drugs for the treatment of alcohol withdrawal are a group of tranquilizers, the diazepines. These drugs are also the most widely used compounds for the treatment of anxiety associated with numerous physical and psychological ailments. The diazepines not only reduce anxiety, but also help reduce tremor, tachycardia (rapid heartbeat), heavy sweating and insomnia that occur during alcohol withdrawal.

The diazepines may be given by mouth or, when necessary, by intramuscular or intravenous injection. These drugs have been shown to be quite safe for the treatment of alcohol withdrawal and, when used judiciously, produce minimal adverse side effects. They can also be used safely in patients with alcohol withdrawal who are suffering from other physical disorders that may be associated with either chronic alcoholism or withdrawal, e.g., heart disease, gastrointestinal disease and liver disease. The decision of whether to administer diazepines or other drugs during withdrawal must be made on the basis of a thorough diagnostic assessment of the patient's physical and mental status. Some patients will require little or no drug treatment from the beginning of withdrawal signs and symptoms.

Every form of drug therapy for virtually every condition, be it anxiety-reducing treatment during alcohol withdrawal, or antibiotic treatment for infectious disease, carries the risk of adverse drug side effects. The decision to use drugs is based upon a risk-benefit ratio and, depending on the situation, uncritical and overzealous use of drugs can be as dangerous as withholding drugs when they are required. There have been repeated admonitions against establishment of a formulary or "cookbook" regimen for drug treatment of alcohol withdrawal. As in all medical conditions that require drug treatment, the dose and duration of drug administration during alcohol withdrawal must be adjusted for each patient. A sound, general principal derived from studies of clinical pharmacology is that the best dose is the lowest which still effectively produces relief or remission.

In summary, during the past decade, opinions about the best treatment modality for alcohol detoxification have changed markedly. The "drunk tanks" in municipal jails were established in part because hospitals and health care facilities once were loath to accept alcoholics for treatment. As the alcohol withdrawal syndromes were better understood, largely through the efforts of dedicated clinical investigators treating alcoholic patients in large, municipal hospitals, it became apparent that the best treatment for alcohol withdrawal was provided by the well-staffed and well-equipped general hospital. Consequently, more and more general hospitals began to accept patients for treatment of alcohol withdrawal. This policy has resulted in an enormous decrease in morbidity and mortality associated with the alcohol withdrawal syndromes. But as hospital costs have risen, policymakers for state, federal and municipal governments have argued that less expensive but equally efficacious programs should replace hospital treatment for alcohol detoxification. These arguments, based largely on fiscal considerations, rather than scientific data, have received approbation from a number of non-medically oriented individuals who subscribe to "a psychosocial detoxification model." The major thrust of this "model" is that detoxification can be carried out in settings other than the hospital and without use of most forms of medical diagnosis and treatment. The popularity of this position has increased among legislatures faced with increasing health care costs, inflation and economic recession. It is regrettable that for a time, custodial care at bargain rates may again replace effective medical treatment of alcohol withdrawal. General awareness of the life-threatening concomitants of alcohol withdrawal may help to alert the socially oriented care providers to the possible need for emergency medical treatment. An alcoholic in alcohol withdrawal is fragile psychologically and medically vulnerable. Denial of either aspect of this acute illness places the alcoholic at unnecessary risk.

TREATMENT OF CHRONIC ALCOHOLISM

There is no single "best" treatment for the condition of chronic alcoholism. As we have seen in earlier chapters (14 and 15), alcoholics do not fit into any homogeneous mold. Anyone from any social, religious, ethnic, or economic background can develop alcohol problems. Consequently, it is not surprising that some types of treatments are more effective for some types of people than for others. In this section, three general approaches to alcoholism treatment are described: psychotherapy, behavior therapy and Alcoholics Anonymous. It should be noted that these are not mutually exclusive and may be effectively combined in a multimodality program with or without pharmacotherapy.

Psychological and Psychiatric Treatment

Psychiatric and psychological treatment programs range from long-term in-patient care in psychiatric facilities to brief weekly individual or group therapy sessions. Some patients who seek psychiatric or psychological assistance for alcohol problems find their way into treatment programs as a function of chance, rather than as a consequence of rational decision-making. If the particular treatment proves unsuitable, the patient may not try to find an alternative. This is unfortunate, since many people who do not respond favorably to one mode of treatment may improve in another therapeutic milieu. This is not meant to encourage "window shopping" or rapid alternation between treatment programs, a strategy certain to postpone any beneficial change in drinking behavior. But, patients who fail in treatment with one form of psychological or psychiatric therapy should be reassured that their case is not hopeless and encouraged to try another type of treatment program. In alcoholism, there is no treatment of last resort!

Patients often ask if there is a difference in the types of treatment that can be provided by psychiatrists and psychologists. Both psychiatrists and psychologists are trained to carry out diagnostic interviews, to assess the type and severity of psychological problems, as well as interpersonal and environmental factors that contribute to alcohol abuse and alcoholism. Both professionals are able to devise and conduct individual, group or family treatment programs within private or institutional settings. Psychiatrists can also prescribe drugs for treatment of anxiety and depression as well as medications which may be helpful during alcohol withdrawal states. Psychiatrists have also been exposed to training in general medicine and thus they may be more likely to identify and treat medical disorders associated with the patient's alcohol problem. Psychologists often are better trained in areas related to social, economic and contextual aspects of alcohol abuse. Psychologists are also more experienced in the administration and interpretation of standardized psychometric instruments for measurement of mood and intellectual and perceptual functions.

During the past three decades, there has been a marked reduction in doctrinaire and parochial positions held by professionals in the fields of psychiatry and psychology. Few psychologists or psychiatrists would identify themselves as belonging to a particular "school" of behavioral science. While no one would deny that great clinicians such as Sigmund Freud, Adolf Meyer, Harry Stack Sullivan, Karen Horney and others made outstanding contributions to psychiatry and psychology, there are few contemporary psychiatrists or psychologists who would identify themselves as Freudians, Sullivanians, Meyerians, etc. Most contemporary psychologists and psychiatrists use eclectic and flexible treatment procedures, and are not limited by the dictates of a single school or theoretical position.

The choice of a psychiatrist or psychologist as psychotherapist should depend in part on the patient's previous treatment history. If a patient has not had any form of assistance before visiting a psychologist or psychiatrist, he or she may be well advised to obtain a general medical evaluation before doing either. Psychologists are not trained to diagnose medical disorders, and most psychiatrists do not carry out physical or laboratory examinations. Since many patients with alcohol problems also have medical disorders associated with the drinking problem, a thorough medical evaluation should always be part of the diagnostic workup that precedes psychiatric or psychological treatment.

Should patients be treated on an individual basis, or in groups with patients who have similar or different problems, or in a setting involving the participation of family members or close friends? The answer to this question depends, in part, upon the personality and preference of the patient and upon the problems engendered by the alcohol abuse pattern. Individuals who are highly sensitive about personal privacy and who ordinarily do not like to share their problems with others probably would not do well in group therapy. On the other hand, persons who ordinarily seek out and encourage opinions and support from others would probably benefit from alcoholism treatment in a group therapy setting. When alcohol problems have a major impact upon family interaction and stability, family therapy programs may be useful.

In general, selection of a mode of psychological or psychiatric assistance should be based upon the specific needs of the patient after a thorough assessment of the advantages and disadvantages of each treatment alternative. Treatment choice should be a process that involves the best clinical judgment of the professional along with an informed evaluation by the patient. Patients should not be told what kind of treatment is best for them until they have had an opportunity to express their enthusiasm or reservations about the various treatment programs available. When this process works well, both the professional provider of health care and the patient consumer enjoy mutual confidence and respect.

Psychiatric and psychological assistance in dealing with an alcohol problem may be especially helpful to those individuals who desire insight into how and why they behave in certain ways. While most experts agree that insight alone is not sufficient for modifying many kinds of self-destructive behavior, including alcohol abuse, insight does facilitate coping with life situations that contribute to alcoholism. For example, during insight oriented therapy, a patient may come to understand how particular circumstances during their early growth and development contributed to the development of a drinking problem. Patients may also better appreciate that problem drinking cannot be blamed on a spouse, children or society generally, but reflects many complex factors within themselves. The need to uncover previous life events in order to facilitate remission of abusive drinking varies greatly from patient to patient. Some individuals find they can successfully modify their drinking behavior with relatively little probing into their psyche, while others may benefit from intensive psychological exploration, including more traditional programs of psychoanalysis.

Psychological and psychiatric treatment programs for alcoholics cannot be arbitrarily categorized as best or worse. The requisites for success of alcoholism treatment are similar to the requirements for success of any kind of treatment. The patient must trust and have confidence in the therapist. The patient must understand that therapists cannot provide easy answers or solutions and that a joint effort by both therapist and patient is necessary. Perhaps the term "therapeutic alliance" has been overused, but it is a practical shorthand for saying that patient and therapist must work together, especially at those times when one partner feels the situation is desperate. The therapist must discuss the rationale for the treatment plan with the patient as it develops. Omnipotence, magic, and obfuscation are never in the patient's best interest.

Most importantly, psychiatric and psychological treatment need not be carried out in isolation from other therapeutic attempts to induce remission of alcoholism. Many patients have responded to the combined efforts of psychiatrists, psychologists, participation in the fellowship of Alcoholics Anonymous, aversion therapy and good general medical care from their family physician.

Psychological or psychiatric treatment should never be considered an action of last resort or required as a prerequisite for other types of treatments. Psychotherapy as practiced by experienced psychiatrists and psychologists is a powerful tool for treatment of alcoholism. When done well, psychotherapy is a combination of art and science analogous to the best therapy that contemporary medicine has to offer.

Behavioral Approaches to Treatment

Behavioral therapy can be a useful adjunct to psychotherapy or an effective alternative. Some patients are intellectually and emotionally unsuited to a psychotherapeutic approach. Others achieve good insight into the psychodynamics of their drinking problem, but may continue to abuse alcohol anyway. In some instances the process of psychotherapy can divert attention from the drinking problem without significantly changing the alcohol abuse pattern. A course of behavioral therapy can effectively reduce drinking in some alcohol abusers.

Behavioral therapists differ from psychotherapists in that they do not discuss development issues and emotional problems in an insight-oriented approach. Rather, they develop targeted strategies to try to modify a particular type of behavior, such as excessive eating in obese patients, excessive smoking in tobacco abusers and excessive drinking in alcohol abusers. The particular behavioral modification strategy may differ for different patients and requires an evaluation of both the severity and pattern of the particular behavioral problem.

Behavioral therapy is not a unitary treatment approach, unified by a standard set of procedures or derived from a single generalized theory of behavior. Yet behaviorally based treatments share one basic premise: that behavior is controlled by its consequences. Therefore, manipulation of these consequences can lead to the modification of ongoing behavioral patterns. It is assumed that the individual patient interacts with the environment, which is one source of consequences often referred to as reinforcers. Any event or consequence that maintains behavior leading to its recurrence is a reinforcer. This deceptively simple premise can be translated into a number of behavioral approaches to treatment of the alcoholic individual. Examples range from manipulating the environmental reinforcers other than alcohol to modifying the reinforcing properties of alcohol itself. The several types of behavioral therapy listed below are not necessarily mutually exclusive and often may be integrated in a comprehensive treatment program.

Expanding the Range of Reinforcers: In chronic alcoholism, the repertoire of available reinforcers seems to have contracted to the point that only alcohol is reinforcing. Some behavioral therapists focus on retraining alcoholic patients in interpersonal skills and vocational skills through practice, role-playing and alternative ways of thinking about their problems and prospects, sometimes called "cognitive restructuring." If successful, the alcoholic has increased access to many more potential sources of reinforcement in the course of daily living. These alternative behaviors (a job, new friends, new interests) may eventually replace reliance upon alcohol. Moreover, as the alcoholic encounters more nonalcohol-related reinforcers, these can become part of a contingency management system designed to further help the alcoholic control drinking.

Analyzing What Reinforcers Maintain Drinking: Some behavioral therapists focus on helping the alcoholic to analyze the environmental factors that reinforce excessive drinking. These can include the alcoholic's perception of the effects of intoxication, as well as the complex social interplay with family and friends. The alcoholic may realize that drinking produces a number of social consequences, which may in turn contribute to the maintenance of drinking. As we learned in Chapter 12, events that would appear negative can still maintain behavior that leads to that consequence. For some alcoholics, the adversarial and emotional interactions with family or friends during intoxication can be one of the important consequences that maintains drinking behavior. A fight during drinking may be followed by reconciliations that are better than the usual interactions. Drunken behavior may produce attention, albeit nagging, from an otherwise indifferent companion. The possibilities are almost endless, but drunkenness usually results in some change in the behavior of significant others. Identification of the reinforcers associated with a self-destructive drinking pattern may be a tentative first step toward controlling drinking by modifying some of the consequences.

Manipulating Reinforcers to Affect Drinking: Some forms of behavioral therapy involve establishing agreements with the alcoholic patient, which clearly specify the target behavior (reduction of drinking) and the consequences (fines, loss of privileges, etc.) if the conditions are not met. This approach, sometimes referred to as "contingency contracting," has been especially effective when the target behavior is maintaining daily intake of disulfiram (Antabuse) medication.

The potential power of environmental contingencies in affecting drinking patterns in alcoholics has been shown repeatedly in clinical studies. When alcoholic patients were confronted with a loss of privileges for drinking beyond an agreed-upon limit, many were able to modulate their drinking to conform to the contract requirements. Other social contingencies have also been shown to affect drinking behavior. When social isolation was contingent on drinking in alcoholics, alcohol intake was reduced. When social privileges were made contingent on moderate drinking, this was also effective in controlling drinking behavior. These techniques have particular value for teaching the patient that control of drinking behavior is feasible.

Relaxation and Behavioral Control Techniques: Some behavioral therapists attempt to identify the critical antecedents of the behavior targeted for change. For example, if a patient believes that anxiety or tension always precedes a drinking spree, it may be useful to try to predict and control such feelings. A number of methods are available to induce relaxation. Various biofeedback measures of heart rate and breathing can help the patient monitor physiological concomitants of anxiety and relaxation. In addition to helping the patient achieve some semblance of self-control, the physical involvement in the process of self-monitoring may reduce drinking. At the least, the patient has an alternative coping strategy when troubled by diffuse anxiety and tension. The power of such procedures is often quite remarkable. Even hypertension in some individuals responds favorably to simple exercises to induce relaxation.

Whether biofeedback or some other procedure is used, the essential element seems to be teaching the patient to control an aspect of behavior that was previously not controlled. This field has been plagued by faddism and occasionally mindless application of monitoring machines to inappropriate problems. But despite the chaos of popularization, the basic principles of control appear to hold considerable promise in several areas of psychiatry and medicine. A particular advantage is that the patient learns skills to modify his or her own behavior outside of the formal therapy setting.

Aversion Therapy: One form of behavioral therapy has been used successfully by two leading proprietary hospital groups for the treatment of alcoholism. Aversion therapy is designed to modify the usual consequences of seeing, tasting, smelling and ingesting alcohol. There has been considerable misunderstanding about this form of behavior therapy, and it has sometimes been incorrectly equated with "punishment" for drinking. In order to better understand how and why aversion therapy works, it is interesting to review the theoretical basis for its application, a basis that has foundations in both modern neurobiology and the pragmatic experiences of everyday life.

Aversion therapy is derived from the classic studies of Ivan Pavlov, an eminent Russian physiologist and winner of the Nobel Prize. Among other distinguished scientific achievements, he conducted some of the first and best-known studies of the "conditioned reflex." The conditioned reflex or response is a physiological or psychological response to a stimulus that does not ordinarily elicit such a response. Perhaps the best-known example of this phenomenon is the conditioning of the salivary response in dogs. When hungry dogs are shown food, salivation is the usual, normal response. Presentation of food, which evokes salivation, may be paired with presentation of another stimulus, for example, a sound that ordinarily does not evoke salivation. If the two stimuli (food and sound) are paired for a long enough period of time, salivation may be elicited by the presentation of a sound alone. In this paradigm, food is called the "unconditioned stimulus" and salivation is the "unconditioned response." When sound is paired with food, it is designated the "conditioned stimulus" and when sound evokes salivation, this response is called the "conditioned response."

In aversion conditioning treatment for alcoholism, the smell and taste of alcohol are paired with the induction of nausea and vomiting, with the goal of inducing a conditioned response of nausea whenever alcohol is smelled or tasted. But before describing in more detail how this treatment is carried out, let us consider a well-known conditioning phenomenon that occurs very frequently. Understanding this process may help to explain the efficacy of aversion therapy for alcoholism.

Most people report an aversion for some kind of food. Some never eat smoked oysters; others avoid olives; my grandmother disdained canned tuna fish and I never eat walnuts. Aversion to many nutritious and otherwise pleasant-tasting foods often results from an unfortunate situation in which a food or beverage was implicated as a cause of illness. In most instances the allegedly tainted food did not actually cause gastrointestinal illness such as nausea and vomiting. The real cause was probably a viral or bacterial infection independent of the quality of a food product. Even in those instances when food poisoning occurs, it is not the nature of the food product per se, but the manner in which food was stored or served.

Even though many persons would accept the notion that a specific food does not induce nausea or vomiting, they nevertheless cannot resume eating the special food and often avoid similar products. Thus, development of a gastrointestinal illness with nausea and vomiting soon after ingesting a specific food or beverage sets the stage for rejecting that food or beverage for many years or even a lifetime.

The strength of a food aversion is very powerful. Experimentally induced food aversion in animals is remarkably persistent. Undoubtedly this process has survival value, because animals need to self-select a balanced and nutritious diet while avoiding potentially lethal substances. This basic biological process also facilitates survival of both plants and animals who are subject to attacks from predators. Many animals innately avoid or learn to avoid plants that contain potentially toxic substances, and both plant and animals develop substances toxic to predators, which helps to increase the survival of their species. Thus, food aversion has its roots in fundamental biology. It is not surprising that if aversion to beverage alcohol is induced, such aversion would be both potent and long-lasting.

The goal of aversion therapy is to induce an uncontrollable sensation of nausea and disdain for the sight, smell and taste of alcoholic beverages. Aversion treatment for alcoholism is usually carried out in specialized hospital settings. The average duration of aversion treatment requires ten consecutive days of hospital treatment following detoxification. Aversion treatments are carried out every other day in special treatment rooms, which have the appearance of a comfortable and appealing barroom setting. The treatment room contains displays of beverage alcohol bottles, mixers and accoutrements usually found in a well-stocked bar. In the midst of this barroom-like environment, the patient is seated in a large, comfortable chair to which is attached a very large and foreboding-looking emesis basin. The patient is then given injections of two drugs. The first drug, emetine, produces a strong and unavoidable sensation of nausea. The second drug, pilocarpine, causes a constriction or closure of the valve at the end of the stomach (the pyloric valve) so that any substance ingested after injection cannot be released from the stomach into the intestinal tract. Thus, anything swallowed by the patient after the pilocarpine injection will remain in the stomach and be regurgitated if the patient becomes nauseated.

After injection of emetine and pilocarpine, the patient is asked to consume rapidly several glasses of his or her preferred alcoholic beverage or beverages. The alcohol available is identical to that ordinarily used by the patient. Although the patient may drink a large volume of his or her favorite brand of beverage alcohol in a manner analogous to his or her usual pattern of drinking, the alcohol never reaches the intestine, and hence very little alcohol is absorbed into the bloodstream. But since a small amount of alcohol is absorbed directly from the stomach, the patient is able to feel a very mild alcohol effect immediately before the onset of nausea induced by the emetine injection. In real-life situations, the early perception of mild alcohol effect usually is a very pleasant event for the alcohol abuser because it heralds the onset of alcohol intoxication. Some believe that one of the significant payoffs for the alcoholic is the anticipation of alcohol intoxication signaled by the initial absorption of alcohol into the blood, albeit in small amounts, from the stomach shortly after drinking.

But, instead of feeling better after drinking in aversion conditioning therapy, the patient promptly begins to feel worse. The onset of nausea is usually followed by vomiting, and most of the ingested alcohol ends up in the emesis basin.

After four or five aversive conditioning sessions, patients report that they have developed a significant degree of aversion to the taste and smell of alcohol. The duration of this effect varies. Some persons maintain a lifelong aversion to alcohol following their initial course of four or five treatments. Others lose the aversion response within weeks or months after the initial program of therapy. Most facilities that provide aversion treatment advise the patient to have periodic "recap" sessions to maintain the maximal strength of aversion. Such recap sessions may be scheduled on a monthly or twice-yearly basis. Usually, patients are encouraged to return any time for further treatment when they feel their degree of aversion to alcohol may be waning. Follow-up studies with patients who have received aversion therapy revealed that many maintained long periods (years) of sobriety.

This illustrates how methods derived from basic behavioral studies can be translated into innovative behavioral therapies.

Alcoholics Anonymous

Probably the most popular and widely advocated treatment for alcoholism is Alcoholics Anonymous. In 1938 it was estimated that over one million people belong to A.A. There are about forty-eight thousand chapters of the organization in one hundred and ten countries. These chapters range in size from six to four hundred persons. The common feature shared by all is that they accept the basic tenets of Alcoholics Anonymous as expressed in the organization's fundamental credo, the Twelve Steps.

1. We admitted we were powerless over alcohol — that our lives had become unmanageable.
2. Came to believe that a Power greater than ourselves could restore us to sanity.
3. Made a decision to turn our will and our lives over to the care of God *as we understood Him* [italics in original].
4. Made a searching and fearless moral inventory of ourselves.
5. Admitted to God, to ourselves, and to another human being the exact nature of our wrongs.
6. Were entirely ready to have God remove all these defects of character.
7. Humbly asked Him to remove our shortcomings.

8. Made a list of all persons we had harmed, and became willing to make amends to them all.

9. Made direct amends to such people wherever possible, except when to do so would injure them or others.

10. Continued to take personal inventory and when we were wrong, promptly admitted it.

11. Sought through prayer and meditation to improve our conscious contact with God as we understood Him, praying only for knowledge of His will for us and the power to carry that out.

12. Having had a spiritual awakening as the result of these steps, we tried to carry this message to alcoholics and to practice these principles in all our affairs.

Spiritualism, peer group association and self-help constitute the critical ingredients of the fundamental philosophy of A.A. But at a pragmatic level, there is more diversity than uniformity in the day-to-day practice of A.A. work by its individual chapters and members.

Bill W., the organization's founder and charismatic leader for over two decades, was a remarkable person. Both supporters and detractors of A.A.'s treatment philosophy may abstract parables from his life story which highlight A.A.'s strengths and weaknesses. The building blocks that Bill W. synthesized into his concept of a fellowship that could help alcoholics were derived from disparate sources: the psychology of Carl Jung, transcendental and existential mysticism, Christian fundamentalism and early notions from American medicine about the role of allergy as a cause of alcoholism. A close friend and confidant of Bill W. was treated by Jung for alcoholism. He told Bill W. that Jung believed that medicine and psychology were of little value in the treatment of alcoholism and a spiritual force was necessary to cure the alcoholic. Spiritualism, mysticism and transcendentalism are embodied in A.A.'s Twelve Steps. The treatment that Bill W. received from Dr. Silkworth, a medical director of an alcoholism treatment hospital, also helped shape his thoughts about the nature of A.A. Dr. Silkworth believed that allergies were important in the cause of alcoholism, a view which is not supported by modern science.

While in treatment for alcohol detoxification and during a period of severe alcohol withdrawal, Bill W. experienced a spiritual revelation that was to change his life course of alcohol abuse and also kindle the creation of Alcoholics Anonymous. Bill W. described this experience as follows:

> My depression deepened unbearably and finally it seemed to me as though I were at the bottom of the pit. I still gagged badly on the notion of a Power greater than myself, but finally, just for the moment, the last vestige of my proud obstinacy was crushed. All at once I found myself crying out, "If there is a God, let Him show Himself! I am ready to do anything, anything!"
>
> Suddenly the room lit up with a great white light. I was caught up into an ecstasy which there are no words to describe. It seemed to me, in the mind's eye, that I was on a mountain and that a wind not of air but of spirit was blowing. And then it burst upon me that I was a free man. Slowly the ecstasy subsided. I lay on the bed, for now for a time I was in another world, a new world of consciousness. All about me and through me there was a wonderful feeling of Presence, and I thought to myself, "So this is the God of the preachers!" A great peace stole over me and I thought, "No matter how wrong things seem to be, they are all right. Things are all right with God and his world."

Whether this experience was a result of spiritual revelation, wish fulfillment, or delusion as a consequence of alcohol withdrawal can never be ascertained with any accuracy. But it is not inconsistent with many descriptions of heightened self-awareness and transcendental experience following ingestion of drugs that affect the brain. If Coleridge could compose "Kubla Khan" during opiate use, surely Bill W. could be inspired to create Alcoholics Anonymous during the delirium of alcohol withdrawal. Later in his life, Bill W. experimented with the hallucinogen LSD, as well as with massive doses of vitamins for the cure of alcoholism. This latter enterprise, he believed, would be a greater achievement in his life than the creation and initial leadership of A.A.

The need for spiritual intervention for the successful mastery of uncontrolled drinking is stressed in the second and third of the Twelve Steps of A.A. But the first step emphasizes the power of alcohol. In some respects this power is analogous to the force accorded alcohol by ancient mythologists. More important, the third step points the way to effective intervention through "God as we understand Him." God is not specifically defined by any theological construct. In the broadest sense, God could be defined within the confines of the most orthodox religion or as no God at all by the atheist. Many believe that this type of flexibility has provided A.A. with one of its most powerful assets. A.A. members who profess atheism, as well as those who hold highly structured religious convictions, have attested to the great value of A.A. in the management of alcoholism.

A.A. has attracted and won the accolades of individuals with wide religious, ethnic, social, cultural and economic backgrounds. The most ardent advocates of the value of A.A. at times appear to border on the grandiose, yet there are few testimonials from those who have failed in A.A. programs. Such testimonials should not be sought after to detract from A.A.'s value but rather to achieve a better understanding of what conditions contribute to success or failure in A.A. treatment.

At present we do not possess scientific data that either support or refute the relative efficacy of A.A. for the treatment of alcoholism. Such scientific data would require well-controlled follow-up studies, a procedure that is largely prohibited by the rightful need to preserve the anonymity of A.A. members. Despite this caveat, A.A.'s strength of fellowship, shared peer experience and an overall philosophy that stresses continued acceptance despite a long chronology of personal failure are formidable therapeutic advantages. However, other therapeutic resources within (doctors) and outside (employer, spouse) of medicine should never coerce patients to accept membership in A.A. as a contingency for receiving other forms of assistance.

WHO SHOULD TREAT ALCOHOLICS AND WHERE?

Advocates of specific modes of alcoholism treatment have attempted to persuade the public of the strengths and advantages of their recommendations. Some argue that since alcoholism is a disease, treatment should be carried out only by qualified medical personnel in institutional settings where the most sophisticated diagnostic and medical treatment procedures are available. Other protest that conventional medical treatment of alcoholism is too costly and that patients may receive as good or perhaps better care in nonmedical institutional settings where treatment is provided by personnel who do not have medical training. These arguments often reflect economic concerns rather than an unbiased appraisal of the needs of the person afflicted with alcoholism. Statistics have been presented to support the contention that the large majority of alcoholic patients do not require the sophisticated services of today's modern hospital for their treatment. Other studies demonstrate that cost-cutting (rechristened cost-effectiveness) can jeopardize chances for recovery. However, as in most strongly polarized debates, there is both merit and weakness in each argument.

The Medical Treatment Model

Those who favor the more medically oriented approach to treatment correctly emphasize that many persons with alcoholism also have a number of serious medical disorders that require prompt and careful medical attention. Repeated surveys of patients treated in hospitals and in private physicians' offices have shown that alcoholics have a disproportionate degree of liver, heart, pancreatic and gastrointestinal disorders. Moreover, recent studies have shown that these medical disorders occur in alcoholics who have otherwise good standards of living and health care. Although physicians are frequently chastised for not recognizing the high incidence of alcohol problems in many patients who are treated routinely for other medical disorders, the general level of care provided to alcoholic patients by their personal doctors is usually of good quality. If this were not the case, the death rate among the estimated ten million persons with alcohol-related problems in the United States would reach epidemic proportions.

Medical treatment of alcoholism often involves some form of pharmacotherapy, and this in itself has become an issue of controversy. Physicians can prescribe drugs such as antidepressants, antianxiety agents and drinking deterrents such as Antabuse. Undoubtedly drug treatment has been of great benefit for many patients with alcohol problems, but it is also likely that overzealous use of drugs has done some harm. Death by suicide has been recorded among alcoholics who have ingested large doses of antidepressants during periods of despondency associated with heavy drinking. Overmedication with antianxiety drugs combined with alcohol abuse also contributes to motor vehicle accidents, and related injuries and death. Drugs such as Antabuse may have adverse side effects in certain patients. One major problem with reliance upon

Antabuse as a treatment for alcoholism is that patients' compliance, that is, willingness to use the drug, often varies from time to time. It should be remembered that Antabuse does not prevent an individual from drinking but does cause physical discomfort if a patient drinks when taking it. It is obvious that any patient who wishes to resume drinking while on Antabuse medication would stop taking the drug for a reasonable period of time before initiating drinking. Yet there are many recorded instances of individuals who consumed alcohol when they were taking regular doses of Antabuse and developed a potentially life-threatening Antabuse-alcohol reaction.

The Nonmedical Approach

Less expensive resources for treatment of alcoholism have existed for a long time. There are numerous halfway houses that provide living accommodations for the alcohol patient, an intermediate between full hospitalization and independent residence. There are programs of day hospitalization or night hospitalization, designed to meet special needs of the partially employed alcoholic or the alcoholic parent with home responsibilities. In addition to Alcoholics Anonymous there are voluntary organizations such as Al-Anon or Ala-Teen, which are designed to provide assistance to spouses or children of alcoholics. There is a cadre of "alcoholism counselors" who provide a range of services extending from concerned friendship and advice to diagnostic decision-making. Alcoholism counselors are employed in both medically oriented and nonprofessional institutional facilities. Although most alcoholism counselors are dedicated, hardworking individuals, it has been argued that their ability and expertise for treating alcoholic patients should not be based upon their own past history as recovered alcoholics. It has been noted that although recovered cancer patients can often provide friendship and compassionate understanding to patients with cancer, they can hardly assume responsibility for treatment solely as a consequence of having had the disease themselves. Recognition and personal experience with the disease does not automatically confer expertise in diagnosis or treatment of that disease. There is a strong movement for developing credentialing procedures for alcoholism counselors to ensure basic standards of education and therapeutic competence. Hopefully, such credentialing procedures for alcohol counselors will be generally available in the near future.

THE PATIENTS' DILEMMA

Given the current state of controversy about who should treat the alcoholic patient and the setting in which treatment should take place, are there any basic guidelines available to the patient seeking help for an alcohol-related problem?

First, as we have pointed out previously, the diagnosis and treatment of alcohol problems must be based upon the assessment of individual needs and individual circumstances. Treatment choices cannot simply be assigned on an a priori basis without adequate information about the patient's physical, psychological and social status. Secondly, since alcoholism is a chronic disorder that may often progress toward greater degrees of debilitation and may ultimately end in death, the patient should seek what he or she believes to be the best available source of treatment. Alcoholism is a serious condition, and the patient who is seeking assistance for this problem should no more compromise the quality of alcoholism treatment than he or she would for the treatment of heart disease or cancer. Stated another way, if a person suspected that he or she had heart disease or cancer, he or she probably would seek assistance from the best-qualified persons who treat those disorders in the community. If the same persons suspected that they had an alcohol-related problem, it would seem logical for them to seek out the most-qualified source of care for alcoholism. It now remains for the advocates of special forms of care, be they professional or nonprofessional, to better demonstrate the efficacy and safety of their various treatments. But it must be remembered that demonstration of quality of care in the treatment of alcoholism may inversely vary with the cost of treatment. Health care maintenance in the United States and in Western Europe is undoubtedly more costly than in other parts of the world. However, we have good reason to believe modern health care is also of superior quality, since it reflects the major advances in science and medicine achieved in this century.

GOALS OF TREATMENT

No topic in the field of alcoholism is more certain to stimulate acrimonious debate than the issue of appropriate treatment goals. The argument, narrowly structured, revolves around social drinking versus complete abstinence from alcohol. The emotionalism that inevitably surrounds debate on this question tends to distract attention from the fact that there are other criteria of treatment effectiveness. Improved health, employability, better family and social relations, renewed sense of personal worth, restored productivity and capacity for enjoying life are irrefutable indications of clinical improvement, each probably more important than any particular pattern of abstinence or alcohol use.

The following comments by Griffiths Edwards, a British psychiatrist and expert on alcoholism, give some flavor of the nature of the problem:

> *Both guess-work and certainty also have their dangers, especially when the two categories are confused. A guess rather easily becomes a received wisdom — and it was never more than an imperfectly informed guess that no alcoholic could ever return to normal drinking. Once the guess has been elevated to received wisdom, no one cared or dared to look for evidence which might refute the conjecture. Moral postures, institutional and professional positions, political expedience, self-comfort, fears and fantasies, all frequently invite premature closure of useful questioning; but there is also the danger of losing touch with our few certainties of gaining and holding new ground. We can be certain that many people who have been vastly beset by drinking have shown themselves capable of dealing with their problem, although we may at present only be able to guess their reasons for recovery.*

Abstinence is the avowed goal of the Alcoholics Anonymous program. For some alcoholics, abstinence may be a very reasonable and realistic goal, but for others it is not. Just as there are vast differences in persons with alcoholism, and in the types of treatment that are likely to be effective for them, so also should there be some flexibility in the goal of treatment.

For some alcoholics, an occasional drink with friends or wine with dinner can be a pleasurable aspect of social rituals, which may enhance rather than detract from their overall adjustment. Forbidding recovered alcoholics ever to taste alcohol again is for some a draconian restriction that distorts rather than assists their overall recovery. Still others may voluntarily choose abstinence as the safest course to avoid returning to alcohol abuse problems. Once recovered, some patients feel that anything that places them at great risk for relapse to alcoholism is too great a risk to take. The decision about maintaining alcohol abstinence or gradually returning to a responsible pattern of social drinking is one that should be made in discussion with the patient and not imposed upon the patient by fiat. It could be argued that occasional controlled drinking helps the patient gain confidence about his or her capacity to control drinking. Thus, the occasional celebratory toast or postprandial liqueur is not necessarily to be viewed as the harbinger of relapse.

Although a choice of social drinking over abstinence, or the converse, is a question for each individual to decide, the alcoholic's expectations will to some extent determine the outcome. The issue of social drinking versus abstinence is one of many issues related to alcoholism that remains enshrouded in fear and unchallenged belief approaching mysticism. The fear is conveyed succinctly in a poem.

> *First the man takes a drink,*
> *Then the drink takes a drink,*
> *Then the drink takes the man!*

The almost animistic belief that alcohol can exert a demonic control over behavior persists almost unchanged from the earliest known writings about alcohol intoxication (see Chapter 1). It is still widely believed that any ingestion of alcohol will unmask an uncontrollable alcohol "craving," and the individual will continue drinking to severe intoxication. It is obvious that phrases such as "loss of control over drinking" are subjective, inferential and difficult to verify objectively. The concept of alcohol craving appears to be defined by the behavior that it is invoked to explain. However, the most significant challenge to the concept of craving has come not from an analysis of its logical inconsistencies or deficiencies as an explanatory concept for alcoholism, but from empirical observations of alcoholics during drinking and clinical studies of alcoholics who resume moderate social drinking.

Clinical studies of alcoholics have consistently shown that alcoholics may interrupt a drinking spree for hours or days, even at the cost of the discomfort of mild withdrawal signs and symptoms. The spontaneous cessation and subsequent resumption of drinking during weeks or months of observation under conditions where alcohol is readily available is inconsistent with the notion that "the drink takes a drink." The fact that many alcoholic individuals are able to reduce their drinking behavior gradually, even after a prolonged drinking episode, in order to avoid the discomfort of abrupt alcohol withdrawal also argues against a "craving" notion. Other evidence that alcoholics have the capacity for controlled drinking comes from behaviorally oriented treatment programs in which alcoholics were paid to drink moderately by making ward privileges or money contingent upon a moderate drinking pattern (defined as five ounces of alcohol per day).

There have been an increasing number of clinical reports that some former alcoholics can drink socially and function well for periods of two and one-half to eleven years. Many clinicians have reported that alcoholics who drink moderately are better adjusted and have better social functioning than ex-alcoholic abstainers. Despite this gradually accumulating data base, the 1976 publication of a national follow-up study of alcoholics treated in federally funded centers, known as the Rand Report, was responded to with outrage by many self-appointed spokesmen for the alcoholism treatment community. The Rand Report contained two "heretical" major findings. First, it was found that some alcoholics were able to return to normal drinking at an eighteen-month follow-up. This group accounted for slightly over one-fifth of the total sample studied. The second finding was that the rates of relapse (i.e., return to serious problem drinking) were no higher among alcoholics who had resumed normal drinking than among alcoholics who had elected to abstain.

When this national sample was followed again after four years, there were no significant differences in relapse rates between alcohol abstainers and nonproblem drinkers. These findings that the relapse rate of long-term abstainers was not significantly lower than the relapse rate for nonproblem drinkers demonstrate that, for some patients, moderate drinking does not increase the rate of relapse. Examination of possible contributing variables such as age, marital status, employment status, and race indicated that no single variable seemed to be the determinant of a differential relapse rate. However, the Rand Report concluded that patients with severe alcohol dependence symptoms have a better prognosis if they abstain.

It is a sad commentary on the politics of alcoholism treatment, and the impact of dogmatic adherence to abstinence as the sole treatment goal, that the main finding of the first Rand Report received very little attention. The major finding was that about 70 percent of the patients treated showed improvement at six- and eighteen-month follow-ups.

The alcoholic patient exposed to the conflicting, sometimes vociferous, claims by the pro-abstinence and pro–social drinking camps will be understandably confused and perhaps resentful that although two decades of clinical research suggest that social drinking may be a viable option, there has not been one or more definitive studies to settle the issue once and for all. Indeed, the area of treatment-outcome evaluation perpetually suffers from generalization based on small or unrepresentative groups of patients, studied with incomplete, imprecise and limited methods, and followed over too short a time. Given current strictures on federal funding for all research, definitive correction of this situation does not seem likely. However, it is of some interest to compare the presumed data base for Jellinek's original formulation of the notion of "craving" and "loss of control" with the basis on which the Rand Corporation reported that social drinking occurred and did not result in different relapse rates than alcohol abstinence. Jellinek was an American pioneer in alcoholism studies. In 1946, Jellinek analyzed responses to a questionnaire circulated by Alcoholics Anonymous and concluded from the ninety-eight responses received that "loss of control means that as soon as a small quantity of alcohol enters the organism, a demand for alcohol is set up which is felt as a physical demand by the drinker . . . the drinker has lost the ability to control the quantity once he has started, but he can still control whether he will drink on any occasion or not."

Scientists at the Rand Corporation chose a representative random sample of fourteen thousand clients treated at forty-four federally funded alcoholism treatment centers. The resulting sample was evaluated with the most sophisticated procedures available and included a large group of geographically and demographically diverse patients. The conclusions of the Rand Corporation Report are somewhat surprising, since the majority of their subjects were those who could not afford to pay (through insurance or private funds) for treatment at a proprietary hospital. Whether these conclusions are also valid for the more affluent middle-class alcoholic remains a question for future studies.

BEYOND TREATMENT — PREVENTION

Are there any effective ways to prevent alcoholism which would be acceptable to a free society? At present, the answer unfortunately is no. However this fact has not decreased the frequency of rhetorical pronouncements that alcoholism can be prevented. National Prohibition was a negative case in point. But today, preventionists still cling to the hope that political action may reduce alcohol problems by increasing alcohol's cost or restricting its availability. The feasibility of such an approach is usually bolstered by elaborate economic and mathematical models. The cost of alcohol probably does influence overall consumption, but it is unlikely to influence alcohol abuse and alcoholism. Alcoholics make enormous economic sacrifices to maintain drinking — loss of job, savings and other resources is not uncommon. Bonded bourbon may be replaced by cheap muscatel, but alcoholism persists. Studies of drug-abusing populations have clearly shown that individuals will expend enormous sums of money on illicit heroin and cocaine.

In the absence of reliable data on efficacy, prevention programs are particularly susceptible to fads and pseudoscientific formulas for action. This lamentable trend appears to be part of a larger process eloquently described in the essays of Dr. Lewis Thomas for the *New England Journal of Medicine*. According to Dr. Thomas:

> *The new theory is that most of today's human illness, the infections aside, are multifactorial in nature, caused by two great arrays of causative mechanisms: 1) the influence of things in the environment and 2) one's personal life-style. For medicine to become effective in dealing with such diseases it has become common belief that the environment will have to be changed and personal ways of living will also have to be transformed and radically.*

Dr. Thomas goes on to say, "These things may turn out to be true, for all I know, but it will take a long time to get the necessary proofs. Meanwhile, the field is wide open for magic."

> *Magic is back again and in full force. Laetrile cures cancer, acupuncture is useful for deafness and low back pain, vitamins are good for anything and meditation, yoga, dancing, biofeedback and shouting one another down in crowded rooms over weekends are specifics for the human condition. Running, a good thing to be doing for its own sake, has ac-*

quired medicinal value formerly attributed to rare herbs from Indonesia.

Dr. Thomas was particularly exercised about one disease prevention message.

There is a recurring advertisement placed by Blue Cross on the op-ed of the New York Times which urges you to take advantage of science by changing your life habits, with the suggestion that if you do so, by adopting seven easy-to-follow items of life-style, you can achieve eleven added years beyond what you'll get if you don't. Since today's average figure is around seventy-two for all parties in both sexes, this might mean going on until at least the age of eighty-three. You can do this formidable thing, it is claimed, by simply eating breakfast, exercising regularly, maintaining normal weight, not smoking cigarettes, not drinking excessively, sleeping eight hours each night, and not eating between meals. The science which produced this illumination was a careful study by California epidemiologists, based on a questionnaire given to about seven thousand people.

Dr. Thomas continues:

The findings fit nicely with what is becoming folk doctrine about disease. You become ill because of not living right. If you get cancer it is, somehow or other, your own fault. If you didn't cause it by smoking or drinking or eating the wrong things, it came from allowing yourself to persist with the wrong kind of personality, in the wrong environment. If you have a coronary occlusion, you didn't run enough. Or you were too tense, or you wished too much, and didn't get a good enough sleep. Or you got fat. Your fault. But eating breakfast? It is a kind of enchantment, pure magic . . . it is hard to imagine any good reason for dying within five years from not eating a good breakfast, or any sort of breakfast.

Dr. Thomas's admonition about uncritical acceptance of disease prevention ideas serves as a good lesson for us all. He cautions:

The popular acceptance of the notion of Seven Healthy Life Habits, as a way of staying alive, says something important about today's public attitudes, or at least the attitudes in the public mind about disease and dying. People have always wanted causes that are simple and easy to comprehend, and about which the individual can do something. If you believe that you can ward off the common causes of premature death — cancer, heart disease, and stroke, diseases whose pathogenesis we really do not understand — by jogging, hopping, and eating and sleeping regularly, these things are good things to believe in, even if not necessarily true. Medicine has survived other periods of unifying theory, constructed to explain all of human disease, not always as benign in their effects as this one is likely to be. After all, if people can be induced to give up smoking, stop overdrinking and overeating and take some sort of regular exercise, most of them are bound to feel better for leading more orderly, regular lives, and many of them surely are going to look better.

Nobody can say an unfriendly word against the sheer goodness of keeping fit, but we should go carefully with the promises.

There is also a bifurcated ideological appeal contained in the seven-life-habits doctrine, quite apart from the subliminal notion of good luck in the numbers involved (7 come 11). Both ends of the political spectrum can find congenial items. At the further right, it is attractive to hear that the individual, the good old freestanding, free-enterprising American citizen, is responsible for his own health and when things go wrong it is his own damn fault for smoking and drinking and living wrong (and he can jolly well pay for it). On the other hand, at the left, it is nice to be told that all our health problems, including dying, are caused by failure of the community to bring up its members to live properly, and if you really want to improve the health of the people, research is not the answer; you should upheave the present society and invent a better one. At either end, you can't lose.

In between, the skeptics in medicine have a hard time of it. It is much more difficult to be convincing about ignorance concerning disease mechanisms than it is to make claims for full comprehension, especially when the comprehension leads, logically or not, to some sort of action. When it comes to serious illness, the public tends, understandably, to be more skeptical about the skeptics, more willing to believe the true believers. It is medicine's oldest dilemma, not to be settled by candor or by any kind of rhetoric; what it needs is a lot of time and patience, waiting for science to come in, as it has in the past, with the solid facts.

In alcoholism-prevention programs and elsewhere, it is often difficult to distinguish absolutely between magic and fact, but program development should proceed with an awareness of the limitations of existing information. Some of the most reasonable, rational and humane disease-prevention programs have been discredited because action was taken without a sufficient basis of "solid facts." One sad example was the creation of well-funded, vigorous drug abuse education programs for primary and secondary schools throughout the United States. Followup studies showed that in many instances, these education programs designed to combat drug abuse actually piqued the children's curiosity about drugs and increased rather than reduced the incidence of drug problems. Such dramatic failures have decreased confidence in the credibility of prevention efforts. Clearly, good intentions are not enough.

If alcohol abuse–prevention specialists are to successfully persuade (and not cajole or coerce) the public to adopt attitudes and behaviors that will ultimately benefit health, the credibility of the health promotion advisor must be untarnished. If experts continue to advocate programs, behaviors and life-styles that have a transient, fad appeal, but are not backed up by solid biomedical or behavioral data, the public will quickly perceive ineptness and hypocrisy. Knowledge, study, and critical appraisal of information are essential components for success in prevention as well as treatment. The accomplishments of the next decade will depend on developing a reliable body of scientific data. Advances in treatment and prevention can evolve only from continuing research in the medical, behavioral and social sciences.

Selected References

Chapter 1: In the Beginning

Campbell, J. *The Masks of God: Primitive Mythology*. New York: The Viking Press, 1959, p. 504.

———. *The Masks of God: Occidental Mythology*. New York: The Viking Press, 1964, p. 546.

Colum, P. *Myths of the World*. New York: Grosset & Dunlap, 1930, p. 327.

Dover, K. *The Greeks*. Austin: University of Texas Press, 1981, p. 146.

Fleming, A. *Alcohol: The Delightful Poison*. New York: Delacorte Press, 1975, p. 138.

Frazer, J. G. *The New Golden Bough*. New York: Criterion Books, 1959, p. 738.

Golding, W. *The Inheritors*. New York: Pocket Books, 1963, p. 213.

Oates, W. J. (ed.) *Seven Famous Greek Plays*. New York: The Modern Library, 1938, p. 446.

Weinhold, R. *Vivat Bacchus*. Watford, Herfordshire, England: Argus Books, 1978, p. 228.

Younger, W. *Gods, Men and Wine*. The Wine and Food Society Limited. Cleveland, Ohio: World Publishing Co., 1966, p. 516.

Chapter 2: Alcohol Use in Colonial Times

Aaron, P., and Musto, D. "Temperance and Prohibition in America: A Historical Overview." In *Alcohol and Public Policy: Beyond the Shadow of Pro-*

hibition (M. H. Moore and D. R. Gerstein, eds.). Washington, D.C.: National Academy Press, 1981, pp. 127–181.

Boyer, L. E. *United States Naval Institute Proceedings* 89 (4, No. 722): 173, 1963.

Brown, S. C. *Wines and Beers of Old New England.* Hanover, N.H.: University Press of New England, 1978, p. 157.

Catton, B., and Catton, W. B. *The Bold and Magnificent Dream. America's Founding Years 1492–1815.* Garden City, N.Y.: Doubleday & Co., 1978, p. 530.

Cherrington, E. H. (ed.). *Standard Encyclopedia of the Alcohol Problem.* Vols. 1–6. Westerville, Ohio: American Issue Press, 1925–1930.

Dorchester, D. *The Liquor Problem in All Ages.* New York: Phillips & Hunt, 1884, p. 656.

Levine, H. G. "The Vocabulary of Drunkenness." *Journal of Studies on Alcohol* 42 (1981): 1038–1051.

Pinson, A. "The New England Rum Era: Drinking Styles and Social Change in Newport, R.I., 1720–1770." Working Papers on Alcohol and Human Behavior. Department of Anthropology, Brown University, 1980, p. 48.

Chapter 3: Loss of Innocence and the Call to Reform

Aaron, P., and Musto, D. "Temperance and Prohibition in America: A Historical Overview." In *Alcohol and Public Policy: Beyond the Shadow of Prohibition* (M. H. Moore and D. R. Gerstein, eds.). Washington, D.C.: National Academy Press, 1981, pp. 127–181.

Blumberg, L. U. "The Institutional Phase of the Washingtonian Total Abstinence Movement: A Research Note." *Journal of Studies on Alcohol* 39 (1978): 1591–1606.

Cassedy, J. H. "An Early American Hangover: The Medical Profession and Intemperance 1800–1860." *Bulletin of the History of Medicine* 50 (1976): 405–413.

Clark, N. L. *Deliver Us from Evil: An Interpretation of American Prohibition.* New York: Norton, 1976.

Dorchester, D. *The Liquor Problem in All Ages.* New York: Phillips and Hunt, 1884.

Gusfield, J. R. "Status Conflicts and the Changing Ideologies of the American Temperance Movement." In *Society, Culture and Drinking Pat-*

terns (D. V. Pittman and C. R. Snyder, eds.). Urbana: University of Illinois Press, 1962, pp. 101–120.

Gusfield, J. *Symbolic Crusade: Status Politics and the American Temperance Movement.* Urbana: University of Illinois Press, 1963.

"J." "Intemperance and disease." *Boston Medical and Surgical Journal* 15 (1836): 261–267.

Krout, J. A. *The Origins of Prohibition.* New York: Knopf, 1925.

Lane, R. *Policing the City: Boston, 1822–1885.* New York: Atheneum, 1977.

Lender, M. E., and Karnchanapee, K. R. " 'Temperance Tales': Antiliquor Fiction and American Attitudes toward Alcoholics in the Late 19th and Early 20th Centuries. *Journal of Studies on Alcohol* 38 (1977): 1347–1370.

Levine, H. G. "The Discovery of Addiction, Changing Conceptions of Habitual Drunkenness in America." *Journal of Studies on Alcohol* 39 (1978): 143–174.

Rorabaugh, W. J. "Estimated U.S. Alcoholic Beverage Consumption, 1790–1860." *Journal of Studies on Alcohol* 37 (1976): 357–364.

———. *The Alcoholic Republic, An American Tradition.* New York: Oxford University Press, 1979, p. 302.

Storer, D. H. "Medical Statistics and Bills of Mortality, for Boston, During Nineteen Years Ending January 1st, 1832." *Medical Magazine* 1 (1833): 516.

Tyrrell, I. R. *Sobering Up: From Temperance to Prohibition in Antebellum America: 1800–1860.* Westport: Greenwood Press, 1979.

Weems, M. O. *The Drunkard's Looking Glass.* 6th edition, 1818.

Woodward, S. B. *Essays on Asylums for Inebriates.* Worcester, 1838.

Chapter 4: Beyond the Civil War: Alcohol, Opium and Cocaine

Byck, R. (ed.) *Cocaine Papers by Sigmund Freud.* New York: Stonehill Publishing Co., 1974, p. 412.

Courtwright. D. T. *Dark Paradise: Opiate Addiction in America Before 1940.* Cambridge, Mass.: Harvard University Press, 1982, p. 270.

Kramer, J. C. "Opium Rampant: Medical Use, Misuse and Abuse in Britain and the West in the 17th and 18th Centuries." *British Journal of Addiction* 74 (1979) 377–389.

Musto, D. F. *The American Disease: Origins of Narcotic Control.* New Haven and London: Yale University Press, 1973, p. 354.

Rorabaugh, W. J. *The Alcoholic Republic: An American Tradition.* New York: Oxford University Press, 1979, p. 302.

Van Dyke, C., and Byck, R. "Cocaine Use in Man." In *Advances in Substance Abuse, Behavioral and Biological Research,* Vol. III (N. K. Mello, ed.). Greenwich: JAI Press, 1983, pp. 1–24.

Chapter 5: The Western Frontier and the Rise of the Saloon

West, E. *The Saloon on the Rocky Mountain Frontier.* Lincoln: University of Nebraska Press, 1979.

Chapter 6: Drinking and The Melting Pot

Blumberg, L. U., Shipley, T. E., and Barsky, S. F. *Liquor and Poverty: Skid Row as a Human Condition.* New Brunswick, N.J.: Rutgers Center of Alcohol Studies, 1979.

Harper, F. D. "Etiology: Why Do Blacks Drink?" In *Alcohol Abuse and Black America* (F. D. Harper, ed.). Alexandria, Va.: Douglass Publishers, 1976, pp. 27–37.

Isetts, C. A. "A Social Profile of the Women's Temperance Crusade: Hillsboro, Ohio." In *Alcohol, Reform and Society: The Liquor Issue in Social Context* (J. S. Blocker, ed.). Westport: Greenwood Press, 1979, pp. 101–110.

Kobler, J. *Ardent Spirits: The Rise and Fall of Prohibition.* New York: Putnam, 1973.

Larkins, J. E. "Historical background." In *Alcohol Abuse and Black America* (F. D. Harper, ed.). Alexandria, Va.: Douglass Publishers, 1976, pp. 13–25.

Paulson. R. E. *Women's Suffrage and Prohibition: A Comparative Study of Equality and Social Control.* Glenview, Ill., Scott Foresman & Co., 1973.

Room, R. "Cultural Contingencies of Alcoholism: Variations between and within Nineteenth-Century Urban Ethnic Groups in Alcohol-Related Death Rates." *Journal of Health and Social Behavior* 9 (1968): 99–113.

Stivers, R. *A Hair of the Dog: Irish Drinking and American Stereotype.* University Park: Pennsylvania State University Press, 1976.

Chapter 7: The New Century and Prohibition

Aaron, P., and Musto, D. "Temperance and Prohibition in America: A Historical Overview." In *Alcohol and Public Policy: Beyond the Shadow of Pro-*

hibition (M. H. Moore and D. R. Gerstein, eds.). Washington, D.C.: National Academy Press, 1981, pp. 127–181.

Binkley, R. C. *Responsible Drinking: A Discreet Inquiry and a Modest Proposal.* New York: Vanguard, 1930.

Blocker, J. S. "The Modernity of Prohibitionists: An Analysis of Leadership Structure and Background." In *Alcohol, Reform and Society: The Higher Issue in Social Context* (J. S. Blocker, ed.). Westport: Greenwood Press, 1979.

————. *Retreat from Reform.* Westport: Greenwood Press, 1976.

Cherrington, E. H. *Evolution of Prohibition in the United States of America.* Westerville, Ohio: American Issue Press, 1920.

Clark, N. L. *Deliver Us from Evil: An Interpretation of American Prohibition.* New York: Norton, 1976, p. 246.

Dublin, L. I. "General Mortality Rates 1900 to 1930, and Death Rates from Alcoholism and Cirrhosis of the Liver." In *Alcohol and Man: The Effects of Alcohol on Man in Health and Disease* (H. Emerson, ed.). New York: Macmillan, 1932, pp. 373–395.

Gussfield, J. *Symbolic Crusade: Status Politics and the American Temperance Movement.* Urbana: University of Illinois Press, 1963.

Kobler, J. *Ardent Spirits: The Rise and Fall of Prohibition.* New York: Putnam, 1973.

Krout, J. A. *The Origins of Prohibition.* New York: Knopf, 1925.

Kyvig, D. E. "Objection Sustained: Prohibition Repeal and the New Deal." In *Alcohol, Reform and Society: The Liquor Issue in Social Context* (J. S. Blocker, ed.). Westport: Greenwood Press, 1979, pp. 211–233.

Morison, S. E., and Commager, H. S. *The Growth of the American Republic,* 3rd Edition. New York: Oxford University Press, 1942.

Paulson, R. E. *Women's Suffrage and Prohibition: A Comparative Study of Equality and Social Control.* Glenview, Ill.: Scott Foresman, 1973, p. 212.

Pollock, H. M. "The Prevalence of Mental Disease Due to Alcoholism." In *Alcohol and Man: The Effects of Alcohol on Man in Health and Disease* (H. Emerson, ed.). New York: Macmillan, 1932, pp. 344–372.

Rorabaugh, W. J. "Estimated U.S. Alcoholic Beverage Consumption, 1790–1860." *Journal of Studies on Alcohol* 37 (1976): 357–364.

Sinclair, A. *Prohibition: The Era of Excess.* Boston: Little, Brown, 1962.

Stoddard, C. F., and Woods, A. *Fifteen Years of the Drink Question in Massachusetts.* Westerville, Ohio: American Issue Press, 1929.

Warburton, C. *The Economic Results of Prohibition.* New York: AMS Press, 1932.

Chapter 8: The Legacy of the Radical Temperance Movement

Aaron, P. and Musto, D. "Temperance and Prohibition in America: A Historical Overview. In *Alcohol and Public Policy: Beyond the Shadow of Prohibition* (M. H. Moore and D. R. Gerstein, eds.). Washington, D. C.: National Academy Press, 1981, pp. 127–181.

Bonnie, R. J. "Discouraging Unhealthy Personal Choices: Reflections on New Directions in Substance Abuse Policy." *Journal of Drug Issues* 8(2) (1978): 199–219.

————. "Discouraging the Use of Alcohol, Tobacco, and Other Drugs: The Effects of Legal Controls and Restrictions." In *Advances in Substance Abuse, Behavioral and Biological Research*. Vol. II (N. K. Mello, ed.). Greenwich: JAI Press, 1981, pp. 145–184.

Matlins, S. M. *A Study in the Actual Effects of Alcoholic Beverage Control Laws*. Vol. I. Washington, D. C.: Medicine in the Public Interest, 1976, p. 258.

Mill, J. S. "On Liberty." In *Utilitarianism, Liberty and Representative Government* (J. S. Mill, 1859). American Edition. New York: E. P. Dutton, 1951, pp. 95–96.

Randall, C. L., and Noble, E. P. "Alcohol Abuse and Fetal Growth and Development." In *Advances in Substance Abuse, Behavioral and Biological Research*, Vol. I (N. K. Mello, ed.). Greenwich: JAI Press, 1980, pp. 327–367.

Rubin, J. L. "The Wet War: American Liquor Control, 1941–1945." In *Alcohol, Reform and Society: The Liquor Issue in Social Context* (J. S. Blocker, ed.). Westport: Greenwood Press, 1979, pp. 236–258.

Chapter 9: The Industry and the Regulators

Annual Report FY 1979. Bureau of Alcohol Tobacco & Firearms, Department of the Treasury. Washington, D.C.: U.S. Government Printing Office, 1980.

Annual Statistical Review 1982. Distilled Spirits Industry. Washington, D.C.: Distilled Spirits Council of the United States, 1983.

The Brewing Industry in the United States. Brewers Almanac 1982. Washington, D.C.: United States Brewers Association, Inc. 1982.

International Survey. *Alcoholic Beverage Taxation and Contol Policies.* Fifth Edition. Ottawa: Brewers Association of Canada, 1982.

Wine Institute Economic Research Report. San Francisco: Wine Institute, 1983.

Chapter 10: The Odyssey of the Alcohol Molecule: To the Brain and Beyond

Freedman, D. X. "Non-pharmacological Factors in Drug Dependence." In *Drug Abuse: Nonmedical Use of Dependence-Producing Drugs* (F. Btesh, ed.). New York: Plenum Publishing, 1972, pp. 25–34.

Kalant, H., Le Blanc, A. E., Wilson, A., and Homatidis, S. "Sensory, Motor and Physiological Effects of Various Alcohol Beverages." *Canadian Medical Journal* 112 (1975): 953–958.

Mendelson, J. H., "Biochemical Mechanisms of Alcohol Addiction." In *The Biology of Alcoholism*, Vol. I, *Biochemistry* (B. Kissin and H. Begleiter, eds.). New York: Plenum Publishing, 1971, pp. 513–544.

Mendelson, J. H., and Mello, N. K. "Biologic Concomitants of Alcoholism." *New England Journal of Medicine* 301 (17) (1979): 912–921.

Chapter 11: Alcohol, Sex and Aggression

Laferla, J. J., Anderson, D. L., and Schalach, D. S. "Psychoendocrine Response to Visual Erotic Stimulation in Human Males." *Psychosomatic Medicine* 40(2) (1978): 166–172.

Marlatt, G. A., and Rohsenow, D. J. "Cognitive Processes in Alcohol Use: Expectancy and the Balanced Placebo Design." In *Advances in Substance Abuse, Behavioral and Biological Research*, Vol. I (N. K. Mello, ed.). Greenwich: JAI Press, 1980, pp. 159–199.

Mello, N. K., and Mendelson, J. H. "Alcohol and Human Behavior." In *Handbook of Psychopharmacology*, Vol. XII, *Drugs of Abuse* (L. L. Iversen, S. D. Iversen and S. H. Snyder, eds.). New York: Plenum Publishing 1978, pp. 235–317.

Mendelson, J. H., Mello, N. K., and Ellingboe, J. "Effects of Alcohol on Pituitary-Gonadal Hormones, Sexual Function and Aggression in Human Males." In *Psychopharmacology: A Generation of Progress* (M. A. Lipton, A. DiMascio, and K. F. Killam, eds.). New York: Raven Press, 1978, pp. 1677–1691.

Wilson, G. T. "The Effects of Alcohol on Human Sexual Behavior." In *Advances in Substance Abuse, Behavioral and Biological Research*, Vol. II (N. K. Mello, ed.). Greenwich: JAI Press, 1981, pp. 1–40.

Chapter 12: Alcohol and Mood: The Illusion of Happiness

Goldman, A. *Ladies and Gentlemen, Lenny Bruce!* New York: Random House, 1971.

Mayfield, D. G. "Psychopharmacology of Alcohol, II: Affective Tolerance in Alcohol Intoxication." *Journal of Nervous and Mental Dis.* 146(4) (1968): 322–327.

Mello, N. K. "A Behavioral Analysis of the Reinforcing Properties of Alcohol and Other Drugs in Man." In *Biology of Alcoholism,* Vol. VII (B. Kissin and H. Begleiter, eds.). New York: Plenum Publishing, 1983, pp. 133–198.

Morse, W. H., McKearney, J. W., and Kelleher, R. T. "Control of Behavior by Noxious Stimuli." In *Handbook of Psychopharmacology,* Vol. VII (L. L. Iversen, S. D. Iversen and S. H. Snyder, eds.). New York: Plenum Publishing, 1977, pp. 151–180.

Morse, W. H., and Kelleher, R. T. "Determinants of Reinforcement and Punishment." In *Operant Behavior,* Vol. II (W. K. Honig and J. E. R. Staddon, eds.). Englewood Cliffs: Prentice-Hall, 1977, pp. 174–200.

Rush, B. "Inquiry into the Effects of Ardent Spirits on the Human Body and Mind. Reprinted in NIAAA, *Alcohol Health and Research World,* 1976, pp. 7–10 (Department of Health, Education and Welfare Publication).

Skinner, B. F. *Science and Human Behavior.* New York: Macmillan, 1953.

Vaillant, G. E. *The Natural History of Alcoholism.* Cambridge, Mass.: Harvard University Press, 1983.

Chapter 13: The Role of Alcohol in Sickness and Health

Alcohol and Health, Third Special Report to Congress, Technical Support Document. Department of Health, Education and Welfare Publication No. (ADM) 79-832. Washington, D.C.: U.S. Government Printing Office, 1978, p. 335.

Bourne, P., and Light, E. "Alcohol Problems in Blacks and Women." In *The Diagnosis and Treatment of Alcoholism* (J. H. Mendelson and N. K. Mello, eds.). New York: McGraw-Hill, 1979, pp. 84–123.

Cicero, T. J. "Common Mechanisms Underlying the Effects of Ethanol and Narcotics on Neuroendocrine Function." In *Advances in Substance*

Abuse, Behavioral and Biological Research. Vol. 1 (N. K. Mello, ed.). Greenwich: JAI Press, 1980, pp. 201–254.

Dreyfus, P. M. "Diseases of the Nervous System in Chronic Alcoholics. In *The Biology of Alcoholism,* Vol. 3, *Clinical Pathology* (B. Kissin and H. Begleiter, eds.). New York: Plenum Publishing, 1974, pp. 265–290.

Feinberg, I. "Effects of Age on Human Sleep Patterns." In *Sleep Physiology and Pathology* (A. Kales, ed.). Philadelphia: J. P. Lippincott, 1969, pp. 39–52.

Greenspon, A. J., Stang, J. M., Lewis, R. P., and Schaal, S. F. "Provocation of Ventricular Tachycardia after Consumption of Alcohol." *New England Journal of Medicine* 301 (19) (1979): 1049–1050.

Hartmann, E. *The Sleeping Pill.* New Haven: Yale University Press, 1978, p. 313.

Knobil, E. "The Neuroendocrine Control of the Menstrual Cycle." *Recent Progress in Hormone Research* 36 (1980): 53–88.

LaPorte, R. E., Cresanta, J. L., and Kuller, L. H. "The Relationship of Alcohol Consumption to Atherosclerotic Heart Disease." *Preventive Medicine* 9 (1980): 22–40.

Lieber, C. S. "Liver Disease and Alcohol: Fatty Liver, Alcoholic Hepatitis, Cirrhosis and Their Inter-relationships. *Annals of the New York Academy of Science* 252 (1975): 63–84.

Mendelson, J. H., and Mello, N. K. "Biologic Concomitants of Alcoholism." *New England Journal of Medicine* 301 (17) (1979): 912–921.

Mendelson, J. H., Mello, N. K., Bavli, S., Ellingboe, J., Bree, M. P., Harvey, K. L., King, N. W., and Sehgal, P. K. "Alcohol Effects on Female Reproductive Hormones." In: *Ethanol Tolerance and Dependence: Endocrinological Aspects* (T. J. Cicero, ed.). NIAAA Research Monograph No. 12, DHHS Publ. No. (ADM) 83-1258, Washington, D.C.: U.S. Government Printing Office, 1983, pp. 146–161.

Ouellette, E. M., Rosett, H. L. Rosman, N. P., and Weiner, A. B. "Adverse Effects on Offspring of Maternal Alcohol Abuse during Pregnancy. *New England Journal of Medicine* 297 (1977): 528–530.

Randall, C. L., and Noble, E. "Alcohol Abuse and Fetal Growth and Development." In *Advances in Substance Abuse, Behavioral and Biological Research,* Vol. I (N. K. Mello, ed.). Greenwich: JAI Press, 1980, pp. 327–367.

Regan, T. J., and Ettinger, P. O. "Varied Cardiac Abnormalities in Alcoholics." *Alcoholism: Clinical and Experimental Research* 3 (1) (1979): 40–45.

Rosenberg, L., Slone, D., Shapiro, S., Kaufman, D., Miettinen, O., and

Stolley, P. "Alcoholic Beverages and Myocardial Infarction in Young Women." *American Journal of Public Health* 71 (1) (1981): 82–85.

Seixas, F. A., Williams, K., and Eggleston, S. (eds.). "Medical Consequences of Alcoholism." *Annals of the New York Academy of Science* 252 (1975): 339.

Van Thiel, D. H., and Gavaler, J. S. "The Adverse Effects of Ethanol upon Hypothalamic-Pituitary-Gonadal Function in Males and Females Compared and Contrasted." *Alcoholism: Clinical and Experimental Research* 6(2) (1982): 179–185.

Victor, M. "Nutrition and Diseases of the Nervous System (Including Psychiatric Disorders)." *Progress in Food and Nutrition Science* 1 (3) (1975): 145–172. (Published by Pergamon Press.)

Chapter 14: Alcoholism: A Search for Origins

Bacon, M. K. "Cross-Cultural Studies of Drinking." In *Alcoholism, Progress in Research and Treatment* (P. G. Bourne and R. Fox, eds.). New York: Academic Press, 1973, pp. 171–174.

Barry, H. "Sociocultural Aspects of Alcohol Addiction. In: *The Addictive States*. Association for Research in Nervous and Mental Diseases. XLVI. Baltimore: Williams and Wilkins Co., 1968, pp. 455–471.

———. "The Correlation between Personality and the Risk of Alcoholism." In *Recent Advances in the Study of Alcoholism, Proceedings, First International Magnus Huss Symposium, Stockholm, 1976*. Amsterdam: Excerpta Medica, 1977, pp. 56–68.

Blume, S. B., and Dropkin, D. "The Jewish Alcoholic: An Under-Recognized Minority?" *Journal of Psychiatric Treatment Evaluation* 2 (1980): 1–4.

Ewing, J. A., Rouse, B. A., and Pellizzari, E. D. "Alcohol Sensitivity and Ethnic Background." *American Journal of Psychiatry* 131 (1974): 206–210.

Goodwin, D. W. *Is Alcoholism Hereditary?* New York: Oxford University Press, 1976.

———. "Genetic Determinants of Alcoholism." In *The Diagnosis and Treatment of Alcoholism* (J. H. Mendelson and N. K. Mello, eds.). New York: McGraw-Hill, 1979, pp. 59–82.

Knupfer, G., and Room, R. "Drinking Patterns and Attitudes of Irish, Jewish and White Protestant American Men." *Quarterly Journal of Studies on Alcohol*, 28 (1967): 676–699.

Kono, H., Faito, F., Shimada, K., and Nakagawa, J. *Drinking Habits of the*

Japanese: Actual Drinking Habits and Problem Drinking Tendencies. Tokyo: Leisure Development Center, Chiyoda-Ku, 1977, p. 15.

Mello, N. K. "Etiological Theories of Alcoholism," In *Advances in Substance Abuse, Behavioral and Biological Research,* Vol. III (N. K. Mello, ed.). Greenwich: JAI Press, 1983, pp. 271–312.

Mello, N. K., and Mendelson, J. H. "Alcohol and Human Behavior." In *Handbook of Psychopharmacology,* Vol. 12, *Drugs of Abuse* (L. L. Iversen, S. D. Iversen and S. H. Snyder, eds.). New York: Plenum Publishing, 1978, pp. 235–317.

———. (eds.). *The Diagnosis and Treatment of Alcoholism.* 2nd edition. New York: McGraw-Hill, 1985.

Pittman, D. J., and Snyder, C. R. (eds.). *Society, Culture and Drinking Patterns.* New York: John Wiley and Sons, 1962, p. 616.

Schuckit, M. A., and Morrissey, E. R. "Alcoholism in Women: Some Clinical and Social Perspectives." In *Alcohol Problems in Women and Children* (M. Greenblatt and M. A. Schuckit, eds.). New York: Grune and Stratton, 1976, pp. 5–35.

Vaillant, G. E. *The Natural History of Alcoholism.* Cambridge, Mass.: Harvard University Press, 1983.

Weissman, M. M., and Klerman, G. L. Sex Differences and the Epidemiology of Depression." *Archives of General Psychiatry* 34 (1977): 98–111.

Wolfe, P. H. "Ethnic Differences in Alcohol Sensitivity." *Science* 175 (1972): 449–450.

Zwerling, I., and Rosenbaum, M. "Alcoholic Addiction and Personality." In *American Handbook of Psychiatry,* Vol. I (S. Arieti, ed.) New York: Basic Books, 1959, pp. 623–644.

Chapter 16: Alcoholism and Polydrug Use Examined

Alcohol and Health, First Special Report to Congress. Department of Health, Education and Welfare Publication No. (HSM) 72-9099. Washington, D.C.: U.S. Government Printing Office, 1971, p. 121.

Alcohol and Health, Third Special Report to Congress. Technical Support Document. Department of Health, Education and Welfare No. (ADM) 79-832. Washington, D.C.: U.S. Government Printing Office, 1978, p. 335.

Alcohol and Health, Fourth Special Report to Congress. Department of Health, Education and Welfare Publication No. (ADM) 81-1080. Washington, D.C.: U.S. Government Printing Office, 1981, p. 206.

Atkinson, J. H., and Schuckit, M. A. "Geriatric Alcohol and Drug Misuse and Abuse." In *Advances in Substance Abuse, Behavioral and Biological Research*, Vol. III (N. K. Mello, ed.). Greenwich: JAI Press, 1983, pp. 195–237.

Bourne, P. G. "Polydrug Abuse — Status Report on the Federal Effort." In *Developments in the Field of Drug Abuse: National Drug Abuse Conference* (E. Senay, V. Shorty and H. Alksen, eds.). Cambridge: Schenkman Publishing, 1975, pp. 197–207.

Brown, J., and Lyons, J. P. "A Progressive Diagnostic Scheme for Alcoholism with Evidence of Clinical Efficacy." *Alcoholism: Clinical and Experimental Research* 5(1) (1981): 17–25.

Chauncey, R. L. "New Careers for Moral Entrepreneurs: Teenage Drinking." *Journal of Drug Issues*, 1 (1980): 45–70.

Diagnostic and Statistical Manual III, American Psychiatric Association, 1980, Washington, D. C., p. 494.

Dorchester, D. D. *The Liquor Problem in All Ages*. New York: Phillips and Hunt, 1884, p. 656.

Drug and Alcohol Abuse: Implications for Treatment. NIDA Treatment Research Monograph. (S. E. Gardner, ed.). Department of Health and Human Services Publication No. (ADM) 80-958. Washington, D.C.: U.S. Government Printing Office, 1981, p. 151.

Freed, E. X. "Drug Abuse by Alcoholics: A Review." *International Journal of Addictions* 8(3) (1973): 451–473.

Jellinek, E. M. *The Disease Concept of Alcoholism*. Highland Park, N.J.: Hillhouse Press, 1960.

Johnston, L. D., Backman, J. G., O'Malley, P. M. *Student Drug Use in America 1975–1980*. Department of Health and Human Services Publication No. (ADM) 81-1066. Washington, D.C.: U.S. Government Printing Office, 1980, p. 113.

Lex, B. W. "Alcohol Problems in Special Populations." In *The Diagnosis and Treatment of Alcoholism*, 2nd edition (J. H. Mendelson and N. K. Mello, eds.). New York: McGraw-Hill, 1985.

Mello, N. K. (ed.). *Advances in Substance Abuse, Behavioral and Biological Research*, Vol. I. Greenwich: JAI Press, 1980.

———. (ed.). *Advances in Substance Abuse, Behavioral and Biological Research*, Vol. II. Greenwich: JAI Press, Inc., 1981.

———. "Some Behavioral and Biological Aspects of Alcohol Problems in Women." In *Alcohol and Drug Problems in Women* (O. J. Kalant, ed.). New York: Plenum Publishing, 1980, pp. 263–298.

Mello, N. K., and Mendelson, J. H. "Clinical Aspects of Alcohol Dependence." In *Handbook of Experimental Pharmacology*, Vol. 45/I (W. Martin, ed.). Berlin: Springer-Verlag, 1977, pp. 613–666.

Mendelson, J. H., and Mello, N. K. (eds.). "Diagnostic Criteria for Alcoholism and Alcohol Abuse." In *The Diagnosis and Treatment of Alcoholism.* New York: McGraw-Hill, 1979, pp. 1–18.

Myers, J. K., Weissman, M. M., Tischler, G. L., Holzer, C. E., III, Leaf, P. J., Orvaschel, H., Anthony, J. C., Boyd, J. H., Burke, J. D., Jr., Kramer, M., and Stoltzman, R. "Six-Month Prevalence of Psychiatric Disorders in Three Communities." *Archives of General Psychiatry* 41 (1984): 959–970.

National Council on Alcoholism. "Criteria for the Diagnosis of Alcoholism." *American Journal of Psychiatry.* 129(2) (1972): 127–135.

National Drug/Alcohol Collaborative Project. *Issues in Multiple Substance Abuse.* NIDA Services Research Monograph (S. E. Gardner, ed.). Department of Health, Education and Welfare Publication No. (ADM) 80-957. Washington, D.C.: U.S. Government Printing Office, 1980, p. 126.

Newsletter, Federation of American Societies for Experimental Biology, 13(6) (1980): 1–8.

Regier, D. A., Meyers, J. K., Kramer, M., Robins, L., Blazer, D. G., Hough, R. L., Eaton, W. W., and Locke, B. Z. "The NIMH Epidemiologic Catchment Area Program." *Archives of General Psychiatry* 41: (1984): 934–941.

Robins, L. N., Helzer, J. E., Weissman, M. M., Orvaschel, H., Gruenberg, E., Burke, J. D., and Regier, D. A. "Lifetime Prevalence of Specific Psychiatric Disorders in Three Sites." *Archives of General Psychiatry* 41 (1984): 949–958.

Scott, J. M. *The White Poppy.* London: Heinmann, 1969.

Victor, M., and Adams, R. D. "The Effect of Alcohol on the Nervous System." *Research Publication of the Association for Nervous and Mental Disease* 32 (1953): 526–573.

Wesson, D. R., Carlin, A. S., Adams, I. M., and Beschner, G. (eds.). *Polydrug Abuse: The Results of a National Collaborative Study.* New York: Academic Press, 1978.

Chapter 17: Health Consequences of Alcohol and Drug Combinations

Abel, E. L. *Marihuana: The First Twelve Thousand Years.* New York: Plenum Publishing, 1980, p. 289.

Atkinson, J. H., and Schuckit, M. A. "Geriatric Alcohol and Drug Misuse and Abuse." In *Advances in Substance Abuse, Behavioral and Biological Research,* Vol. 3 (N. K. Mello, ed.). Greenwich: JAI Press, 1983, pp. 195–238.

Berridge, V. "Opium in the Fens in Nineteenth-Century England," *Journal of the History of Medicine and Allied Sciences* 34(3) (1979): 293–313.

Bourne, P. G. (ed.). *A Treatment Manual for Acute Drug Abuse Emergencies.* National Commission on Drug Abuse Issues Publication No. 16. Washington, D.C.: U.S. Government Printing Office, 1974, p. 178.

Brecher, E. M. (ed.). *Licit and Illicit Drugs.* Mount Vernon, N.Y.: Consumers Union, 1972, p. 623.

Byck, R. (ed.). *Cocaine Papers by Sigmund Freud.* New York: Stonehill, 1974, p. 402.

Byck, R. "Drugs and the Treatment of Psychiatric Disorders." In *The Pharmacological Basis of Therapeutics,* fifth edition (L. S. Goodman and A. Gilman, eds.). New York: Macmillan, 1975, pp. 152–200.

Cohen, P. J. "History and Theories of General Anesthesia." In *The Pharmacological Basis of Therapeutics,* fifth edition (L. S. Goodman and A. Gilman, eds.). New York: Macmillan, 1975, pp. 53–59.

Cohen, S. "The Effects of Combined Alcohol/Drug Abuse on Human Behavior." In *Drug and Alcohol Abuse: Implications for Treatment* (S. E. Gardner, ed.). NIDA Treatment Research Monograph Series, Department of Health and Human Services Publication No. (ADM) 80-958. Washington, D.C.: U.S. Government Printing Office, 1981, pp. 5–17.

————. "The Intentional Inhalation of Volatile Substances." In *Advances in Substance Abuse, Behavioral and Biological Research,* Vol. 2 (N. K. Mello, ed.). Greenwich: JAI Press, 1981, pp. 123–143.

Csontos, K., Rust, M., and Hollt, V. "The Role of Endorphins During Parturition." In *Problems of Drug Dependence, 1980* (L. S. Harris, ed.). NIDA Research Monograph, 34, Department of Health and Human Services Publication No. (ADM) 81-1058. Washington, D.C.: U.S. Government Printing Office, 1981, pp. 264–271.

Douglas, W. W. "Histamine and 5 Hydroxytryptamine (Serotonin) and Their Antagonists." In *The Pharmacological Basis of Therapeutics,* sixth edition (A. G. Gilman, L. S. Goodman and A. Gilman, eds.). New York: Macmillan, 1980, pp. 609–646.

Ellinwood, E. H., and Kilbey, M. M. (eds.). *Cocaine and Other Stimulants.* New York: Plenum Publishing, 1977, p. 721.

FDA Drug Bulletin. *Alcohol Drug Interactions* 9(2) (1979): 10–12.

Gardner, S. E. (ed.). *National Drug/Alcohol Collaborative Project: Issues in Multiple Substance Abuse.* NIDA Research Monograph, Department of Health, Education and Welfare Publication No. (ADM) 80-957. Washington, D.C.: U.S. Government Printing Office, 1980, p. 126.

Gerbino, L., Oleshansky, M., and Gershon, S. "Clinical Use and Mode of Action of Lithium." In *Psychopharmacology: A Generation of Progress* (M. A. Lipton, A. DiMascio and K. F. Killam, eds.). New York: Raven Press, 1978, pp. 1261–1275.

Gottschalk, L. A., McGuire, F. L., Heiser, J. F., Dinovo, E. C., and Birch, H. *Drug Abuse Deaths in Nine Cities: A Survey Report.* NIDA Research Monograph 29, Department of Health, Education and Welfare Publication No (ADM) 80-840. Washington, D.C.: U.S. Government Printing Office, 1980, p. 172.

Greenblatt, D. J., Shader, R. I., Weinberger, D. R., Allen, M. D., and MacLaughlin, D. S. "Effect of a Cocktail on Diazepam Absorption." *Psychopharmacology* 57 (1978): 199–203.

Griffiths, R., Bigelow, G., and Liebson, I. "Human Drug Self-administration: Double-blind Comparison of Pentobarbitol, Diazepam, Chlorpromazine and Placebo." *Journal of Pharmacology and Experimental Therapeutics* 210(2) (1979): 301–310.

Harvey, S. C. "Hypnotics and Sedatives." In *The Pharmacological Basis of Therapeutics,* sixth edition (A. G. Gilman, L. S. Goodman and A. Gilman, eds.). New York: Macmillan, 1980, pp. 339–375.

Hayter, A. *Opium and the Romantic Imagination.* London: Faber and Faber, 1968, p. 388.

Hoefer, H. J., Black, S., Morgan, A., and Cross, T. B. *Singapore.* Hong Kong: APA Productions Ltd. and Times Publishing Berhad, 1980, p. 240.

Jaffe, J. H. "Drug Addiction and Drug Abuse." In *The Pharmacological Basis of Therapeutics,* sixth edition (A. G. Gilman, L. S. Goodman and A. Gilman, eds.). New York: Macmillan, 1980, pp. 535–584.

Jaffe, J. H., and Martin, W. R. "Opioid Analgesics and Antagonists." In *The Pharmacological Basis of Therapeutics,* sixth edition (A. G. Gilman, L. S. Goodman and A. Gilman, eds.). New York: Macmillan, 1980, pp. 494–534.

Jasinski, D. R., Martin, W. R., and Haertzen, C. A. "The Human Pharmacology and Abuse Potential of N-Allylnoroxymorphone (Naloxone)." *Journal of Pharmacology and Experimental Therapeutics* 157(2) (1967): 420–426.

368 SELECTED REFERENCES

Jefferys, D. B., Flanagan, R. J., and Volans, G. W. "Reversal of Ethanol-Induced Coma with Naloxone." *Lancet* 1:8163: 1980, pp. 308–309.

Katz, M. M. and Hirschfeld, M. A. "Phenomenology and Classification of Depression." In *Psychopharmacology: A Generation of Progress* (M. A. Lipton, A. DiMascio and K. F. Killam, eds.) New York: Raven Press, 1978, pp. 1185–1195.

Kessler, K. A. "Tricyclic Anti-depressants: Mode of Action and Clinical Use." In *Psychopharmacology: A Generation of Progress* (M. A. Lipton, A. Di-Mascio and K. F. Killam, eds.). New York: Raven Press, 1978, pp. 1289–1302.

Kissin, B. "Interactions of Ethyl Alcohol and Other Drugs." In *The Biology of Alcoholism*, Vol. 3, *Clinical Pathology* (B. Kissin and H. Begleiter, eds.). New York: Plenum Publishing, 1974, pp. 109–161.

Mello, N. K. "A Behavioral Analysis of the Reinforcing Properties of Alcohol and Other Drugs in Man." In *The Biology of Alcoholism*, Vol. VII (B. Kissin and H. Begleiter, eds.). New York: Plenum Publishing, 1983, pp. 133–198.

Mello, N. K., and Mendelson, J. H. "Self-administration of an Enkephalin Analog by Rhesus Monkey." *Pharmacology, Biochemistry and Behavior* 9(5) (1978): 579–586.

Mello, N. K., Mendelson, J. H., Kuehnle, J. C., and Sellers, M. L. "Operant Analysis of Human Heroin Self-Administration and the Effects of Naltrexone." *Journal of Pharmacology and Experimental Therapeutics* 216(1) (1981): 45–54.

Meyer, R. E., and Mirin, S. M. *The Heroin Stimulus.* New York: Plenum Publishing, 1979, p. 254.

Milne, G. M., Johnson, M. R., Wiseman, E. H., and Hutcheon, D. E. (eds.). "Therapeutic Progress in Cannabinoid Research." *Journal of Clinical Pharmacology*, Vol. 21 Supplement (Nos. 8 and 9), 1981, p. 494.

Moss, L. M. "Naloxone Reversal of Non-Narcotic Induced Apnea." *Journal of the American College of Emergency Physicians* (Jan./Feb. 1973): 46–48.

Musto, David F. *The American Disease: Origins of Narcotic Control.* New Haven and London: Yale University Press, 1973, p. 354.

Noble, E. P., Alkana, R. L., and Parker, E. S. "Ethanol Induced CNS Depression and Its Reversal: A Review." In *Proceedings, Biomedical Research in Alcohol Abuse Problems, Non Medical Use of Drug Directorate, National Health and Welfare.* Canada, 1975, pp. 308–367.

Nuotto, E., Mattila, M. J., Seppälä, T., and Konno, K. "Coffee and Caf feine and Alcohol Effects on Psychomotor Function." *Clinical Pharmacology and Therapeutics* 31 (1) (1982): 68–76.

Petersen, R. C. (ed.). *Marijuana Research Findings: 1980.* NIDA Research Monograph 31, Department of Health and Human Services Publication No. (ADM) 80-1001, Washington, D.C.: U.S. Government Printing Office, 1980, p. 221.

Petersen, R. C., and Stillman, R. C. (eds.). *Cocaine: 1977.* NIDA Research Monograph 13, Department of Health, Education and Welfare Publication No. (ADM) 77-432. Washington, D.C.: U.S. Government Printing Office, 1977, p. 221.

————. *Phencyclidine (PCP) Abuse: An Appraisal.* NIDA Research Monograph 21, Department of Health, Education and Welfare Publication No. (ADM) 78-728. 1978, p. 313.

Pizzetti, I., and Crocker, H. *Flowers: A Guide for Your Garden,* Vol. II. New York: H. W. Abrams, 1975, p. 1477.

S-2013. Temporary Program on Availability of Heroin for Cancer Patients. Congressional Record — Senate. January 25, 1982, S40-S42.

Scott, J. M. *The White Poppy.* London: Heinemann, 1969, p. 205.

Sharp, C. W., and Brehm, M. L. (eds.). *Review of Inhalants: Euphoria to Dysfunction.* NIDA Research Monograph 15, Department of Health, Education and Welfare Publication No. (ADM) 77-553. Washington, D.C.: U.S. Government Printing Office, 1977, p. 345.

Snyder, S. H. *Uses of Marijuana.* New York: Oxford University Press, 1971, p. 127.

Snyder, S. H. "Brain Peptides as Neurotransmitters." *Science* 209 (1980): 976–983.

Sorensen, S. C., and Mattison, K. W. "Naloxone as an Antagonist in Severe Alcohol Intoxication." *Lancet* ii (1978): 668–689.

Stillman, R. C. and Willette, R.E. (eds.). *The Psychopharmacology of Hallucinogens.* New York: Pergamon Publishing, 1978.

Tallman, J. F., Paul, S. M., Skolnik, P., and Gallagher, D. W. "Receptors for the Age of Anxiety: Pharmacology of the Benzodiazepines." *Science* 207 (1980): 274–281.

Van Dyke, C., and Byck, R. "Cocaine Use in Man." In *Advances in Substance Abuse, Behavioral and Biological Research,* Vol. 3 (N. K. Mello, ed.). Greenwich: JAI Press, 1983, pp. 1–26.

Willette, R. E. *Drugs and Driving.* NIDA Research Monograph 11, Department of Health, Education and Welfare Publication No. (ADM) 77-432. Washington, D.C.: U.S. Government Printing Office, 1977, p. 13.

Woodbury, D. M., and Fingl, E. "Analgesic-Anti-Pyretics, Anti-Inflammatory Agents and Drugs Employed in the Therapy of Gout." In *The Pharmacological Basis of Therapeutics,* fifth edition (L. S. Goodman and A. Gilman, eds.). New York: Macmillan, 1975, pp. 325–358.

Woody, G. E., O'Brien, C. P., and Greenstein, R. "Misuse and Abuse of Diazepam: An Increasingly Common Medical Problem." *International Journal of Addictions,* 10(5) (1975): 843–848.

Chapter 18: Alcoholism Treatment and Prevention: Rhetoric, Faith and Science

Armor, D. J., and Meshkoff, J. E. "Remission among Treated and Untreated Alcoholics." In *Advances in Substance Abuse, Behavioral and Biological Research* (N. K. Mello, ed.). Greenwich: JAI Press, 1983, pp. 239–269.

Armor, D. J., Polich, J. M., and Stambul, H. B. *Alcoholism and Treatment.* Santa Monica: The Rand Corporation, 1976.

————. *Alcoholism and Treatment.* New York: J. Wiley and Sons, 1978, p. 348.

Bean, M. H., and Zinberg, N. E. (eds.). *Dynamic Approaches to the Understanding and Treatment of Alcoholism.* New York: The Free Press, 1981, p. 214.

Benson, H. *The Relaxation Response.* New York: William Morrow, 1975.

Benson, H. "Systemic Hypertension and the Relaxation Response." *New England Journal of Medicine* 296 (1977): 1152–1156.

Edwards, G., and Grant, M. (eds.). *Alcoholism Treatment in Transition.* London: Croom Helm, 1980, p. 327.

Griffiths, R. R., Bigelow, G. E., and Liebson, I. "Relationship of Social Factors to Ethanol Self-administration in Alcoholics." In *Alcoholism: New Directions in Behavioral Research and Treatment* (P. E. Nathan, G. A. Marlatt, and T. Loberg, eds.). New York: Plenum Publishing, 1978, pp. 351–359.

Jellinek, E. M. "The Phases of Alcohol Addiction." *Quarterly Journal of Studies on Alcohol* 13 (1952): 673–684.

Keller, M. "On the Loss of Control Phenomenon in Alcoholism." *British Journal of Addiction* 67 (1972): 153–166.

Kissin, B., and Begleiter, H. (eds.). *Treatment and Rehabilitation of the Chronic Alcoholic.* New York: Plenum Publishing, 1977, p. 631.

Kurtz, E. (ed.) *Not-God, A History of Alcoholics Anonymous.* Center City: Hazelden Educational Services, 1973, p. 363.

Marlatt, G. A. and Nathan, P. E. (eds.). *Behavioral Approaches to Alcoholism.* New Brunswick: Publications Division, Rutgers Center of Alcohol Studies, 1978, p. 222.

Mello, N. K. "A Semantic Aspect of Alcoholism." In *Biological and Behavioural Approaches to Drug Dependence* (H. D. Cappell and A. E. Leblanc, eds.). Ontario: Addiction Research Foundation of Ontario, 1975, pp. 73–87.

Mendelson, J. H., and Mello, N. K. (eds.). *The Diagnosis and Treatment of Alcoholism.* New York: McGraw-Hill, 1979, p. 405.

Mendelson, J. H., Miller, K. D., Mello, N. K., Pratt, H., and Schmitz, R. "Hospital Treatment of Alcoholism: A Profile of Middle Income Americans." *Alcoholism: Clinical and Experimental Research* 6 (1982): 377–383.

Pattison, E. M., Headley, E. B., Gleser, G. C., and Gottschalk, L. A. "Abstinence and Normal Drinking: An Assessment of Changes in Drinking Patterns in Alcoholics after Treatment." *Quarterly Journal of Studies on Alcohol* 29 (1968): 610–633.

Polich, J. M., Armor, D. J., and Braiker, H. B. *The Course of Alcoholism: Four Years after Treatment.* Santa Monica. The Rand Corporation, 1979.

Pomerleau, O. F. "Behavioral Medicine: The Contribution of the Experimental Analysis of Behavior to Medical Care." *American Psychologist* 34(8) (1979): 654–663.

Thomas, L. "Notes of a Biology Watcher: Magic in Medicine." *New England Journal of Medicine* 299(9) (1978): 461–463.

Index